CONNECTIVE TISSUE IN MEAT AND MEAT PRODUCTS

CONNECTIVE TISSUE IN MEAT AND MEAT PRODUCTS

ALLEN J. BAILEY

AFRC Institute of Food Research, Langford, Bristol, UK

and

NICHOLAS D. LIGHT

Devro Ltd, Chryston, Glasgow, UK

ELSEVIER APPLIED SCIENCE
LONDON and NEW YORK

ELSEVIER SCIENCE PUBLISHERS LTD
Crown House, Linton Road, Barking, Essex IG11 8JU, England

Sole distributor in the USA and Canada
ELSEVIER SCIENCE PUBLISHING CO., INC.
655 Avenue of the Americas, New York, NY 10010, USA

WITH 17 TABLES AND 106 ILLUSTRATIONS

British Library Cataloguing in Publication Data

Bailey, Allen J.
Connective tissue in meat and meat products.
1. Meat. Meat products. Quality. Effects of collagen in connective tissues in muscles.
I. Title II. Light, N. D. (Nicholas D.)
664'9

Library of Congress Cataloging-in-Publication Data

Bailey, Allen J.
Connective tissue in meat and meat products.

Bibliography: p.
Includes index.
1. Meat—Composition. 2. Meat—Quality. I. Light, N. D. (Nicholas D.) II. Title.
TX556.M4B35 1989 641.3'6 88-24602

ISBN 1-85166-284-7

Printed by The Universities Press (Belfast) Ltd.

Preface

Connective tissue comprises a small part of muscle, but plays a significant role in the ultimate physiological function of that muscle and, more importantly, as we have tried to show in this book, plays an important role in determining the eating quality of meat. The major component of this connective tissue is the structural protein collagen, whilst proteoglycans and some glycoproteins form the physically and functionally minor part of the tissue. As meat scientists we feel that collagen, although not exactly neglected, has generally been relegated to a minor role in determining quality. This has been due partly to an understandable emphasis on the myofibrillar proteins together with a more detailed understanding of their structure and function compared to that of intramuscular collagen, but primarily a lack of understanding of the complexity of the parameters determining the properties of native and denatured collagens in raw and cooked meat. As the answers to these questions are unravelled we are beginning to build an overall view of the part played by collagen in meat. Many meat scientists 'measure' one or two ready parameters of collagen, e.g. amount or solubility, but the role of collagen must be multifactorial. It is this rather simplistic view of the treatment of meat collagen which we have sought to redress in this book. Much information is now available on the structure and function of collagen but, unfortunately, recourse to in depth reviews by collagen specialists is seldom helpful to the meat scientist. We therefore felt that what was needed was an attempt to simplify and to put into context the properties of collagen and other connective tissue macromolecules relevant to meat. We have not tried, therefore, to produce another specialised review of the literature, but have made our own judgement on the relevance of a mass of published data in an attempt to produce an hypothesis on the role of collagen in meat quality.

In this book, therefore, we have tried both to redress the balance of emphasis on the myofibrillar proteins and provide a clearer understanding of the role of collagen in meat quality. We have not attempted to write a meat science text but have concentrated on connective tissue. It is for the reader to determine whether we have been successful or one-sided. However, we feel that if the book generates questions leading to further research on the role of collagen and other connective tissue components in meat and meat products, then the writing of this book will have been well worthwhile.

Allen J. Bailey
Nicholas D. Light

Contents

6 STRUCTURE AND LOCALISATION OF CONNECTIVE TISSUE COMPONENTS IN MUSCLE

7 MECHANICAL AND THERMAL PROPERTIES OF CONNECTIVE TISSUES

8 THE ROLE OF CONNECTIVE TISSUE IN DETERMINING THE TEXTURAL QUALITY OF MEAT

9 THE EFFECTS OF ANTE- AND POST-MORTEM FACTORS ON MEAT CONNECTIVE TISSUE: GROWTH RATE, HANDLING AND CONDITIONING

14 FUTURE PROSPECTS

Chapter 1

General Introduction

Connective tissues have been well known to man for a long time. They were described by the early anatomists as the pale tissues which appeared to bind together organs and help retain the structural integrity of the body. They were known to meat-eating man from pre-history as the tough and chewy parts of meat which formed a network throughout the food. They were equally well known to the medical world as being those tissues often involved in disease states, especially those involving fibrosis. However, despite the well-known and ubiquitous nature of connective tissues, their apparent simplicity and the amount of research that has been committed to this field, it has only been in relatively recent years that our understanding of the structure and function of these diverse and important tissues has really begun to develop. We should therefore start by asking the question—what are connective tissues?

Animal bodies are extraordinarily complex structures made up of a wide range of cell types and functions. Even very simple animal bodies such as sea urchins, sea cucumbers, worms and molluscs are highly differentiated machines with strictly segregated parts and highly co-ordinated functions. These bodies require structure and order to separate various parts of the organism within the body and allow those parts to carry out their multitudinous jobs. This structure needs to be organised and strong, and for these reasons an extracellular framework of connective tissues evolved.

Connective tissues appear throughout the animal kingdom from the most primitive beings through to man, but in every organism they are found in many different forms, each one adapted to its own physiological function. The three main roles of connective tissues, as we shall explore below, are to give mechanical strength to organs, to provide a framework for movement and to promote the right environment for cell growth and proliferation. In the first two cases there is a need to provide physical strength and so connective tissues may be viewed as biomechanical structures. The physical properties of the various connective tissues are determined by their composition and specific macromolecular organisation. The forces which the different connective tissues have to withstand vary tremendously, depending on their site in the body and the functions of that particular tissue. The differences are manifest in, for example, the virtual inextensibility and high tensile strength of tendon to ensure transmission of muscular contraction without loss of energy, and the flexibility and elasticity of dermis to permit movement of appendages. At the other extreme is found the fine intercellular framework holding together the highly cellular liver, and the selective filtration properties of the non-fibrous basement membranes of the kidney. In other words, the structures of the range of connective tissues found in higher animals have to be both varied and specialised to carry out the functions required of them.

There are two main classes of connective tissue in higher animals, the skeletal tissues and the soft connective tissues. The first class, the skeletal tissues, consist of the bones, teeth and cartilages of the body. Bone and the dentine of the teeth are unusual amongst the connective tissues because they are mineralised, i.e. they contain high levels of calcium salts. The soft connective tissues are the second class of connective tissues, i.e. skin, tendons, ligaments, blood vessel walls, intra- and extra-organ connective tissues, the specialised tissues of the eye (e.g. lens and cornea) and the heart (e.g. heart valves and ligaments) and the highly specialised linear tissues upon which many types of cell grow—the basement membranes.

The biomechanical functions of the various connective tissues are both varied and specialised. For example, bone needs to be rigid and non-elastic with compressive and tensile strength, whilst on the other hand the cartilages which sheath the ends of the long bones must be elastic and shock absorbant and yet be resistant to wear. Skin needs to be tear and shear resistant as well as being waterproof whilst having a degree of elasticity, the latter property being bestowed by the random organisation of the connective tissue fibres of the dermis. A different elasticity is required in the connective tissues of artery walls where a recovery from the high, regularly pulsating forces of the blood pressure within the artery is necessary. In this case a high proportion of elastic protein is present with the matrix. Ligaments and tendons have to have very high tensile strength, although the former generally need a degree of elasticity whilst the latter do not. The elastic ligaments contain elastic fibres whilst the tendons are largely made up of strong inelastic fibres in a parallel array. Intra-organ connective tissues must bind the organ together and generally support it with fine fibres. The basement membranes must support and separate cell layers and often function as filters or barriers, therefore requiring a unique organisation of the connective tissue components.

The insoluble elastic and inelastic fibres of connective tissue are embedded or laid down in an amorphous material with high water content, known traditionally as the ground substance. This material may be present in varying quantities and may have different chemical compositions depending on the function of the particular connective tissue in which it is found. For example, in cartilage, which has a shock absorbing function as mentioned above, the content of ground substance, and therefore water, is high in relation to the content of the fibres. In tendon tensile properties are required, hence the major component is fibrous protein and ground substance content is low. The biochemical components of the ground substance are large, complex and contain high proportions of carbohydrates. These carbohydrates are often charged, as we shall see later. Basement membrane does not contain ground substance as we have defined it here, but does contain limited amount of proteoglycans and glycoproteins.

The two types of connective tissue of most concern in this book are the intra-organ connective tissues of the muscles and the basement membranes which surround each individual muscle cell. In the following sections we will consider the general structural characteristics of these, but before doing so it would be useful to review briefly the molecular components of connective tissues.

MOLECULAR COMPOSITION OF CONNECTIVE TISSUES

All connective tissues have different compositions and contain a mixture of several different proteins and complex polysaccharides. However, one major protein is common to all connective tissues in all higher animals (and even in a remarkable range of primitive organisms)—this protein is collagen. Indeed, so ubiquitous is this protein that it has been estimated that the bodies of mammals contain as much as 25–30% of their total protein as collagen. The main, functional property of this macromolecule is that it is able to form fibrous, filamentous and network structures (Fig. 1.1) in the extracellular milieu which are cross-linked in a fashion analogous to that induced in synthetic polymers like nylon and dacron. This singular chemical feature ensures extraordinary strength in collagen matrices, endowing its fibres, filaments and networks with mechanical properties which are invaluable and indispensable to the individual organism and species in which it is found.

Such is the importance of collagen that much of the research carried out on connective tissues over the last fifty years has centred on this one protein. Because of this, in terms of the bulk of knowledge accumulated, most of what we currently know and understand about connective tissues revolves around collagen and its properties. Indeed, although considerable progress has been made in recent years in understanding the role of collagen in the connective tissue of meat, very little research has been carried out to date on the other components of this important tissue. For this reason the major portion of this book will centre on collagen, its biochemistry, mechanical properties and role in meat connective tissues.

The other common components of connective tissues are elastin (a protein which forms large insoluble elastic fibres), the family of mixed macromolecules called the proteoglycans, and a limited number of large polysaccharides. The latter two components, together with water, form the ground substance. A number of large and small glycoproteins are also present, especially in basement membranes.

Each different connective tissue contains varying quantities of collagen and these other components. In many connective tissues some or nearly all of the non-collagenous compounds mentioned above may be absent, or present only in minute quantities. For example, some tendons contain 95% or more (as a percentage of dry weight) collagen, with only very small amounts of elastin and proteoglycans, whereas the connective tissue of major arteries like the aorta contain very high proportions of both elastin and collagen. As explained above, cartilages, with their high content of ground substance, contain extremely large amounts of proteoglycans with small amounts of collagen. Table 1.1 shows the comparative collagen contents of some of the main connective tissues of the mammalian body.

(a)

(b)

(c)

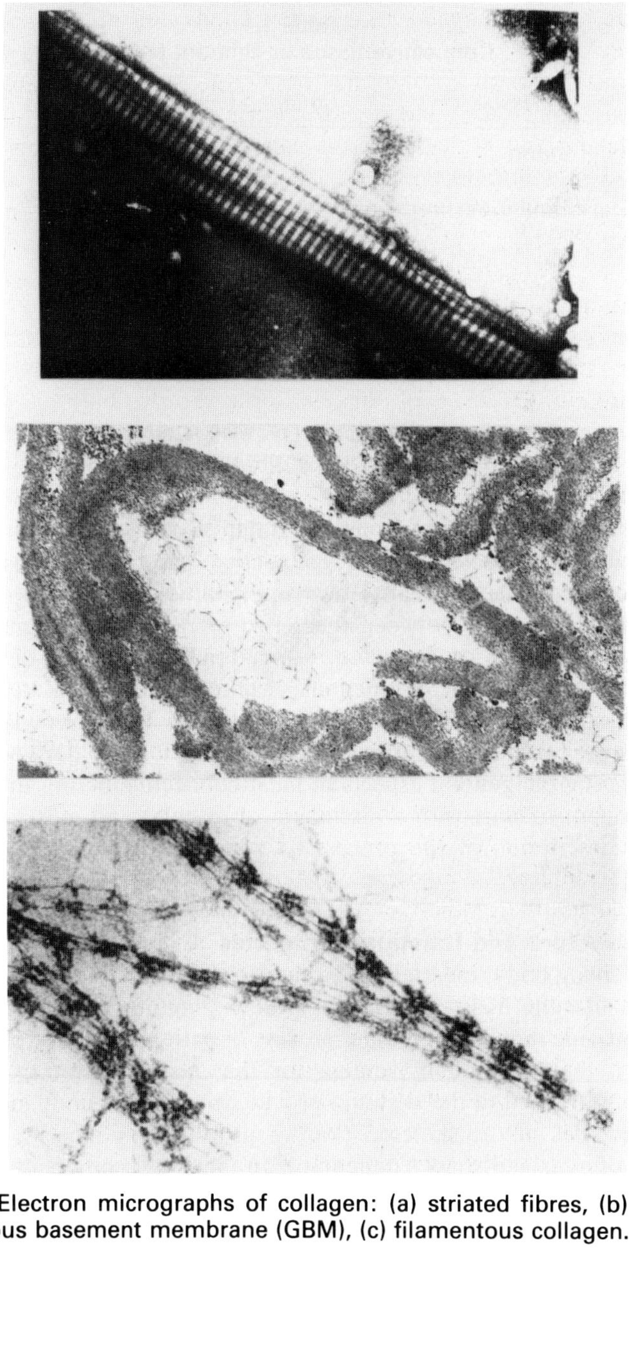

Fig. 1.1 Electron micrographs of collagen: (a) striated fibres, (b) non-fibrous basement membrane (GBM), (c) filamentous collagen.

Table 1.1
Connective tissue in different tissues

Tissue	*Collagen* (%)	*Elastin* (%)
Tendon	~95	1
Skin	~50	5
Vascular system	~40	50
Bone	25	—
Lung	15	20
Kidney	4	—
Liver	2	—
Muscle	2	0·1

GENERAL ASPECTS OF THE CONNECTIVE TISSUE OF MEAT

In this book we are concerned with the physical, morphological and biochemical characteristics of connective tissues in terms of their functions and role in meat. However, meat is a derivative of muscle, and the relationship between these two tissues is somewhat complex. Muscles are part of the living animal and have well-defined and controlled physiological functions, whereas meat is a dead tissue, having some biochemical functions but with little physiological relationship to the living tissue from which it was derived. Before considering the general aspects of meat connective tissue, therefore, it is pertinent to start with a discussion of muscles and muscle structure and a description of the process by which muscles are converted to meat.

The structure and function of muscles

In the living body, muscles are primarily responsible for movement. In organs like the heart, muscles provide a pumping action; in the face they provide movement for the process of eating and for expression; in the gut they provide movement for the mixing of food; and those muscles attached to the skeleton provide, in general, body locomotion. The muscles of the skeleton (known as voluntary muscle) exist in a range of sizes and shapes depending on their function, as illustrated in Fig. 1.2.

The semitendinosus muscle of the hind limb possesses muscle fibres that run the whole length of the muscle, about 50 cm, and attach at

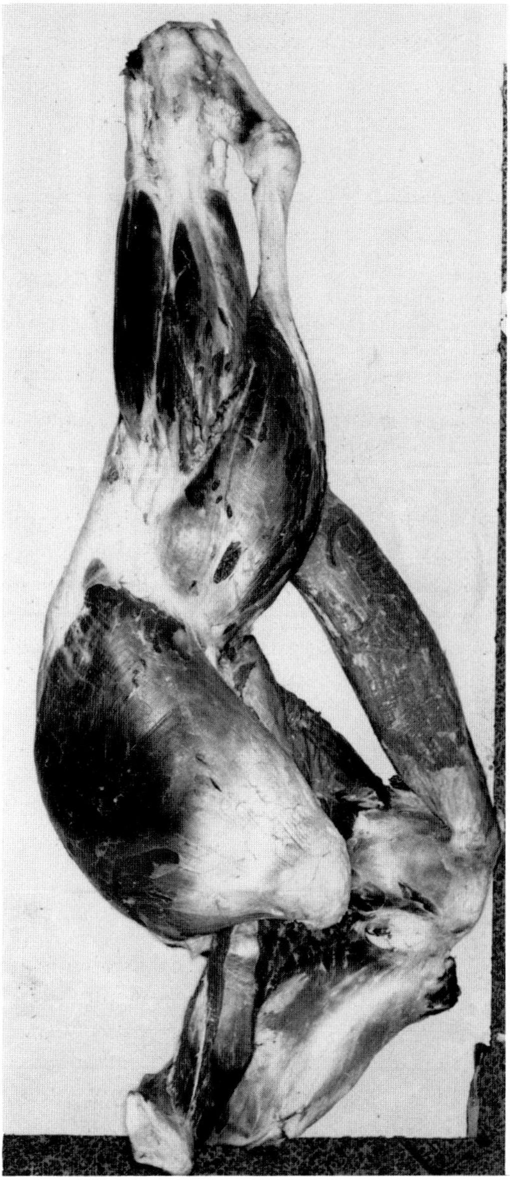

Fig. 1.2 Variation in muscle shape and size as demonstrated by some of the muscles of the bovine hind leg.

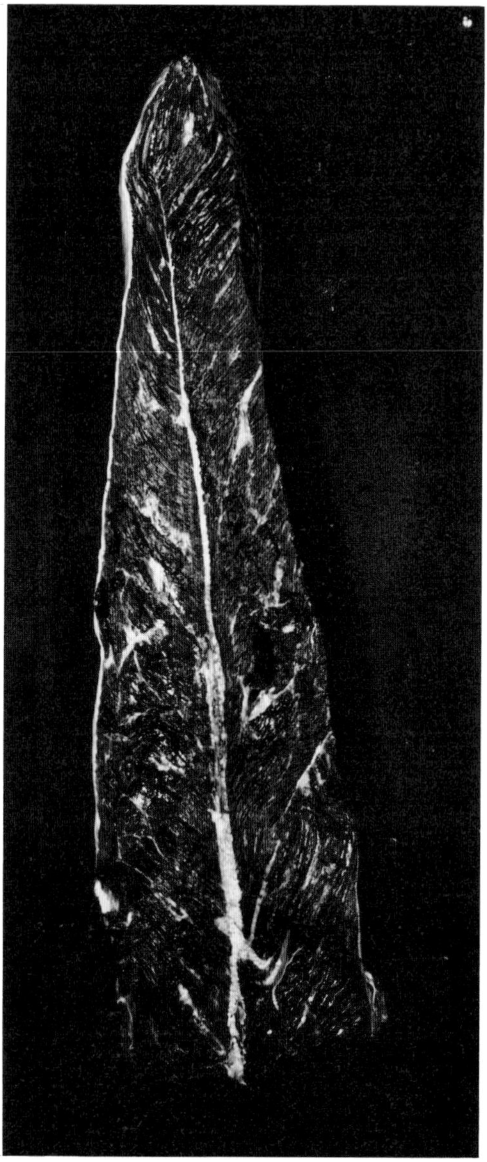

(a)

Fig. 1.3 Section of two pennate muscles. (a) Pennate muscle (Infraspinatus).

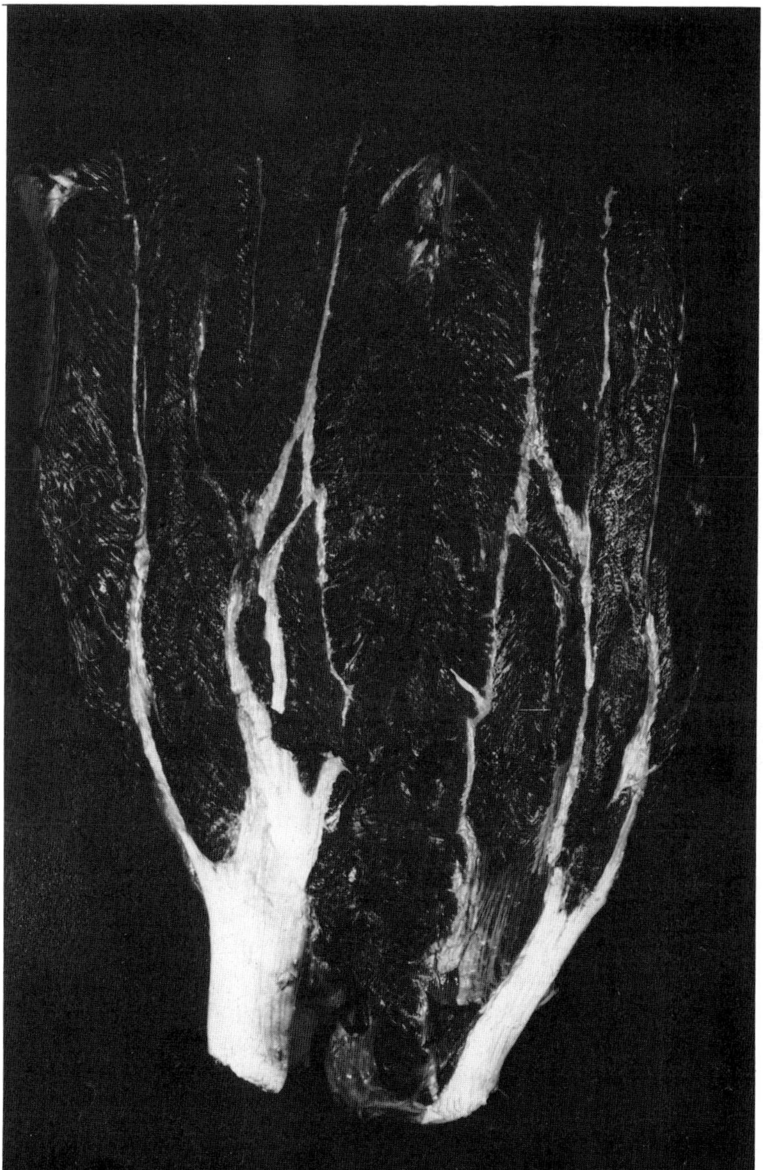

(b)

Fig. 1.3—*contd.* (b) Multipennate muscle (Flexor digitorum).

tendons at either end. However, most muscles are more complex (Fig. 1.3). The longissimus dorsi is the longest muscle of the body but the fibres do not run along its length, rather, they run from the spinous processes along the backbone to thick collagenous sheets forming the outer covering of the muscle, i.e. only about 10 cm. In the fusiform type of muscle the muscle fibres attach at an angle to the upper and lower outer collagenous sheaths. The bipennate muscles contain an inner sheet of connective tissue as an extension of the tendon and the muscle fibres run from this sheet to the outer connective tissue sheath. In the extreme case, the multipennate muscle, the internal connective tissue sheets divide the muscle into three or four parts. The muscle fibres extend from these inner sheets to the external sheath and are therefore very short in comparison to the overall length of the muscle.

Muscles, therefore, are complex cellular structures made up of a mixture of biochemical components organised into supramolecular structures, whose major functions are contraction and relaxation to bring about movement. The individual muscle cells, the specialised units which form the vast majority of the muscle structure, are highly differentiated and fuse during development to form extremely long, multinucleate tubes. These muscle cells (myofibrils) are packed with the protein-based contractile apparatus (built up into segmentally arranged fibrils from actin, myosin and a range of minor protein components) and are arranged in groups known as fibre bundles. Segments of muscle are made up of regularly arranged fibre bundles (Fig. 1.4.).

The coherent structure of the bundles of muscle fibres is maintained by the connective tissue of muscle, known as intramuscular connective tissue. Intramuscular connective tissue may be defined as all those connective tissues found in the body of any muscle which do not compose the main bulk of any tendon, the tendon being the thick, fibrous rope-like structure which binds skeletal muscles to bones. The force of contraction of the living muscle is transferred to the skeleton with a minimum loss of energy to ensure efficient locomotion, by the tough, inelastic fibrous material, primarily collagen, of the tendon. Whilst these properties are so vital in the living muscle for movement and for preventing the weak myofibrils stretching too far and rupturing, in meat, as we shall see, these tough fibres can dramatically affect eating quality.

Intramuscular connective tissue is composed of an extremely well developed network of sheaths and layers of films, both thick and thin,

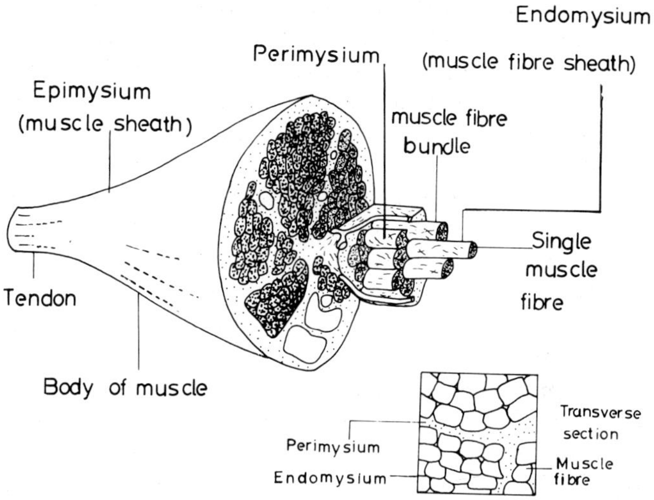

Fig. 1.4 Exploded diagram of muscle showing general muscle structure, epimysium, perimysium and endomysium.

which appear macroscopically and microscopically in the muscle as a white or pale yellow mosaic between and within the red muscle tissue itself. Each muscle is surrounded by a fibrous sheath of connective tissue which encloses the whole body of the muscle (Fig. 1.4). This sheath may be heavy, thick and durable (in bovine muscles it can be up to several millimetres in thickness), or it may be gossamer thin, translucent and almost impossible to see with the human eye, depending on the site and function of the particular muscle. This tissue is called the epimysium. The epimysium in most muscles is continuous with the tendons which link the muscle to bones.

The next level of intramuscular connective tissue is known as the perimysium and surrounds the muscle fibre bundles and groups of bundles (Fig. 1.4). This tissue varies from large tendon-like tracts penetrating the body of some muscles to extremely fine sheaths and films surrounding small fibre bundles which are barely visible to the human eye. The perimysium is contiguous with the tendons and forms the soft skeleton of each muscle, a well-articulated and proliferative lattice, the structural integrity of which is of great importance.

The final, microscopic level of intramuscular connective tissue is the basement membrane which surrounds each and every muscle cell (myofibril). This structure is a non-fibrous sheath which itself is associated with fine connective tissue fibrils that surround it (Fig. 1.4). The sheath, the basement membrane proper, is known as the endomysium. The associated fine fibrils have long been known as reticulin (although this is now a somewhat dated term as it has been shown to be a type of collagen, the intense silver staining being due to glycoproteins on the surface of the fibres). The endomysium is as important as the perimysium in terms of muscle function.

Muscle contracts *in vivo* at constant volume, and this could clearly affect the orientation of the collagen fibres. Several authors have reported that initially the crimps in the collagen fibres are straightened out, and then the angle of the collagen fibres between layers of the perimysium decreases as the muscle lengthens. Similarly changes must occur in the fibres of the endomysium. The mechanism of transferring the *in vivo* myofibrillar contraction to the inelastic collagen fibres of the tendon has not been clarified. Two alternative hypotheses are available.

Muscular contraction could result in the transfer of that energy via the endo- and perimysial fibres and ultimately to the tendon. Certainly following contraction the tension on these elements exterior to the muscle would tend to exert a pulling force on the tendon. However, both the perimysium and endomysium networks are by nature compliant because of the collagen fibre orientation and the fibre crimp. Therefore it is unlikely that the small changes in muscle length on contraction could be translated into movement of the tendon. The immediate response to muscular contraction would be difficult to achieve by this mechanism unless the myofibrils were closely bound to the endomysium. Unfortunately, there is little evidence for such a cross-linking, indeed the ease of removal of myofibrils from the endomysium to leave an intact 'ghost' suggests that a strong covalent link is unlikely to be present. On the other hand the collagenous network would certainly provide protection against overstretching of the muscle fibre.

A more attractive hypothesis is that the force of muscle fibre contraction is transmitted directly through the muscle–tendon junction. Again there is little evidence for any mechanism of attachment of the myofibril to the tendon, but some detailed studies are being carried out (Chapter 6). This should prove an important research area for both muscle biologists and meat scientists.

Conversion of muscle to meat

In life the muscles are active components of the body structure under strict control of the central nervous system and well supplied with nutrients and oxygen by the blood. After death, muscles become isolated structures, no longer under any control and with no supply of nutrients or oxygen, i.e. anaerobic. As a consequence the conversion of glycogen to lactic acid continues by the same mechanism as *in vivo*, but in the absence of oxygen the process is irreversible and continues until the ultimate pH (pH_u) is reached, about pH 5·4–5·5. The rate and extent of pH fall are determined by the species, the type of muscle and the pre-slaughter environment.

The contractile apparatus, deprived of its store of ATP, goes into a state of excessive and uncontrolled contraction called rigor mortis (the stiffness of death). This stiffness is caused by the interaction of the myosin heads and the actin, now irreversible due to the loss of ATP, to form a rigid structure of actin and myosin filaments.

The muscle does not remain in this 'stiffened' state but becomes flaccid due to a series of degradative enzymic processes on the integrated structure. This breakdown and resolution of rigor is achieved by cytoplasmic (neutral) and lysosomal (acidic) hydrolases (particularly proteases). As the pH falls the relative importance of the neutral proteinases decreases and the importance of the acid hydrolases increases. These acid cathepsins are released from the intracellular lysosomes to membranes which become unstable as the pH falls. The degradative changes brought about by these enzymes include the partial breakdown of the integrated structure of the contracted actomyosin complex, its anchoring proteins and other intracellular proteins, and damage to the sarcolemma. However, the precise mechanism is unknown. It is unlikely that the actomyosin interactions are cleaved, as this would require extensive breakdown, but limited damage to the Z-line structure could have a considerable effect on the ability to stretch the muscle fibres and could account at least in part for the 'resolution' of rigor. The degradation of desmin, a protein which aligns the myofibrils, could also affect the stiffness, possibly by allowing the muscle fibres to slide past each other. The cathepsins also damage the connective tissue surrounding the muscle cells, but it is not known if this would affect the stiffness generated by the actomyosin. It is possible that the rigor condition puts the connective tissue under considerable tension, in which case limited proteolytic damage could release this tension and contribute to the resolution phenomena.

The continued resolution of rigor, carried out by hanging the carcass at chill temperatures for several days, is known as conditioning. Analysis of meat conditioned for 2–3 weeks certainly demonstrates rupture of both the Z-lines and the connective tissue (Chapter 8).

The product obtained after resolution of rigor is meat, and the longer the conditioning period the more tender the meat. Muscle eaten before the onset of rigor mortis (normally within 8 h of slaughter) is particularly tender, but muscle in rigor mortis is extremely tough. The conditioning process makes the post-rigor muscle edible.

The whole process of conditioning needs to be closely controlled as time and temperature have been shown to have a profound effect on the quality of the final meat. Low temperature storage is used (normally about 4°C) although cooling the carcass too quickly after death can result in accentuated contraction of the myofibrils, a condition known as cold-shortening (see Chapter 9), in which state the meat is very tough and tends to remain so throughout conditioning. Carcasses, then, are cooled slowly to chilled storage temperatures and held at that temperature for 14–21 days, depending on the origin of the carcass (species, breed, etc.) and the country in which the conditioning is being done (various cultural differences exist). The details of these processes, not strictly relevant here, are discussed in a number of excellent general texts.

Meat and meat connective tissue

The term meat is generally taken to refer to all 'red' or skeletal muscle (otherwise known as striated or voluntary muscle). Some smooth or non-striated muscle (also known as involuntary muscle) is included within the definition of the term but, in general, this type of muscle is regarded as a separate commodity. Smooth muscle comprises only a small proportion of meat, generally being confined to the gastrointestinal tract, the vascular system and the reproductive tract. Smooth muscles vary in size and shape from flattened to triangular or polyhedral in cross-section. The smooth muscle fibres have a single nucleus and there are no striations along the fibre, despite their having about the same proportion of actin and myosin as skeletal muscle fibres. The fibres form membrane-to-membrane contact bridges with neighbouring fibres, and these are supported by a delicate network of reticular fibres.

Non-muscle animal tissues are referred to as offal, e.g. liver, kidney, tripe, brain and heart. With the exception of the brain, all these tissues contain varying amounts of connective tissue. The liver contains a very delicate network of fibres amounting to about 0·1% of its weight, the heart contains connective tissue in the cardiac muscles and tendons, whilst tripe contains a high proportion of collagen.

All of these various tissues obviously contain and require connective tissue. However, because the major part of our knowledge is about skeletal muscle, this book will concentrate on the connective tissue of or related to that tissue. The meat animal carcass contains over six hundred skeletal muscles, each with its own shape, size and function (Fig. 1.5), which taken together account for between 35% and 65% of the carcass weight. Although the general appearances of all of these diverse muscles are very different, they do share major characteristics in common.

All skeletal muscles are connected to bones by tendons. Both bones and tendons, as well as skin, play an important role in the meat processing industry and are important sources of collagen, generally as gelatin, the denatured and partly degraded form of collagen (Chapter 10). Skin and tendons are also used to some degree as extenders in comminuted meats (Chapter 10). Skin, bones and tendons are all important and major connective tissues but will not be dealt with individually in this book.

The general topographical and compositional aspects of meat connective tissue are much the same as those of muscle. However, the function of the connective tissue in meat, although still involved in maintaining the structural integrity of the tissue, does not relate to its function in the living muscle. Furthermore, the properties of the connective tissue in cooked meat are very different from those in raw meat. These changes in property are extremely important and will be discussed extensively in the following chapters. Finally, the connective tissue content of different muscles of the carcass is quite variable, as can be seen by the illustration in Fig. 1.6. Related to this and other factors such as the amount of work the muscle has to do in life, different muscles of the carcass show different quality aspects when eaten in that some are tougher than others. This can be illustrated by the differences in their mechanical properties (Fig. 1.7) which, as we shall see later (Chapter 7), are related to eating quality.

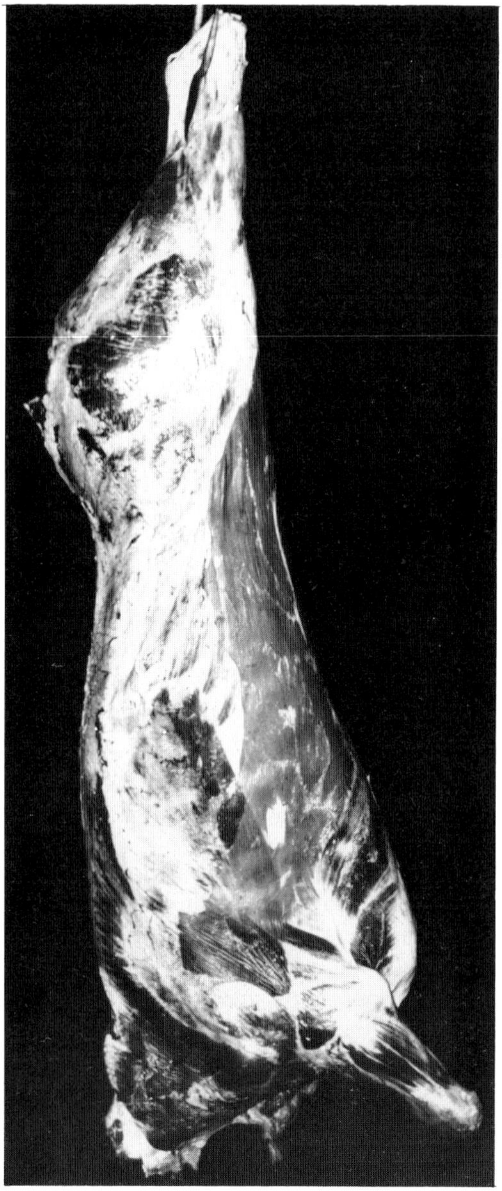

Fig. 1.5 The bovine carcass.

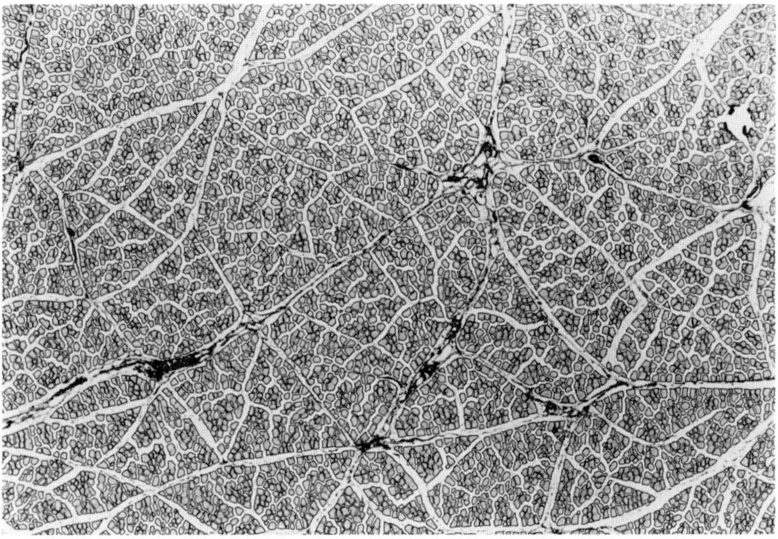

(a)

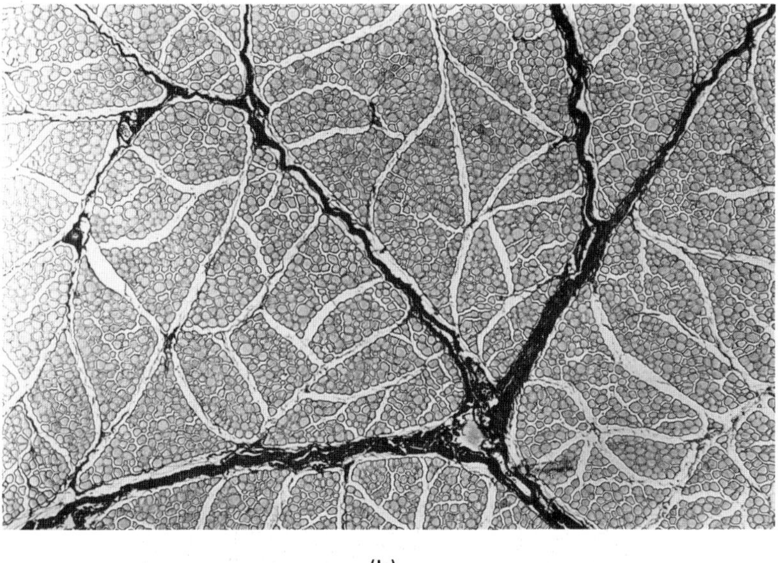

(b)

Fig. 1.6 Variation in visible intramuscular connective tissue in bovine muscles. (a) Psoas major, (b) sternomandibularis.

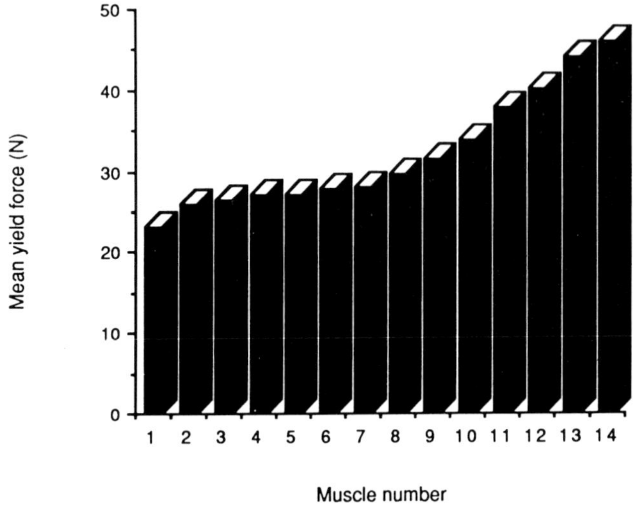

Fig. 1.7 Histogram showing differences in yield strength as an indicator of eating quality (higher yield strengths being associated with lower eating quality) of a range of bovine muscles.

MORPHOLOGICAL ORGANISATION OF COLLAGENOUS TISSUES

It would be appropriate at this point to examine the main topographical and structural states in which collagens are found. These are in the form of fibres, filaments and networks. The fibres are best described by consideration of the structure of the fibres of the tendon. The filaments, a rarer form, are illustrated by the fine filamentous structures found in cartilage and aorta and the networks have been described in the basement membrane. As all these three structures have now been found in muscle, a brief description will be given now, but the detailed molecular structure will be considered later in Chapter 2.

The fibres formed by collagen vary in size from tissue to tissue, and even within one tissue there may be a wide range of fibre diameters. In meat, the vast majority of collagens are in the fibrous form, i.e. fibrous collagen may represent as much as 95–97% by weight of the total collagen with network, basement membrane collagen making up the remaining 3–5%. Examination of collagen fibres in the electron microscope reveals a characteristic 67 nm axial repeat pattern of banding appearing as closely placed stripes (Fig. 1.1), the origin of which is discussed in the next chapter. The smallest measurable fibril has a diameter of 8 nm whilst the largest fibres (in skin and tendons) have been measured at 400 nm. In some tissues the collagen fibres show remarkable uniformity in diameter. For example, in human cornea, which contains highly regular arrays of collagen fibres to allow transparency of the tissue, all the collagen fibres have a diameter of 32·5 nm. The fibres in the cornea of the rabbit have diameters of 20 nm and in catfish they are 10 nm. Similarly, in foetal rat flexor tendon the fibres are all 16 nm in 15–16 day foetuses, 20–25 nm in 28 day foetuses and 40 nm at birth. In other tissues, however, the fibres show a wider and less specialised variation in diameter. Studies of tendon and other connective tissues showed that the distribution of sizes for fibre diameter was either unimodal or bimodal, depending on the tissue. Unfortunately very little work has been carried out to date specifically on the diameters of fibres in meat connective tissues.

At the next level of structure it is the arrangement of fibres with respect to each other that determines the integrity of each particular connective tissue. Collagen fibres form many different types of structure. In cornea, the fibres are arranged extremely regularly in anti-parallel sheets, i.e. each sheet contains parallel fibres but is orientated at 90° with respect to the next sheet. In skin the collagen fibres form larger fibre bundles which vary in size depending on their depth below the skin surface and are randomly orientated to form a flexible but mechanically stable matrix. In tendons the fibres can be extremely large and the vast majority are always orientated in the direction of the tendon axis. The fibres are arranged in large bundles, each bundle surrounded by a specialised thin connective tissue sheath, the endotendinium.

The collagen fibres of intramuscular connective tissue have not been extensively studied, although limited electron microscopy has revealed something of their macromolecular structure. The epimysial fibres are usually large in diameter but vary considerably with the muscle type. The fibres of the perimysium are larger in general than those found in the endomysium, being of the order of 50–95 nm compared with 40 nm for the fibres associated with the endomysium. Although the fibres in the tendons are contiguous within the perimysium, the organisation of the intramuscular collagen matrix appears quite different from the elongated cylindrical structures of the tendon themselves. In the perimysium, the fibres are arranged in sheets. The orientation of sheets with each other appears to be random, although there may be some degree of order imposed by stretching when the muscle is under normal load. This aspect will be more fully covered in the chapter dealing with mechanical properties (Chapter 7).

The organisation of the basement membrane collagen of the endomysium is quite different from that seen in fibrous collagen. The membrane itself is very thin, about 20–50 nm, and even at the highest magnification in the electron microscope appears as an amorphous tissue (Fig. 1.1). Collagen represents about 40% of its dry weight, the remainder being proteoglycans and glycoproteins. Unlike the fibrous collagen, the molecular organisation cannot be visualised, but recent studies have revealed an open network structure of polymerised molecules, which is discussed in Chapter 2.

The third type of collagenous structure in which there is a loose packing of the microfibrils to form a filamentous structure has only recently been studied. These filamentous collagens occur in cornea, aorta and cartilage and have been shown to be present in tendon and intramuscular collagen to a small extent. However, little is currently known about their function.

NON-COLLAGENOUS COMPONENTS OF MEAT CONNECTIVE TISSUE

Elastin

An important component of connective tissues is the protein, elastin, which is present in intramuscular connective tissue in very small amounts and is observed in the electron microscope as non-banded fibres of varying diameter. The elastic nature of the elastin fibres is in complete contrast to the inelastic semi-crystalline structure of collagen. The structure/function relationship of these components will be discussed in Chapter 5.

Proteoglycans

Other muscle connective tissue components include the large modified polysaccharides, like hyaluronic acid, the proteoglycans such as chondroitin, dermatan and heparan sulphate, and the specialised minor glycoproteins, like fibronectin, laminin and nidogen. The polysaccharides (mucopolysaccharides) and proteoglycans form a minor component of the fibrillar network of the perimysium and epimysium which is generally referred to as the ground substance, as

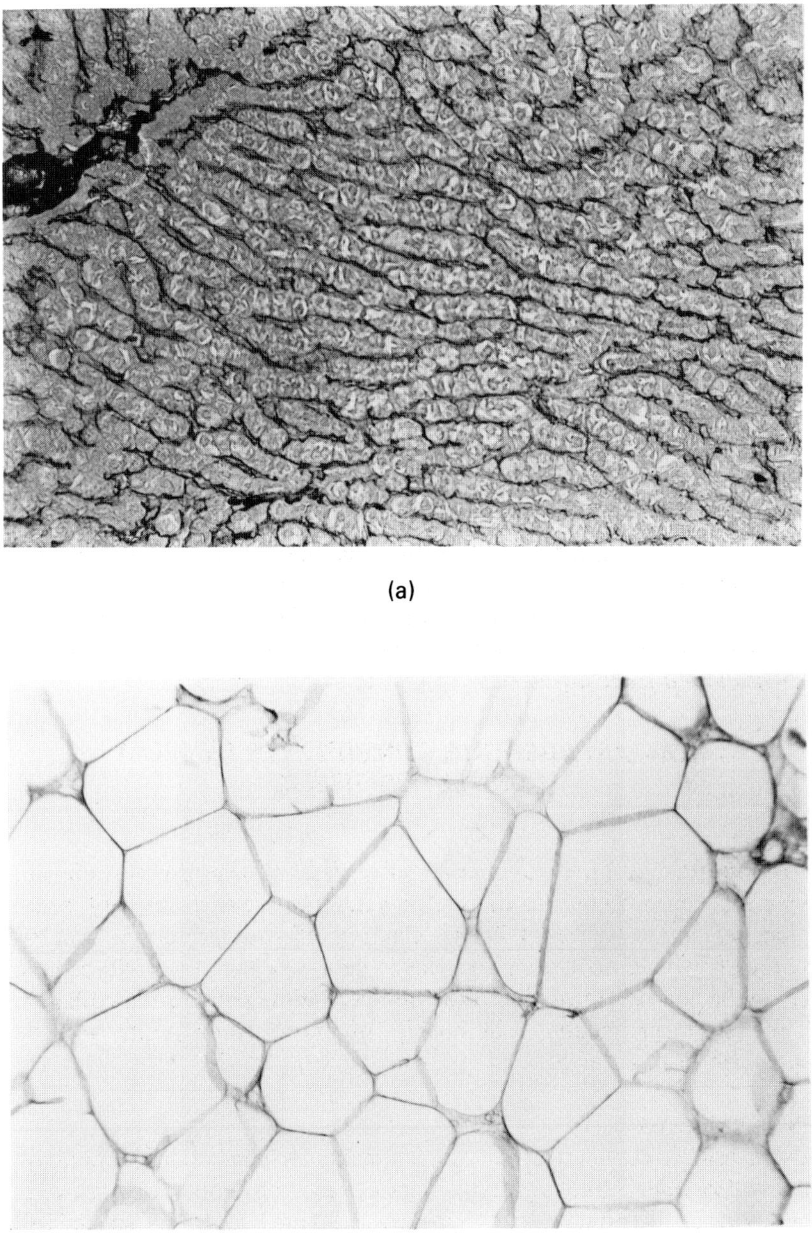

(a)

(b)

Fig. 1.8 Micrograph of the distribution of connective tissue in (a) liver and (b) adipose tissue.

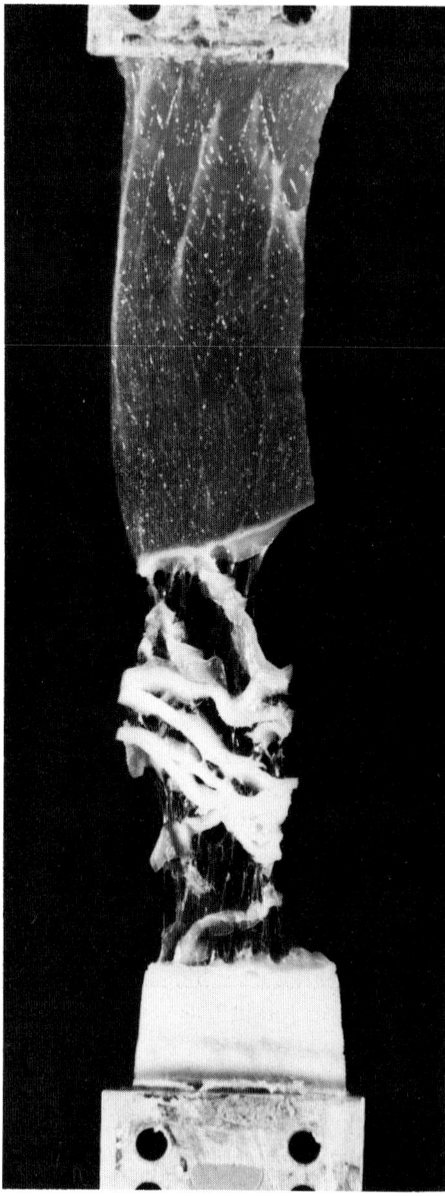

Fig. 1.9 Mechanical stretching of subcutaneous adipose tissue illustrating the connective tissue layers.

discussed earlier. It is also possible that such components may be involved in interfibre interactions or in the function of the endomysial basement membrane. The structure and function of proteoglycans and glycoproteins will be discussed in detail later (Chapter 5), as will their possible function and role in meat (Chapter 8).

ADIPOSE TISSUE AND OFFAL

Adipose tissues are aggregates of specialised cells (adipocytes) which are capable of storing large amounts of fat as an energy reserve. The cells are held together and supported in the tissue mass by a loose connective tissue often referred to in biological texts as areolar tissue (Fig. 1.8). This is largely made up of collagen but also consists of some of the other major components of the 'ground substance' mentioned above, e.g. proteoglycans and glycoproteins. Areolar tissues can also contain elastin depending on their location and function in the body, as well as a small number of cells such as fibroblasts and some macrophages. Adipose tissue is loosely associated with a large number of muscles in the animal carcass and forms, with intramuscular fat, the major source of meat lipids.

Organs used in the meat industry such as liver, kidney, heart, tongue, intestines, spleen, brain, etc. are known collectively as offal. All bodily organs, except the brain, are masses of packed cells bound together by connective tissue and arranged in well-ordered arrays whose organisation determines the function of the particular organ (Fig. 1.9). These cell masses are held together by tight intercellular junctions known as gap junctions and by the loose connective tissue. They are often also surrounded by a protective connective tissue sheath, normally of a thin, gauze-like appearance. This areolar tissue is generally similar in composition to adipose tissues. Some organs have rather specialised structures like those seen in the lungs, heart and kidneys, and these are often based on very specialised connective tissue structures (e.g. the valves of the heart, the tubules and glomeruli of the kidney). Overall, the connective tissue contents of these offal organs is very low.

SUMMARY

Connective tissues are ubiquitous in the bodies of animals and have an important role in providing a structural basis for the body and its organs. They account for a large proportion of the dry weight of animal tissue and their major protein component, collagen, accounts for about 25% of the total bodily protein complement. Muscles have a well developed connective tissue system which helps to support the tissue and also to transmit the force of muscle contraction to the skeleton. Connective tissues, therefore, have considerable tensile strength. Muscle connective tissue can be conveniently separated morphologically into the three domains of epimysium (muscle sheath), perimysium (intramuscular connective tissue) and endomysium (individual muscle fibre sheath). The main component of these domains is collagen. Collagen forms fibres, filaments and networks, all of which are present in meat connective tissue. The other components of connective tissue include elastin, the large modified polysaccharides, proteoglycans and specialised minor glycoproteins. Non-muscle organs also contain these elements in their loose connective tissues, which is sometimes referred to as areolar tissue.

FURTHER READING

1. Lawrie, R. A. (ed.), *Meat Science,* 4th edn. Pergamon Press, Oxford and New York, 1985.
2. Sims, T. J. & Bailey, A. J., Connective tissues. In *Developments in Meat Science* Vol. 2, ed. R. A. Lawrie. Applied Science Publishers, London and New Jersey, 1981.
3. Bailey, A. J. (ed.), *Recent Advances in the Chemistry of Meat.* Royal Society of Chemistry, London, 1984.
4. Bendall, J. R., *Muscles, Molecules and Movement.* American Elsevier, New York, 1971.
5. Briskey, E. J., Cassens, R. G. & Trautman, J. C. (eds), *The Physiology and Biochemistry of Muscles.* The University of Wisconsin Press, Madison, WI, 1966.
6. Fawcett, D. W., *A Textbook of Histology.* W. B. Saunders, Philadelphia, PA, 1968.
7. Forrest, J. C., Aberle, E. D., Hedrich, H. B., Judge, M. D. & Merkel, R. A., *Principles of Meat Science.* W. H. Freeman & Co., San Francisco, CA, 1975.
8. Wainwright, S. A., Biggs, W. D., Curry, J. D. & Gosline, J. M., *Mechanical Design in Organisms.* Edward Arnold, London, 1976.

Chapter 2

Molecular and Fibre Structure of Collagen

INTRODUCTION

Collagen was once regarded as a single protein common to most connective tissues and was always described as a long, rigid, rod-like molecule. Although much of this definition still holds at least approximately true, with the advances made in recent years in our knowledge of the genetic family of collagen and its more subtle structural properties, a more exacting description of this important connective tissue protein is now needed.

Firstly, collagen is not merely one single type of extracellular protein but a whole family of closely related molecules each coded for in the DNA by different genes which are often located on different chromosomes. To date, twelve different forms of collagen have been isolated and identified, each having its own unique sequence of amino acids, molecular weight, associated carbohydrate molecules and physico-chemical characteristics (see Table 2.1). Clearly, what is required here is a generalised definition which will fit all of these various forms of the molecule.

Collagen, then, is a large extracellular protein which is composed of three polypeptide chains (called α-chains) each of which contains long sequences of repeating tripeptides based on the general structure Gly–X–Y (where X is commonly proline and Y can represent any amino acid but is often the modified amino acid hydroxyproline).

–$\underline{\text{Gly}}$–Pro–Lys–$\underline{\text{Gly}}$–Asp–Thr–$\underline{\text{Gly}}$–Pro–Arg–$\underline{\text{Gly}}$–Pro–Arg–$\underline{\text{Gly}}$–

All collagens also contain varying quantities of the amino acid hydroxylysine which is formed by the enzyme-catalysed hydroxylation of lysine residues in newly synthesised α-chains (see Chapter 3).

Because of this primary sequence the collagen polypeptide α-chain adopts a left-handed polyproline type helix. These individual chains are unstable but when three chains are wrapped round each other they form the very stable right-handed triple helix characteristic of collagen. Variations in the primary amino acid sequences and in post-translational modifications (Chapter 3) of the α-chains lead to differences in the properties of the molecule, and consequently in their self-assembly process into fibres and their interaction with other matrix molecules.

The general chemical characteristics of the different collagen types therefore include the amino acid composition of one-third glycine, significant amounts of hydroxyproline and hydroxylysine, and the presence of an invariant repeating sequence (Gly–X–Y), such that Gly occurs every third residue. Arising from this highly characteristic structure, all collagen types share the property of an X-ray diffraction pattern revealing the 0·28 nm meriodonal characteristic of the residue spacing of the triple helix.

This provides a broad definition of the class of protein molecules called collagens. However, in order to understand fully the structural and functional role of these connective tissue molecules in animal tissues, and in meat in particular, it is necessary to take a closer view of the molecule itself.

MOLECULAR STRUCTURE

Primary structure

The polypeptide α-chains which make up the collagen molecule vary in size and composition from collagen type to collagen type. However, there is a structural similarity between all the types and in many cases a high degree of sequence homology, as might be expected for closely related genetic products. The twelve known types of collagen share a general structural formula in that the molecules of all collagen types consist of three polypeptide chains (α-chains) in close association. These three α-chains form long, extended triple helical regions and shorter, unextended globular domains often referred to as non-triple helical domains (see Table 2.1 for summary).

The extended triple helical regions of the molecules consist of long, unbroken sequences of the repeating tripeptide Gly–X–Y, where X and Y are often proline and hydroxyproline but can be any other amino acid except tryptophan. This latter amino acid is not present in any known collagen type. All collagens contain a relatively high proportion (about 14% by weight) of the modified amino acid hydroxyproline within this highly conserved repeating tripeptide region. This secondary amino acid is derived from proline by post-translational enzymic hydroxylation (see Chapter 3) and has the structure shown below. The hydroxyl group is generally found at the 4-position and it is this group which accounts for the high degree of intramolecular hydrogen bonding in collagen, as will be seen later. 3-Hydroxyproline is present in small quantities but its function is unknown. Because the only other proteins in which it occurs are the membrane-bound enzyme acetylcholinesterase, the complement component Clq and elastin (in very small quantities), hydroxyproline is considered a unique identifying amino acid for collagen.

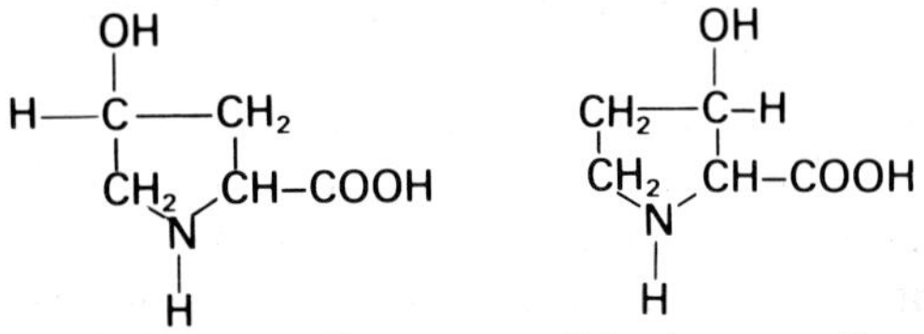

4-hydroxyproline 3-hydroxyproline

The triple helical regions are long rod-like structures which are relatively rigid. In non-fibrous collagens the triple helical domains may be interrupted by short sequences in which glycine does not occur every third residue, thus leading to a more flexible and structurally adaptable molecule. Figure 2.1 shows a generalised structural model for both the fibrous and non-fibrous collagen types.

Because of the unusual tripeptide repeating unit, the amino acid composition of collagen is itself unique. Analyses of amino acid content for several α-chains of the more common types is shown in Table 2.2. As can be seen, each contains about one-third glycine, one-quarter proline and one-seventh hydroxyproline. Consideration of the amino acid composition of all twelve collagen types shows that the incidence of the large hydrophobic residues phenylalanine and tyrosine is low, the latter being commonly present in some of the fibrous forms at levels below 10 residues/1000 residues. Histidine too is relatively

Fig. 2.1 Diagrammatic representation of helical and globular domains of the collagen molecule.

low in abundance, occurring, like tyrosine, at levels below 10 residues/1000 residues. In the fibrous collagens, both tyrosine and histidine are located predominantly in the terminal non-helical domains. In Type I collagen, for example, tyrosine is found at residues 15, 1035, 1043 and 1044 only (note that the C-terminus of the

Table 2.1
Collagen types and their location in animal tissues

Type	*Molecular length (nm)*	*Retained terminal globular regions*	*Molecular composition*	*Tissue location*
I	300	Processed	$\alpha 1(I)[\alpha 2(I)]_2$	Intramuscular, skin, tendon, bone, dentine
II	300	Processed	$\alpha 1(II)_3$	Cartilage, disc, vitreous humour
III	300	Processed	$\alpha 1(III)_3$	Intramuscular, skin, vascular, intestine
IV	420	C-terminal	$\alpha 1(IV)[\alpha 2(IV)]_2$	Basement membranes
V	300	N-terminal	Various combinations of $\alpha 1(V)[\alpha 2(V)]_2$ and $\alpha 3(V)$	Intramuscular, skin, embryonic tissues
VI	105	N- and C-terminal	$\alpha 1(VI)\alpha 2(VI)\alpha 3(VI)$	Vascular system
VII	450	Not known	Not known	Skin, amniotic membrane
VIII	Not known	Not known	Not known	Aortic endothelium
IX	200	N-terminal	$\alpha 1(IX)\alpha 2(IX)\alpha 3(IX)$	Cartilage
X	150	N- and C-terminal	$\alpha 1(X)_3$	Cartilage
XI	300	Not known	$\alpha 1(XI)\alpha 2(XI)\alpha 3(XI)$	Cartilage

polypeptide consists of a dityrosine unit). As mentioned above, tryptophan is not found in collagen.

The small hydrophobic amino acid alanine is as abundant as hydroxyproline, accounting for about one-seventh the total amino acid content. Cysteine is absent from many collagen types altogether. Of the most common forms of the molecule, only Types III and IV contain any cysteine, and therefore disulphide bonds, and these are confined to the non-helical terminal regions. The other sulphur-containing amino acid, methionine, is also relatively low in abundance in all collagen types. The distribution of amino acids in the X and Y positions of the tripeptide is not entirely random. As stated earlier proline and hydroxyproline prefer the X and Y position respectively. Glutamine, leucine and phenylalanine usually occur in the X position, whilst threonine, lysine and arginine are found predominantly in the Y position.

The amino acid lysine is important in collagen even though its incidence is somewhat low compared with globular proteins. This amino acid is distributed along the length of the polypeptide chains, and a proportion of the lysine residues in each collagen chain are hydroxylated to form hydroxylysine during post-translational modification (Chapter 3).

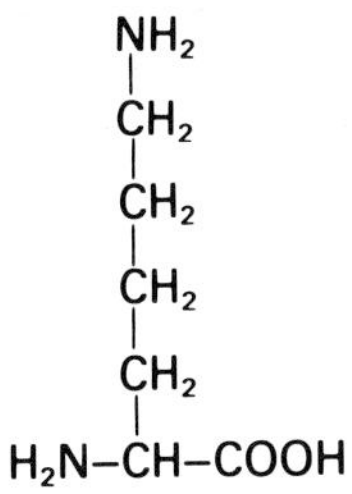

lysine

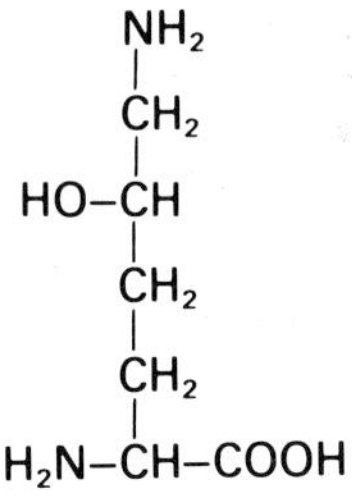

hydroxylysine

Many of the hydroxylysine residues are subsequently glycosylated during intracellular processing and contain O-linked β-D-galactopyranoside or β-D-glucopyranosyl-β-D-galactopyranoside. These glycosylated hydroxylysine residues are located exclusively in the triple helical domains of collagen and, in the fibrous forms, are the only glycosylated amino acids. Finally, one or two hydroxylysine residues, which, like lysine, are distributed along the chains, are also known to be involved, with lysine, in the formation of intermolecular cross-links (Chapter 4).

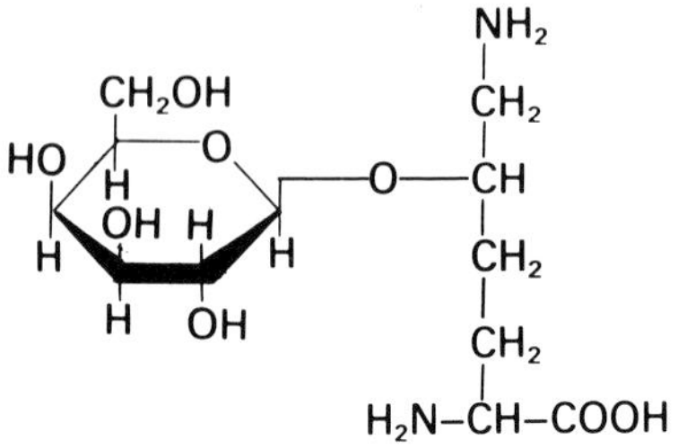

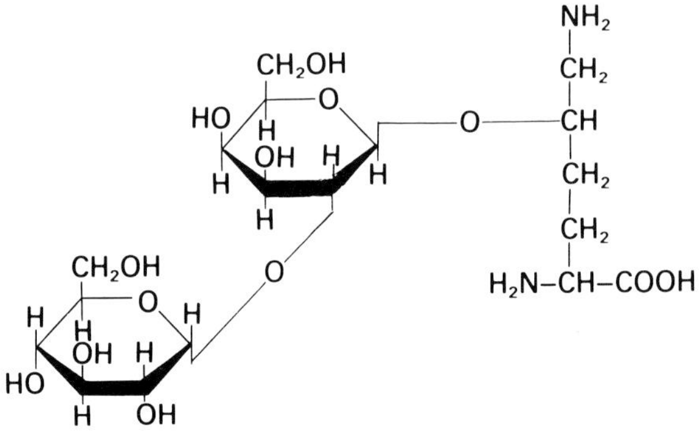

The incidence of hydrophobic and ionic amino acid side chains along the collagen α-chains generates a charge profile which controls the ability of collagen molecules to self-associate in a highly specific manner. Studies using computer simulations from known sequences have shown that hydrophobic and ionic regions in different molecules are in registration when molecules are brought together in specific alignments. Studies in which ionic side chains were chemically blocked supported the view that these residues were important in the association of collagen molecules in fibres, since molecules with blocked side chains could not form fibres.

The relatively small number of methionine residues randomly distributed along the α-chains of collagen types mentioned above has provided an excellent technique for identifying the various collagen types, as each has its unique distribution of this amino acid. Cyanogen bromide is used to cleave specifically the chains at the methionine residues and the small number of fairly large peptides released (Fig. 2.2) readily produce a characteristic peptide map on gel electrophoresis.

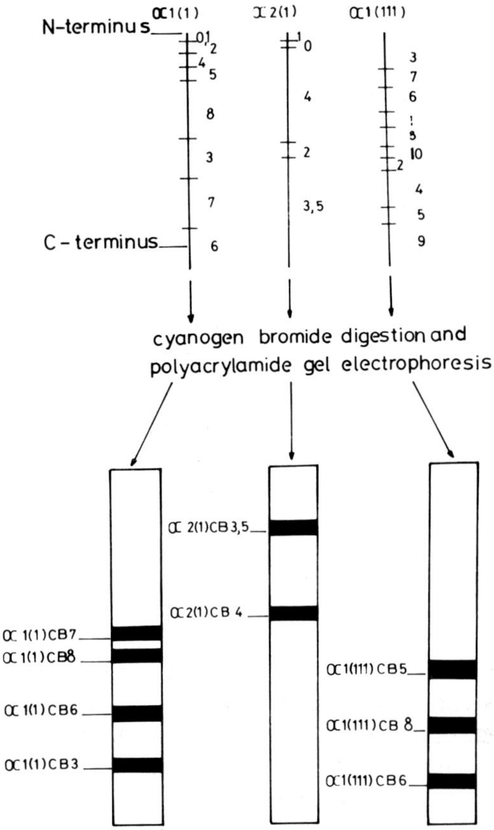

Fig. 2.2 Cyanogen bromide (CNBr) cleavage points in the α1(I), α2(I) and α1(III) chains and the characteristic peptide banding pattern seen after polyacrylamide gel electrophoresis of CNBr digested collagens.

Structure of the triple helix

The triple helices in the major fibrous (or interstitial) collagens are all very similar, being semi-rigid rods about 300 nm long and approximately 1·5 nm in diameter. Each α-chain in the triple helix contains about 1000 amino acids with the conformation of a polyproline-like helix having about three amino acids per turn. This means that each turn of the helix contains about one triplet of the Gly–X(Pro)–Y type as described above. The triple helix is right-handed and is formed by

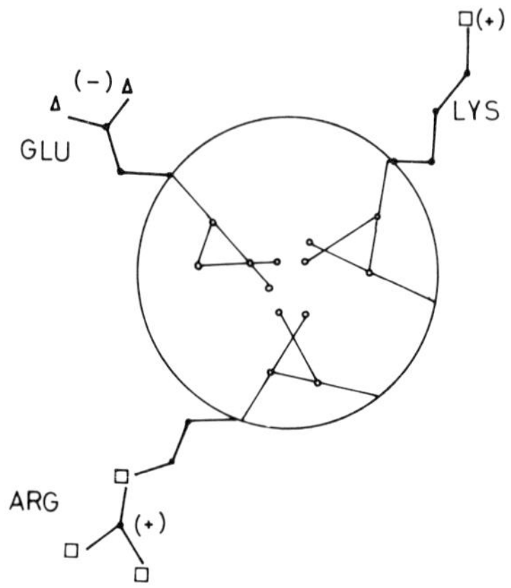

Fig. 2.3 Diagrammatic representation of the 'end-on' view of the collagen triple helix showing the inter-relationship of three amino acid residues.

the three α-chains which are wound around a common axis (Fig. 2.3). These rod-like triple helices can readily be visualised in the electron microscope. The physical properties of the helical molecules in solution fit the concept of a long rigid asymmetrical rod.

The three α-chains are able to pack very tightly together due to the occurrence at every third residue of glycine, whose small side chain (a hydrogen atom) is orientated into the centre of the helix. Any other amino acid in the third position would have a side chain too large to be incorporated into the helix structure and would therefore sterically hinder triple-helix formation. There is therefore an absolute requirement for glycine in every third position for coherent triple helical structure. The triple helix itself is stabilised by one hydrogen bond per turn from the glycine residue to the carbonyl oxygen of the peptide backbone. The ring structure of proline which prevents rotation of the N—C bond in the peptide chain is important in stabilising the helix. Further stabilisation occurs via the hydroxyl group of the hydroxyproline and water. The water molecules can form a hydrogen-bonded water-bridge from one hydroxyl group of hydroxyproline to another and to the carbonyl oxygen of the peptide backbone. Collagen that is under-hydroxylated in respect of proline is considerably less stable, the thermal stability being reduced by 20%.

Amino acids in the X and Y positions are located on the outer surface of the helix and have their side chains orientated outwards. It is these side chains which are involved in ionic and hydrophobic interactions between molecules of collagen and are involved in fibril formation, as we shall see later. From the above considerations it is easy to see that the minimum requirements for stable triple helix formation are that glycine should occur at every third residue in the sequence of the α-chains, and that hydroxyproline should be present at a minimum level.

The amount of proline and hydroxyproline is directly related to the thermal stability of the helix (Fig. 2.4). The thermal stability of collagen is an important property, not only in its physiological function, but also in respect of the properties of meat. The thermal denaturation of the molecule in solution can be monitored by measuring viscosity or optical rotation, both of which decrease dramatically as the helix collapses to a globular form. The melting temperature (T_m) range is reasonably sharp and the mid-point of the transition for mammals is just above body temperature i.e. about 39°C. Examination of the T_m for a wide range of animals reveals a correlation between the T_m and the environmental temperature at which the particular species exists. Collagens of fish which live at very

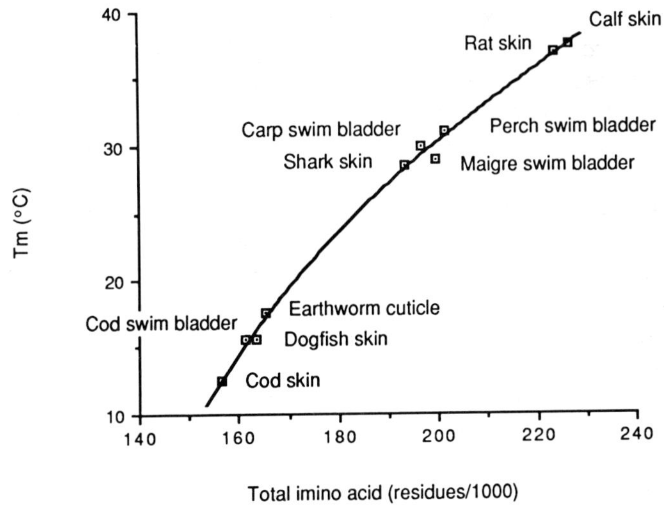

Fig. 2.4 Relationship between the total imino acid content (proline + hydroxyproline) of various collagens from different species and the molecular melting temperatures (T_m) of these collagens.

low temperatures contain reduced quantities of hydroxyproline as a lower T_m is important in its metabolism, for example in solubilised cod skin collagen the T_m is about 12°C.

In addition to the total number of proline and hydroxyproline residues, the location of these two imino acids is also important to helix stability. Thus, in studies using synthetic peptides, $(Gly\text{–}Pro\text{–}Y)_n$ was shown to be more stable than $(Gly\text{–}X\text{–}Pro)_n$ and $(Gly\text{–}Pro\text{–}Hyp)_n$ more stable than $(Gly\text{–}Pro\text{–}Pro)_n$. Under-hydroxylated collagen synthetic peptides, like the fish collagen mentioned above, had a denaturation temperature of up to 15°C less than normal collagen. The distribution of triplets along the α-chains also appears to be related to the overall stability of the collagen molecule in that, near the C-terminal end of the helix, there is a region of high stability. In Types I and II collagen this is caused by the presence of a group of Gly–Pro–Y and Gly–Pro–Hyp triplets (six in Type I and five in Type II). In Type III this region is thought to be stabilised by intramolecular disulphide bridges as well as two such triplets.

The triple helix is so well stabilised by the internal non-covalent bonding structure that it shows a remarkably high resistance to proteolytic attack. At physiological temperatures, isolated collagens in solution or in the form of a matrix are eventually totally solubilised and reduced to small fragments by pepsin- and trypsin-like proteases, but at lower temperatures (0–25°C) the triple helix remains undigested even after prolonged exposure (72 h). In the latter case whole molecules of collagen can thus be liberated from the insoluble matrix by controlled proteolysis with pepsin or an equivalent enzyme. Such molecules, which lack the N- and C-terminal non-helical peptides, are remarkably stable and show all the general characteristics of the whole collagen molecule, even in their ability to form fibres. Because of this many studies have centred on pepsin-solubilised collagen.

Non-triple helical domains

All genetic forms of collagen share not only the characteristic triple helical structure described above which forms the larger portion of each molecule, but also shorter non-helical regions. These domains are, in general, globular in structure and contain an amino acid composition which is more like that of the majority of animal globular proteins (Table 2.2). The classes of non-helical domain can be divided into two: those which make up the identifiable regions of the procollagen extra peptides at the N- and C-termini of all newly

Table 2.2
Amino acid composition of the α-chains of the main collagen types in meat

Amino acid	Type I		Type III	Type IV		Type V	
	α1(I)	α2(I)	α1(III)	α1(IV)	α2(IV)	α1(V)	α2(V)
Hyp	102	87	127	146	127	107	108
Asp	43	46	43	48	52	55	51
Thr	17	19	13	18	28	27	22
Ser	39	36	39	34	26	34	26
Glu	74	69	72	77	62	90	99
Pro	139	123	109	79	69	92	119
Gly	352	350	354	318	313	318	320
Ala	121	113	97	31	49	59	46
Val	21	34	14	30	24	30	19
Cys	—	—	2	—	—	—	—
Met	7	5	8	15	12	10	7
Ile	7	15	13	34	42	18	20
Leu	21	33	22	54	59	37	40
Tyr	2	4	3	5	6	—	1
Phe	13	12	8	24	37	12	12
Hyl	5	9	5	61	42	25	36
His	2	11	6	6	8	10	7
Lys	31	23	30	5	6	18	19
Arg	51	53	47	20	45	57	48

synthesised collagens (see Chapter 3), and those small regions which are left attached to the N- and C-termini following processing of the larger propeptides. As will be seen later (Chapter 4), the covalent cross-links which stabilise the fibre and network structures of collagen form between helical and non-helical domains in adjacent molecules. These regions are more susceptible to proteolytic attack (as are all non-helical collagenous domains) and so, in these collagens, the triple helical molecule can be released from fibre or network matrices by controlled proteolysis at the terminal sites (see Chapters 3 and 12).

In the interstitial, fibrous collagens the molecular structure is remarkably consistent between collagen types with a triple helix of about 300 nm and short non-helical regions at both the N- and C-terminal ends. In all cases these non-helical domains are very short, being between only 18 and 25 amino acids. The N-terminal amino acid is always pyro-glutamic acid and all of the N-terminal non-helical sequences contain a lysine residue at position 9 (or the equivalent of position 9), as well as containing relatively high proportions of hydrophobic amino acids including between one and two residues of tyrosine.

N-terminal sequences

```
                         5N                      *   10N
α1(I) calf   pGlu–Leu–Ser–Tyr–Gly–Tyr–Asp–Glu–Lys–Ser–Thr–Gly–
                                                 *
α2(I) calf   pGlu  —   —  Phe  —   —  Asp–Ala–Lys–Gly  —  Gly–
                                                 *
α1(II) calf  pGlu–Met–Ala–Phe–Gly–Glx–Asp–Glu–Hyl–Gly–Ala–Gly–
                                                 *
α1(III) calf pGlu–Tyr–Glu  —  Ala–Tyr–Asp–Val–Lys–Ser–Gly–Val–
```

*Lysine (or hydroxylysine) residue 9 involved in intra- and intermolecular cross-linking.

The C-terminal non-helical domain also contains higher proportions of hydrophobic amino acids than the triple helix, notably tyrosine, and also contains a lysine residue very near to the C-terminus at the equivalent of residue 1045. This lysine and the one present in the N-terminal non-helical region is the substrate for enzymatic oxidative deamination and becomes involved in the formation of intra- and intermolecular cross-links (see Chapter 4).

C-terminal sequences

```
                                 15C       *          20C
α1(I) calf     –Glu–Pro–Pro–Gln–Glu–Lys–Ala–His–Asp–Gly–Gly–Arg–Tyr–Tyr–Arg–Ala
α2(I) chick    –Asp–Gly–Gly–Gly–Tyr–Glu–Val–Gly–Phe–Asp–Ala–Glu–Tyr–Tyr–Arg–Ala
                                           *
α1(II) chick   –Leu–Gly–Gln–Thr–Glu–Lys–Gly–Pro–Asp–Pro–Ile–Arg–Tyr–Met–Arg–Ala
```

*Lysine residue 17 involved in cross-linking.
N.B. No lysine residue is available for cross-linking in the α2(I) C-terminal sequence.

It is not necessary in this book to describe in detail the molecular structure of all the known collagens; only those known to be involved in the structure of muscle and meat will be discussed in detail.

FIBRE-FORMING COLLAGENS

The collagen molecules of this group self-assemble to form fibres possessing a periodic band pattern with a repeat of 67 nm. This pattern arises from a molecule 4·4 D long quarter-staggered by D and end-overlapped by 0·4 D with respect to the nearest neighbour

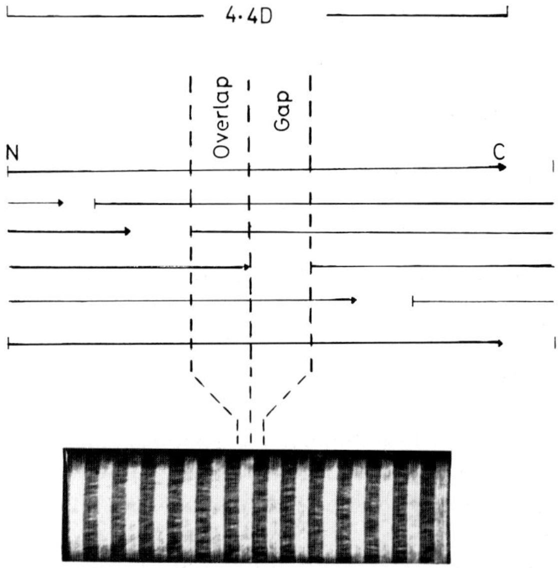

Fig. 2.5 Illustration of the 4·4 *D* over-lap structure of the collagen fibre demonstrating the derivation of the banding pattern of a negatively stained collagen fibre as seen in the electron microscope.

molecules, where *D* is the repeat period 67 nm. This type of alignment gives rise to overlap and gap regions which can be readily identified in negatively stained collagen fibres in the electron microscope (Fig. 2.5). The detailed packing structure of the fibre is described later.

Type I collagen

Type I collagen is the most common type in the collagen family. Each molecule is a co-polymer (heteropolymer) of two α1 chains (called α1(I)-chains) and one α2 chain (called the α2(I)-chain). The α1(I)- and α2(I)-chains are products of different genes located on different chromosomes. Type I collagen contains one-third glycine, is composed of a triple helix of approximately 1050 amino acids in length (300 nm), has short non-triple helical sequences at either end of the triple helix, contains no tryptophan, very low tyrosine and histidine and no cysteine (and therefore no disulphide bonds). It is found in all major connective tissues including skin, bone, tendon, dentine, and all intra-organ connective tissues. In common with the other fibrous forms of collagen, Type I produces fibres of high tensile strength due to the comprehensive network of intermolecular cross-links laid down within the matrix (see Chapter 4).

Type I collagen, because of its ubiquity, was the first of the collagen family to be fully investigated and consequently there is a wealth of information about this type. Several Type I collagens from different species have been sequenced both as the protein and from the DNA sequence of the gene coding for the molecule, as have the propeptides of the biosynthetic precursor form of the molecule. Throughout this book, where generalised information is given about fibrous collagen, this is based on what we know about Type I collagen unless otherwise stated.

Type II collagen

Type II collagen is found largely in cartilaginous tissue and was initially characterised in the early 1970s. Although it is another fibre-forming collagen, unlike Type I it is a homopolymer of three identical chains (called α1(II)-chains). The hydroxylysine content is three times greater than in Type I and a higher proportion of these residues are O-glycosylated with galactose and galactose–glucose. The fibres of Type II collagen are much finer than those of Type I and this has been linked to the higher carbohydrate content. The methionine content is greater than Type I, therefore leading to the production of much smaller peptides on digestion with cyanogen bromide. Apart from these differences it is similar in structure and properties to Type I collagen. The triple helix of Type II shows about 75% amino acid sequence homology with Type I, although there is considerably less similarity in the non-helical regions. As it is not present in meat connective tissue we shall not consider Type II collagen further in this book.

Type III collagen

First characterised in the early 1970s, this fibrous collagen has been located in embryonic tissue, scar tissue, skin, arteries, heart valve and many intra-organ connective tissues. It is normally physically associated with Type I collagen and is prominent in tissues which require a degree of flexibility. The monomer, like Type II collagen, is composed of three identical chains, the α1(III)-chains. However, unlike Types I and II, it contains cysteine residues located at the C-terminal end of the triple helix which form intra- and, possibly, intermolecular disulphide bonds. Also, examination of SLS crystallites (see Chapter 13) of Type III collagen show that its helix is longer (by two triplets) than that of Type I but that, like Type II, it shows a high degree of homology with Type I, especially within the helix.

Type V collagen

This unusual collagen is probably a fibrous type but it has proved extremely hard to demonstrate that it is able to form fibres *in vivo*. It has been isolated from tissues rich in basement membranes and appears, by histological methods, to be associated with these structures. It has a high hydroxyproline and hydroxylysine content like Type IV collagen and also contains 3-hydroxyproline, but it has no cysteine. In most tissues it exists as a heteropolymer of two chains, denoted $\alpha1(V)$ and $\alpha2(V)$, with the ratio of the two chains 2:1 respectively, as seen in Type I collagen. However, a third chain has been isolated and given the nomenclature $\alpha3(V)$. This chain may form trimeric molecules or dimeric co-polymers with either of the other two chains in a limited number of tissues.

STRUCTURE OF THE FIBRE

The early proposals for a quarter-stagger alignment of collagen molecules in the fibres were subsequently modified to include an end overlap, and this is now generally accepted (Fig. 2.5). The molecular length of 300 nm is represented by $4{\cdot}4\,D$ where D is the 67 nm periodicity observed in the electron microscope, and hence $0{\cdot}4\,D$ is the end overlap distance. This organisation has been confirmed by electron microscopy, amino acid sequence and cross-linking studies. In a computer study the known amino acid sequence of the $\alpha1(I)$ chain was overlaid upon itself with both N-termini in line. One sequence was then 'moved' one residue at a time with respect to the other sequence. The interactions of the two sequences in each position were calculated using a formula for hydrophobic or ionic interactions and the 'best-fit' in terms of maximal interactions between the two sequences was determined. This was found to be at a position where the two chain sequences were overlapped (or staggered) relative to each other by 234 residues, which is equivalent to the 67 nm observed in the electron microscope. The amino acid sequence of the molecule therefore determines the subsequent arrangement of the molecules in the self-assembly process to produce a fibril. However, the precise

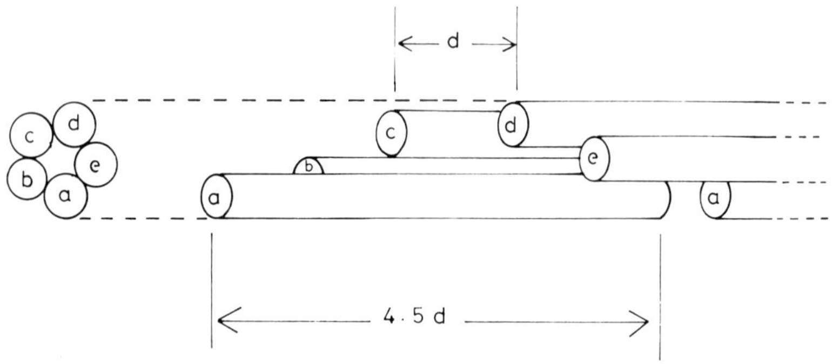

Fig. 2.6 Diagram of the classical pentafibril model of collagen fibre structure.

organisation of the fibrils within a large fibre is still a matter of debate.

In the 1960s Smith suggested a three-dimensional cylindrical microfibril based on a five molecule repeat (Fig. 2.6). This repeating unit was called a pentafibril, and the fibre was assumed to be built up of arrays of pentafibrils. This was a very favourable model since it could be readily related to the build-up of pentafibrils (or microfibrils) into large diameter collagen fibres. Support for specific interactions occurring to form an ordered microfibril was obtained by analysing the charged and hydrophobic residue distribution along the surface of the molecule, which demonstrated that one edge of the molecule was hydrophobic and the other ionic. The hydrophobic edge was thought to act between the microfibrils and the ionic edge within the microfibril.

Some recent detailed X-ray diffraction studies have modified this simple model. These studies have indicated that a more accurate explanation of the available data is that collagen molecules are organised in a quasi-hexagonal lattice, in which there are disordered areas and crystalline areas where there is a hexagonal packing of the molecules (Fig. 2.7). In this model there is also a limiting cell of molecules which is based on a five-fold molecular alignment, but in this case it is an entirely theoretical system. The single unit of the quasi-hexagonal model is the individual collagen molecule, not a five molecule unit as suggested by the older pentafibril model. This gives a considerable fluidity to the structural integrity of the model and also to the mechanism by which more molecules may be accreted to increase fibre diameter. It should be remembered that X-ray diffraction can

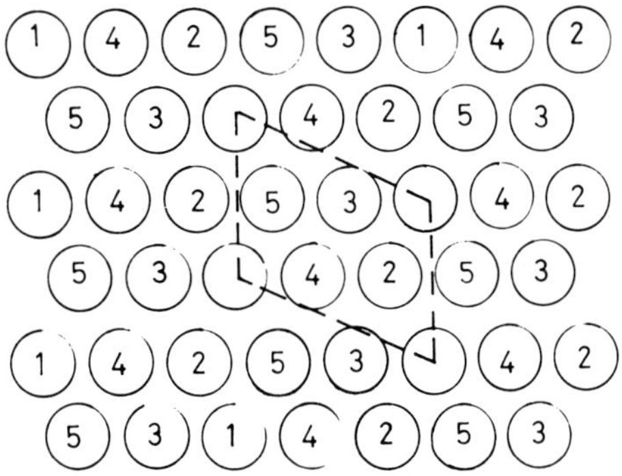

Fig. 2.7 Cross-sectional view of the molecular packing of collagen molecules in a fibre based on the quasi-hexagonal model.

only be used to confirm proposed models; it does not provide derivative data. It may be that both the structured microfibril and the quasi-hexagonal models are correct. The basic building block may indeed be the microfibril, but in the confines of a large fibre distortion may take place. The interactions between molecules are flexible enough to allow other stable arrangements, leading to areas of crystallinity and disorder. Indeed it may be biologically advantageous for there to be some disorder in a large fibre. Evidence from nuclear magnetic resonance (NMR) spectroscopy indicates that the molecules are rotating through approximately 30° about their axes; in other words contacts between molecules in the fibre are fluid, the interactions between molecules constantly being broken and reformed. This would clearly allow for flexibility in fibre organisation.

The end-overlap of 0·4 D (26·8 nm) has been supported by electron microscope studies, but direct evidence has been obtained by chemical studies on the location of the intermolecular cross-links (Chapter 4). Residue 87 in the helix is 4 D from the C-terminal region and residue 930 is 4 D from the N-terminal region and both these residues have been shown to be involved in the formation of a cross-link. Thus the molecules which are 4·4 D in length overlap 0·4 D at each terminal region. Similar evidence exists for Types II and III collagens.

Cross-link studies have also been useful in limiting the number of possible models based on the X-ray data. The formation of cross-links in mature tissue is believed to be facilitated through molecules aligned in register, i.e. at 4 D (Chapter 4). This is theoretically possible for both the pentafibril model in which adjacent pentafibrils would be in register, and in the quasi-hexagonal model through 1:5 interactions.

Collagen fibres range in size from just 8 nm in diameter to very large fibres in tendons and skin with diameters of over 400 nm. They are extremely long in most tissues and it is impossible to measure their length in the microscope. In many tissues the size is age dependent, the fibre growing in diameter with age until development is complete. Large fibre diameter is associated with tissues which need a high tensile strength such as mature tendons. It is the organisation of these fibres in the tissue that determines its mechanical properties, but the mechanism by which this organisation takes place is still unknown. In skin the fibres are orientated in different directions to provide flexibility in resisting force in several directions, whereas in tendon the fibres are unidirectional in order to transmit force in a single direction. In bone the fibres are woven in different directions and then into lamellae to resist compression and stretching, whereas in cartilage the fibres appear randomly distributed to resist compressive forces.

The fibres observed in most tissues are seen to be in some sort of woven form, referred to as a planar crimp. This crimping has been shown to be highly regular in tissues with a periodicity of about 100 μm, in for example, rat tail tendon (Fig. 2.8). The crimp therefore extends over several *D* periods. It has been suggested that it acts as a sort of shock absorber to sharp tensile forces. The crimp is readily straightened out in physiological situations.

The features of collagen fibres discussed here are characteristic of the epi-, peri- and endomysial fibres of muscle (see later chapters). Fibre diameter varies between these groups and between different muscles. Similarly, the crimp has been observed in the perimysium and is seen to disappear under contractile or extensile forces.

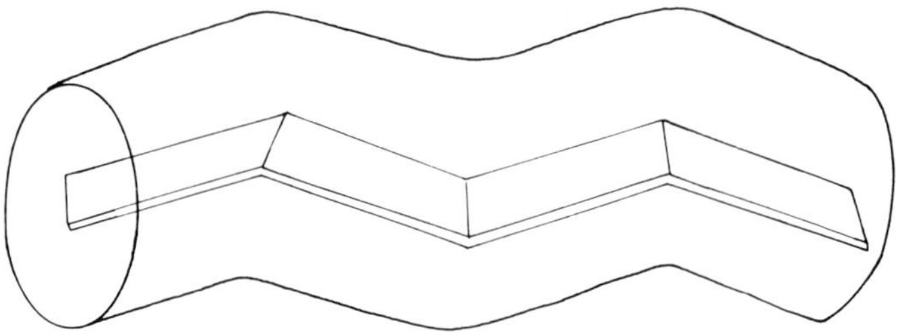

Fig. 2.8 Diagrammatic representation of the planar crimp within a collagen fibre.

NON-FIBRE-FORMING COLLAGENS

At the present time Type IV is the only characterised member of this group and is the major structural component of the amorphous basement membranes.

Type IV collagen

The Type IV collagen molecule contains two dissimilar α-chains, α1(V) and α2(IV), and the molecule is considerably longer than those of Types I, II and III, being about 400 nm. The triple helix is flexible and susceptible to proteases due to non-triple helical interruptions in the Gly–X–Y sequences. Approximately twelve such interruptions of varying length occur in the long triple helical portion of the molecule.

The distribution of the amino acids does not produce the asymmetric charge profile seen in the fibre-forming collagens, suggesting that lateral interactions of the molecules are less important. The Type IV molecule contains a high number of hydroxylysine residues which are almost exclusively O-glycosylated in the form of the glucosylgalactoside. The level of hydroxyproline is slightly increased compared with fibrous forms and about 10% of these are 3-hydroxyproline residues.

Type IV collagen is synthesised as a procollagen molecule, as are all other collagen types so far investigated (see Chapter 3), and is secreted as such from the cells which produce it. However, once outside the cell, Type IV procollagen is not processed by having its extra terminal peptides removed by specific proteases (Chapter 3) to give a triple helical collagen molecule similar to Types I, II and III collagen. Rather, the procollagen is incorporated directly into the basement membrane in which the non-collagenous regions as well as the collagenous domains play an important part in stabilising the matrix through intra- and intermolecular cross-linking.

The N-terminus of the molecule consists of a short helical sequence, about 30 nm long (known as the 7S domain due to its sedimentation characteristics in the analytical centrifuge), a short non-collagenous region (called NC2), then a long triple helical domain equivalent in length to the normal triple helix of the fibrous types. At the C-terminal end of the molecule is a large globular, non-collagenous region called the NC1 domain. Tetramers of these molecules have been observed in extracts from basement membrane, in which the four N-termini are joined in an anti-parallel fashion. This tetrameric unit has been proposed as the building block for a network structure throughout the membrane in which the C-terminal regions of the tetramers interact to form a unique open network (Fig. 2.9).

Although the tetramer unit has been observed in the electron microscope following extraction from the basement membrane, as yet there is no direct evidence that these tetramers polymerise through their C-terminal domains to form an open network scaffolding for the

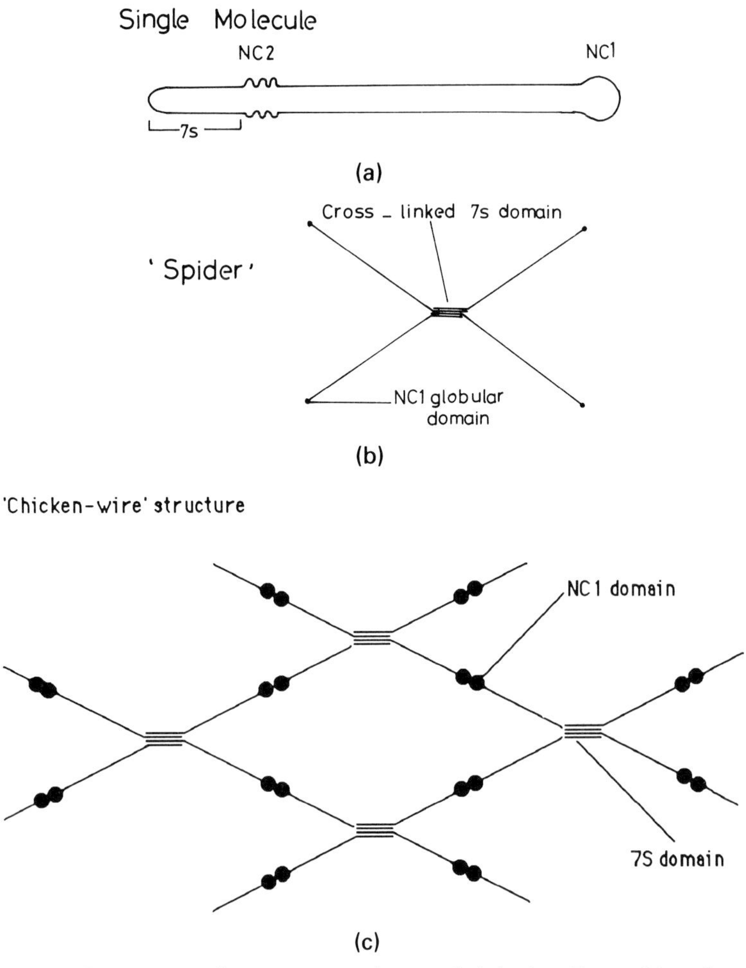

Fig. 2.9 Diagrammatic representations of (a) the Type IV collagen molecule, (b) the Type IV collagen tetramer and (c) the proposed 'chicken-wire' network.

membrane. The proposed network structure is a monomolecular sheet and as such would be expected to be rather fragile. Consequently other models for the basement membrane framework have been proposed involving lateral association of the molecules. This lateral association is unlikely to be tight or specific compared to the fibrous collagens since no '*D*' repeat pattern has been discerned along the amino acid sequence of the molecule. Most of the models are hypothetical or based on the *in vitro* aggregation studies of Type IV molecules. These studies indicate not only that two monomers aggregate to forms dimers via either the N- and C-termini, but also aggregate laterally through the NC1 domain binding at four points along the triple helix. Evidence for lateral packing of about four molecules *in vivo* has been obtained from X-ray diffraction studies, and from the location and age-related changes of the cross-links. Additional lateral packing could occur by the layering of any type of network structure. Clearly the structure of basement membrane is not yet fully understood but the mechanism of the self-assembly of Type IV collagen molecules, as well as the other main types, provides a fascinating insight into the structure/function relationships of proteins.

FILAMENTOUS FIBRES

A number of the recently identified collagens appear to belong to this class, e.g. Types VI, VII, IX and X. In the context of this book we are only interested in Types VI and VII, but the general characteristics of these filamentous collagens will be briefly discussed here.

Type VI collagen

Type VI collagen is observed in tissue as a filamentous structure of loosely packed microfibrils with an axial repeat of 100 nm. This collagen has a wide distribution in the extracellular matrix, e.g. the vascular system, cartilage and cornea. In the vascular system the microfibrils are generally perpendicular to the fibrous collagen and elastin and may act as an interconnecting network. Rotary shadowing of Type VI monomers extracted from aortic intima with pepsin revealed a rod-like helical section 105 nm long with a large globular domain on each end. The globular regions constitute two-thirds of the molecule, the triple helix portion being relatively short.

Molecular aggregation studies based on *in vitro* reconstitution experiments indicate that dimers are first formed through the anti-parallel and lateral association of the two molecules with an overlap of 30 nm. The structure is stabilised by disulphide bonds. The dimers then aggregate by lateral association with their ends in register to form a tetramer. The filaments are then formed by the end-to-end alignment of the tetramers to give a microfibril with a 110 nm repeat (Fig. 2.10).

Type VI collagen is unusual in being readily extracted and analysis indicates that the filaments are not cross-linked by the usual collagen cross-links. Type VI has been identified in the perimysium of muscle, but as yet its role has not been determined.

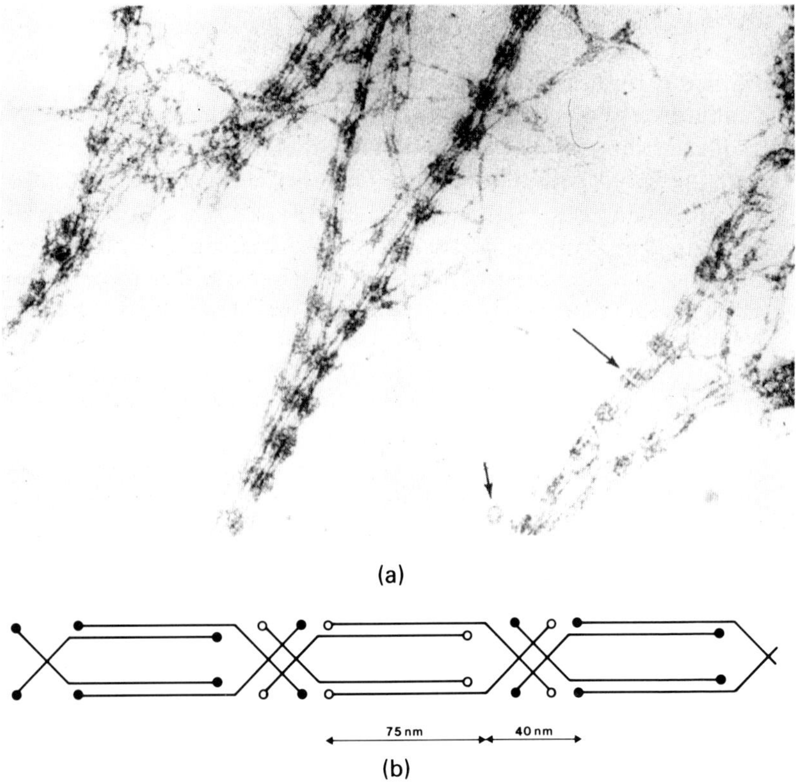

Fig. 2.10 Type VI collagen. (a) Electron micrograph of a Type VI filament and (b) representation of molecular organisation of the filament.

Type VII collagen

The microfibrils underlying the placental and dermal basement membrane are believed to be Type VII collagen. These microfibrils, which have been referred to as anchoring fibrils, do not occur in all basement membranes, but may occur in the endomysium. The Type VII molecule as isolated has been shown to be 1·5 times longer than the fibrous collagens, that is about 450 nm. Little is known about its macrostructure but *in vitro* reassociation results in the formation of anti-parallel dimeric structures showing a 60 nm overlap. Using monoclonal antibodies and immuno-electron microscopic techniques Type VII appears to form the anchoring fibrils that are believed to link basement membrane to the underlying matrix. At the present time no studies have been carried out on the presence of Type VII collagen in endomysial basement membrane, but if present its fibrils could be important in determining texture in meat (see Chapter 8).

OTHER COLLAGEN TYPES

Several other gene products of the collagen family have been identified in recent years, these being termed Types VIII, IX, X and XI. These have been found in various tissue sources including epithelia, arteries and cartilage. All of them contain the basic triple helical structure with greater or lesser proportions of non-triple helical peptides. One Type (IX) even appears to be a part of a proteoglycan. All of these types, including Type I trimer which is a co-polymer of $\alpha 1(1)$-chains, are minority collagens in that they exist in very small quantities in the tissues in which they have been located to date. This does not diminish their importance and some have already been postulated to have special roles. However, these rarer forms of the collagen species will not be discussed in detail in relation to meat.

SUMMARY

The family of collagen molecules varies in size and structure, but all the collagens are composed of three polypeptide chains, called α-chains, with a simple repeating primary structure Gly–X–Y, where X is often proline and Y hydroxyproline. These α-chains form a polyproline helix, and three of these chains wrap around each other to form a triple helix.

The triple helix is right-handed, about 1000 residues (300 nm) in length in fibrous collagens with three amino acids per turn, and the molecule is rod-shaped. The hydrogen atom of the glycine residues is orientated inside the helix giving a tight compact structure stabilised by hydrogen bonds. The stability is further facilitated through hydroxyproline residues and bound water and thus variation in the melting temperature of the helix is determined by the number of hydroxyproline residues. The tight packing of the helix provides it with a strong resistance to proteolytic attack. The non-triple helical or globular domains vary from one collagen type to another and can be small residual regions (15–20 residues) left after procollagen processing is complete or large unprocessed procollagen domains.

The incidence of ionic and hydrophobic side chains along the length of the α-chains of the triple helix contributes to the self-association of molecules into supramolecular structures. The collagens can therefore be roughly classified based on their functional form as fibrous collagen, non-fibrous collagen and filamentous collagen.

The common fibrous collagens are Types I, II, III and V, which are distributed ubiquitously throughout the bodily connective tissues with Type I being the predominant type. It has a molecular structure of $[\alpha1(I)]_2\alpha2(I)$ and forms fibres of high tensile strength. Type II is located in cartilage and vitreous humour and has the structure $[\alpha1(II)]_3$. Type III collagen is associated with more elastic tissues such as skin, intestine and blood vessel walls and has the molecular structure $[\alpha1(III)]_3$. Type V collagen is thought to be fibrous and has a variable molecular structure derived from its three chains. Collagen fibres are banded when negatively stained with a repeat of 67 nm. This is due to the molecules being aligned with a 234 residue overlap and are thus in staggered array (known as the quarter stagger and end-overlap array). The fibres probably have a quasi-hexagonal array of molecules making up their structure based on a five molecule cell.

The non-fibrous collagen group currently only includes Type IV. It is found in basement membranes and has a molecular structure composed from its three chains of $[\alpha1(IV)]_2\alpha2(IV)$. Type IV collagen was thought, until recently, to form open networks and be completely non-fibrous, but may in certain membranes form non-striated microfibrils.

The filamentous collagens include Types VI and VII which form loosely packed molecular arrays in a filamentous form. In Type VI collagen these are thought for form from laterally associated anti-parallel molecules with a repeat period of 100 nm. Type VII forms a filamentous dimer. Other collagen types include VIII, IX and X, but their supramolecular organisation has not yet been elucidated.

FURTHER READING

1. Piez, K. A. & Reddi, A. H. (eds), *Extracellular Matrix Biochemistry*. Elsevier, New York & Amsterdam, 1984.
2. Fleischmajer, R., Olsen, B. R. & Kuhn, K. (eds), *Biology, Chemistry and Pathology of Collagen*. New York Academy of Science, New York, 1985, pp. 1–537.
3. Viidik, A. & Vuust, J. (eds), *Biology of Collagen*. Academic Press, London, 1980.
4. Parry, D. A. D. & Creamer, L. (eds), *Fibrous Proteins, Scientific, Industrial and Medical Aspects*. Vols I & II, Academic Press, London, 1979.
5. Squire, J. M. & Vibert, P. J. (eds), *Fibrous Protein Structure*. Academic Press, London, 1987.
6. Pearson, A. M., Dutson, T. R. & Bailey, A. J. (eds), *Advances in Meat Research*. Vol. 4: *Collagen as a Food*, Van Nostrand Reinhold, New York, 1987.
7. Burgeson, R. E. & Mayne, R. (eds), *Structure and Function of Collagen Types*. Academic Press, Orlando, FL, 1987.

Chapter 3

Genes, Biosynthesis and Degradation of Collagen

INTRODUCTION

The collagens are a family of closely related proteins each coded for by a separate gene, some of which exist on different chromosomes. The synthesis of collagen is basically similar to that of other secreted proteins, but is characterised by extensive post-translational modifications, the majority of which are unique to the collagen family. The final product of synthesis, procollagen, is secreted through the Golgi complex in the same way that other extracellular proteins are secreted. Following secretion the procollagen may be fully or partially processed before being incorporated into the extracellular matrix as fibres, filamentous structures or non-fibrous networks. The supramolecular structures finally produced are then stabilised by the formation of intermolecular cross-links. The degradation of these stable polymers is achieved by the concerted action of several specific collagenases extracellularly and completed by intracellular digestion within the lysosomes.

GENES

The twelve genetically distinct types of collagen identified over the past few years are composed of at least fifteen different α-chains, hence there must be at least fifteen different genes coding for these chains. As other types of collagen are identified this number of presumptive or identified genes will increase further. The genes may all be located on different chromosomes, even the genes coding for α-chains of the same type of collagen. For instance, for Type I collagen it has been shown that each haploid genome contains only one copy of the gene for α(1)-chains located on chromosome 17 and one for α2(I)-chains located on chromosome 7. This is not unusual—many well characterised protein hetero-polymers have genes for their constituent polypeptide chains on different chromosomes, e.g. the α and β globin chains of haemoglobin.

Several genes for collagen α-chains have been completely or partially sequenced to date. These include the genes for mouse α(I), chick α2(I), chick α1(II) and chick α1(III) chains. All these genes are very similar, as might be expected from the very close homology of the protein chains for which they code. The genes so far isolated appear to be many times longer than the size of the corresponding mRNA and their expected size based on the number of amino acids in the collagen α-chains. The coding regions, known as exons, are distributed in 50 short regions; the major portion of the gene is however composed of the non-coding regions or introns which interrupt the exons. The genes are complex and large, but also vary considerably in size. For example, the gene for chick α2(I) has 39 000 bases or 39 kilobases (kb) whilst the gene for α2(I) has 18 kb. As three bases code for each amino acid a procollagen chain of 1500 amino acids would be coded for by a total of only 4500 bases, i.e. 4·5 kb. This illustrates the large preponderance of non-coding intron sequences in collagen DNA.

The coding sequences, or exons, which amount to only 10–30% of the gene, are remarkably conserved. They are made up of units containing 54 base pairs (bp) with some dimers of 108 bp. Slight variations can occur with the loss of 9 bp, i.e. exons of 45 or 99 bp. It has therefore been proposed that the triple helical region of collagen evolved by the duplication of an ancestral 54 bp unit. This unit codes for $(Gly–X–Y)_6$. However, recent studies on the Type IX collagen DNA failed to reveal this ancestral 54 bp unit. Hence this mechanism may not be the evolutionary basis of all collagens, but perhaps only of the fibrous collagens.

The size of the introns is highly variable, ranging from 80–2000 bp. As expected, the exons coding for the propeptide are not multiples of nine, but surprisingly the exons continue into the triple helix and are not interrupted by an intron at the junction of propeptide and triple helix.

BIOSYNTHESIS

Before examining the actual components and mechanisms of collagen synthesis, it is worthwhile briefly to review the process of protein synthesis in general.

The first step in protein synthesis is for the code for that protein contained in the large DNA molecules of the chromosomes to be transcribed into a mobile version which can then travel from the nucleus to the protein synthetic machinery in the cytoplasm (Fig. 3.1). This process, called transcription, involves the synthesis of a new nucleic acid molecule (ribonucleic acid—RNA) from the DNA template in the nucleus. This RNA is called the precursor messenger RNA (pre-mRNA) and contains copies of both the exons and introns. The

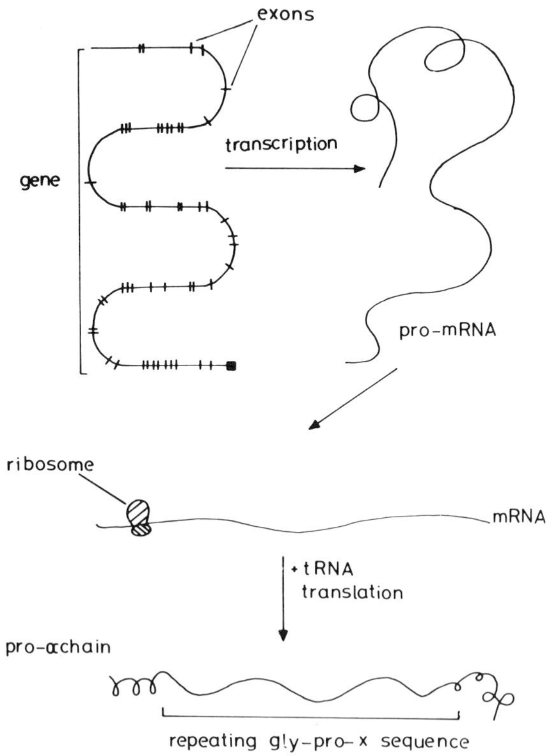

Fig. 3.1 Flow diagram showing the structure of a collagen gene and the pathway for transcription and translation of the DNA code.

pre-mRNA is subsequently modified by enzymic excision and splicing of exons to form the messenger RNA (mRNA), the component which carries the code from the DNA to the cytoplasm.

The mRNA is transported from the nucleus to the external surface of the rough endoplasmic reticulum where it couples with ribosomes to form a polysome (Fig. 3.2). The ribosomes are complex organelles which contain enzymes capable of catalysing the series of reactions involved in the production of the polypeptide chain from the coded message. A further nucleic acid species is now bound to the ribosome/mRNA complex (polysome) which is known as transfer RNA (tRNA). The cell contains a range of these small molecules,

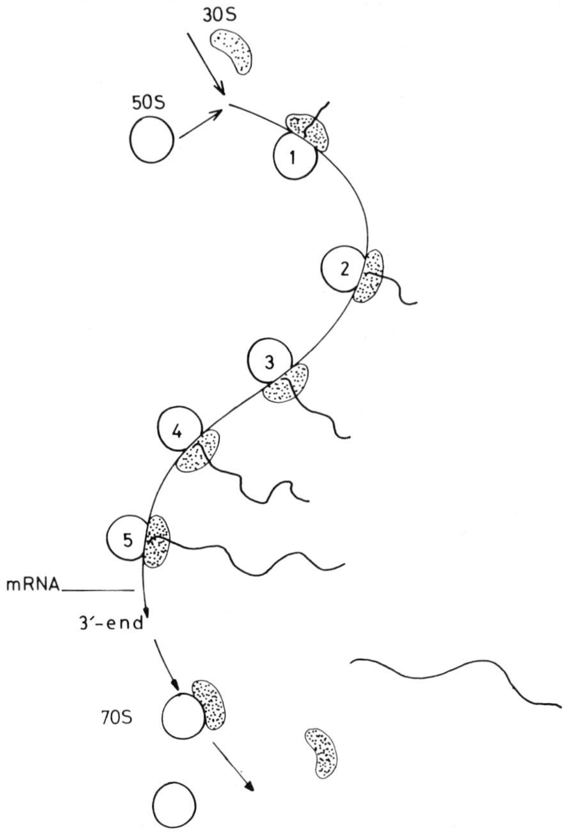

Fig. 3.2 Formation of a polysome.

each of which is able to bind a different amino acid. Thus, as each of the two tRNA binding sites on each ribosome is made specific for only one amino acid by the small region of three bases (the codon) in the nucleic acid (mRNA) located at that point, one amino acid-bearing tRNA molecule is bound to each site. An amino acid is then transferred from one tRNA to the other and a peptide bond is formed, thereby completing the synthesis of a dipeptide, i.e. the first two amino acids of the new polypeptide chain. The ribosome then moves one codon along the mRNA molecule, thus revealing a new binding site for another tRNA molecule carrying the next amino acid. The dipeptide is then transferred to the new amino acid and another peptide bond is formed. In this way the nucleic acid code is translated into protein and the whole polypeptide is synthesised. The process, therefore, is called translation (see Fig. 3.3).

The fully translated polypeptide is then released from the ribosome and passes into the rough endoplasmic reticulum. Here it can be further modified, if necessary, by local enzymes and can then be

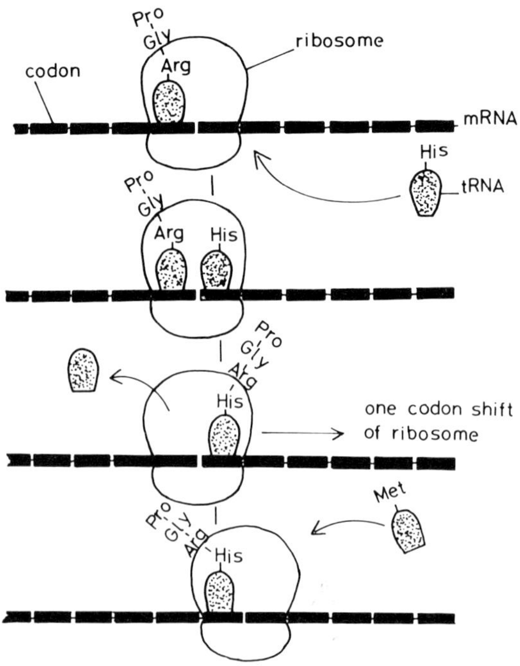

Fig. 3.3 Flow diagram showing the details of translation.

formed into a native molecule either alone, if that native molecule consists of only one chain, or by the association of several chains as in the case of polymers. The native molecule is then transported to the Golgi apparatus if it is to be glycosylated and then, if it is an extracellular glycoprotein, it is transported to the cell surface and secreted into the extracellular space.

Transcription of mRNA and translation of procollagen chains

DNA coding for collagen is fully transcribed to form a pre-RNA which corresponds to the full intron and exon complement of the gene. The large complex nature of the collagen genes means that processing of the pre-mRNA to yield the final mRNA involves at least 50 excision and splicing events. In addition, other reactions such as capping, polyadenylation and methylation are involved. During processing of the pre-mRNA the introns form large loops (called R-loop), allowing close association of the exons in a linear structure, thus facilitating the synthesis of mRNA on the exon template when the double-stranded DNA unwinds. These R-loop structures on the genes have been 'mapped' by electron microscopical techniques which have assisted in the interpretation of biochemical data on the structure of these very large genes. The transcription of a gene for collagen can give rise to several mRNAs of different lengths. In the case of human pro-α1(I) and pro-α2(I), two and three mRNAs, respectively, have been identified.

It should be remembered that genes and, therefore, each mRNA molecule, only code for one polypeptide chain and not whole collagen molecules. The newly synthesised mRNA contains the necessary code for the whole of one procollagen polypeptide chain. mRNAs for procollagen α-chains are very large due to the size of the polypeptides they code for. Because of this, procollagen mRNAs have some secondary structure established through limited base pairing between guanidine and cytosine residues. This secondary structure may be the reason for the rather slow rate of translation of procollagen polypeptides (see below), although the high demand for glycyl-tRNA (due to the high glycine content of the procollagen α-chains) and the slow passage of the very large newly synthesised polypeptide through the rough endoplasmic reticulum may also be causes.

After the translatable mRNA is synthesised in the nucleus it is transported to the cytosol where it forms polysomes (Fig. 3.2) in association with ribosomes and procollagen chains are synthesised.

Subsequently, the chains pass into the cisternae of the rough endoplasmic reticulum where they are modified and assembled to form procollagen molecules. As in the case of other secretory proteins, the peptide chain is synthesised with a hydrophobic N-terminal sequence, referred to as a signal peptide, of about 100 amino acids. This peptide is believed to bind to the rough endoplasmic reticulum to facilitate transport of the chain across the membrane. The signal or pre-peptide is then enzymically removed to give procollagen chains.

Post-translational modification of procollagens

In common with many other proteins, especially extracellular proteins, collagen is modified by specific enzymes after translation. Procollagen is unusual in this respect, however, in that it is modified by a total of eight enzymes. These changes include the hydroxylation of proline residues, hydroxylation of certain lysine residues, glycosylation of some of these hydroxylysine residues, and cleavage of amino terminal (pN) and carboxyterminal (pC) procollagen propeptides.

The first of these modifications, which is required for stable triple helix formation, is the hydroxylation of proline. Proline residues in the newly synthesised procollagen polypeptide are hydroxylated by the enzyme 4-prolyl hydroxylase. This step is required before stable triple helices can be formed because of the involvement of hydroxyproline in the formation of non-covalent bonds within the helix. It is thought that the enzyme, which is bound to the cisternae of the endoplasmic reticulum, hydroxylates nascent procollagen polypeptides while they are still attached to the polyribosome and as they pass through into the cisternae of the endoplasmic reticulum (Fig. 3.4).

Only proline residues in the Y position of the Gly–X–Y triplet structure can act as a substrate for the enzyme. The nature of the amino acid at position X can have an effect on the rate of hydroxylation; the highest activity for the enzyme is when proline is in the X position, but activity is twenty times slower when glutamic acid occupies this site. The enzyme is not active on triple helical procollagen. Ascorbic acid is required for the reaction to occur, although there is some confusion about the role of the vitamin. It seems most likely that it binds to the enzyme and retains its reduced state, which is required for the full activity of the enzyme. Vitamin C deficiency certainly leads to defects in connective tissues, indicating the important role of this co-factor in the reaction.

The second post-translational modification of the newly formed

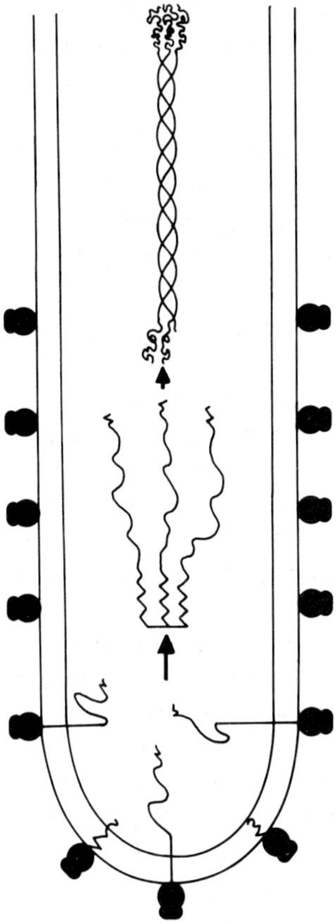

Fig. 3.4 Diagrammatic representation of the passage of collagen chains into the rough endoplasmic reticulum and the formation of the triple helix.

procollagen polypeptides is the hydroxylation of certain lysine residues. This hydroxylation reaction is very similar to that for proline in that molecular oxygen, non-haem iron and ascorbate (or a similar reducing agent) are required, but a separate enzyme, also located in the rough endoplasmic reticulum, is involved (lysyl hydroxylase). Fewer lysines than proline residues are hydroxylated, 10–20 compared with 100. As with 4-prolyl hydrolyase, the enzyme requires the substrate lysine to be in the polypeptide at the Y position. The activity of the enzyme, and therefore the number of hydroxylysine residues modified, appears to be controlled by the rate of triple helix formation in the endoplasmic reticulum, as the enzyme is not active on native procollagen. Hydroxylation of the lysine in the terminal non-helical regions of the collagen molecule is critical in that this will determine the type of bifunctional cross-link formed after fibre formation and, therefore, the stability of the tissue (see Chapter 4). It has been suggested that there may be a specific lysyl hydroxylase for the non-helical collagen lysines, as the extent of hydroxylation of these lysines compared to that of helical lysines varies with age. However, no separate enzyme has yet been demonstrated.

The other modifications which must occur before the procollagen molecule is complete are glycosylation of some of the hydroxylysine residues and formation of the triple helix. There are two enzymes involved in this last step before triple helix formation, galactosyltransferase and galactosylhydroxylysineglucosyltransferase. These two glycosylating enzymes add galactose to hydroxylysine and glucose to galactosylhydroxylysine, respectively. This means that the carbohydrate-bearing parts of the main repeating Gly–X–Y triplet section of the procollagen α-chain are hydroxylysine residues and that only two types of glycosylated hydroxylysine exist. These are *O*-β-D-galactosylhydroxylysine and 2-*O*-α-D-glucosyl-*O*-β-D-galactosylhydroxylysine (see Chapter 2). Few of these residues are formed in most collagen types although the amount present is both tissue and collagen type-dependent. Type I collagen contains less (10–20%) of these residues than Type II collagen (60–70%), which also contains less than type IV collagen (approximately 90%). On the other hand, some Type I collagens contain more carbohydrate than others, e.g. corneal and bone Type I contain more than skin Type I collagen.

A great deal is now known about the two hydroxylysine glycosylating enzymes of procollagen, for example the galactosyl transferase requires a divalent cation and UDP-galactose for activity and the galactosylhydroxylysineglucosyltransferase also requires a divalent cation for activity and has a sulphydryl group at its active site. However, the most important fact about both enzymes in relation to this review is that they have both been shown to be located in the rough endoplasmic reticulum. Thus, there is now little doubt that, like the two hydroxylating enzymes, they act on the hydroxylated, newly synthesised pro-α-chains.

Glycosylation of the procollagen molecules is not confined to the triple helical hydroxylysines, oligosaccharide units being transferred to the asparagine residues in the non-helical propeptides. These oligosaccharides are preformed units of mannose and *N*-acetylglycosamine, the size probably varying between collagens. The function of these asparagine-linked carbohydrates is unknown.

The final step within the rough endoplasmic reticulum is the formation of the completed triple helix. Its formation from the pro-α-chains appears not to require the aid of an enzyme but seems to occur spontaneously between the modified procollagen α-chains in the cisternae of the rough endoplasmic reticulum. Free α-chains of native collagen will reassociate after denaturation into triple helices *in vitro*, but the rate of re-formation is very slow and is dependent on random collisions of the molecules. This mechanism is too slow to be physiologically meaningful. The most likely process, favoured by researchers at present, is that the modified pro-α-chains in the rough endoplasmic reticulum associate initially at the C-terminal end through binding of the pC-propeptides and formation of intrachain disulphide bonds after release of the polypeptide chain from the ribosomes (Fig. 3.4). After association and bonding of the pC-propeptides the three pro-α-chains wind up to form the triple helix, and the last step is the binding together of the pN-propeptides. Certainly, disulphide bond formation seems to be the rate-limiting step, as inhibition of this step with disulphide reducing agents drastically reduces the rate of helix formation. The formation of these disulphide bonds may be spontaneous or involve an enzyme, protein disulphide isomerase. This enzyme is present in embryonic tissues and decreases with growth in line with reduction of prolyl hydroxylase activity, but although highly likely to be involved there is no direct evidence for its involvement in collagen synthesis.

Structure of procollagens

As we have seen, all of the major types of collagen are synthesised as larger precursor molecules. These molecules are called procollagens and consist of the triple helical oligomer with short non-helical (telopeptide) domains at either end plus large mainly globular regions at both the N- and C-terminals (Fig. 3.5).

Procollagen peptides are represented by the prefix p- so that the extra non-helical domain at the N-terminal end of the triple helical part of a newly synthesised procollagen molecule is known as the pN region. The extra peptide at the N-terminal end of procollagen is called the pN propeptide, and that at the C-terminal end is called the

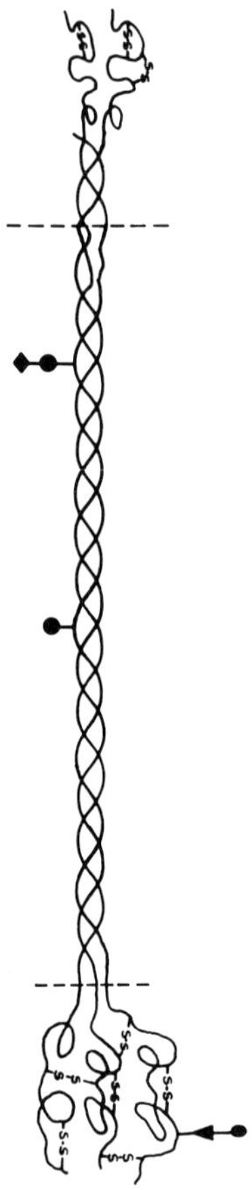

Fig. 3.5 The generalised structure of procollagen.

pC propeptide. Also, a procollagen α-chain which has lost its pC region and only contains the extra peptide at the N-terminal end is known as pN $\alpha 1(x)$ (where x is a Roman numeral representing the particular collagen type). Similarly, a procollagen α-chain which only contains the pC propeptide is known as pC $\alpha 1(x)$.

The illustration in Fig. 3.5 shows the generalised structure of the family of procollagen molecules. The pN propeptide is attached to the main body of the molecule through a peptide bond between a proline and a glutamine residue. On the C-terminal side of the dipeptide is the short non-helical structure which separates the pN propeptide proper from the triple helix and which remains after cleavage of the propeptide as the non-helical domain of the complete collagen molecule. The propeptide consists of a collagenase-sensitive collagen-like region of about 40 residues at its C-terminal end and a non-collagenous, collagenase-insensitive sequence of approximately 100 residues at its N-terminal end. Collagenase, a highly specific bacterial protease, only cleaves polypeptides with the triplet Gly–Pro–Y sequence and so is a useful diagnostic agent for triple helical domains of collagen.

The non-collagenous domain of the pN propeptide of Type I collagen has a high content of aspartic acid and glutamic acid which comprise about one-third of the total residues. It also has a high cysteine content, with about five internal disulphide bonds, low proline and glycine and no hydroxylysine. The equivalent peptide in Type III procollagen is very similar in structure; the non-collagenous domain contains 97 residues and the collagenous domain contains 33 amino acids.

The pC propeptide, like the pN propeptide, has a very different amino acid composition from the collagen triple helix. It is resistant to collagenase attack and is therefore non-triple helical. It is relatively poor in its content of glycine and has no hydroxyproline or hydroxylysine, but is high in glutamic acid, aspartic acid, cysteine, tyrosine and tryptophan. Each pC propeptide also contains two *N*-acetylglucosamine and nine mannose residues. The pC propeptide from chick tissues contains 246 amino acids, with seven methionines, eight cysteines and has a molecular weight of 100 000. The propeptide from rat tissues is very similar with regions of high homology. It contains 245 amino acids, three of which are methionines and seven cysteines. The cleavage point for this peptide from the main body of the collagen molecule has been shown to be between an alanine and an aspartic acid residue.

Secretion of the procollagen molecule

The route for the secretion of the completed procollagen molecule has been shown to be similar to that taken by all other extracellular secreted proteins. It travels from the rough endoplasmic reticulum to the smooth endoplasmic reticulum; from there to the Golgi apparatus and then, in secretory granules, to the cell surface. Again, the exact mode of the transport mechanism is not yet known and any further action on the protein in the smooth endoplasmic reticulum and Golgi complex has not been elucidated. The molecules may form mini-fibres in the secretory granules in which they are transported to the cell surface. Several researchers have reported the presence of banded aggregates in these granules indicating that the procollagen molecules do, in fact, form precursor fibrils before secretion, but the observed formation of these aggregates is limited and needs confirmation (see for example Fig. 4.1).

The secretion of procollagen involves the microtubular microfilament system of the cell. However, whatever their conformation within the granules or the exact mechanism of their secretion, the procollagen molecules are liberated at the cell surface into the extracellular space by exocytosis, where the next step in their modification takes place.

Extracellular modifications: procollagen propeptidases

Collagen molecules are produced from the precursor procollagen molecules in the extracellular space by the action of two specific proteolytic enzymes which cleave off the pN and pC propeptides. Both enzymes act independently of each other and show different rates of cleavage for propeptides from $\alpha1$(I)-chains as compared to $\alpha2$(II)-chains. The pC-propeptidase appears to act in a stepwise manner and it has been suggested that more than one version of each of the propeptidase enzymes may exist for the different collagen types. It has been shown that in the case of Type I collagen the pN-propeptidase acts before the pC-propeptidase. Thus, the two propeptides are cleaved in a stepwise manner, the pN region first and the pC region second. This sequence is not obligatory and Types III and V collagen appear to have a reversed mechanism of processing.

The site of cleavage for both enzymes has been investigated and good data is available on the peptide bond involved at the N-terminal end of the procollagen molecule. The bond between a proline and glutamine is hydrolysed N-terminus to the glutamine, which then becomes the new N-terminus of the partially processed procollagen molecule. Other –Pro–Glu– bonds exist in the collagen molecule but it is thought that this one is recognised by the pN-procollagen peptidase because of the low glycine content in the area and high concentration of large hydrophobic amino acids.

The more complex stepwise action of the pC-propeptidase is not yet understood, although it is thought that this mechanism may be required in Type I collagen in order to cleave one pro-$\alpha1$(I) and the pro-α2(I) so that the second pro-α1(I)-chain is revealed for hydrolysis. This mechanism has yet to be fully resolved. There has been a suggestion that, in the fibrous collagens, the pN-propeptide may remain uncleaved from the collagen molecule after fibre formation has started as a control of fibre growth. If this is the case then the control mechanism of propeptide cleavage is far from resolved. This issue will be discussed further in the next chapter.

Partially processed collagens

As outlined in Chapter 2, not all collagens are processed by extracellular propeptidases in the same way as are the fibrous types (Types I, II and III). Types IV and V collagens are examples of these types, with Type IV being perhaps the best understood at present. These collagens require the extra propeptides to remain an intact part of their structure in order to form the insoluble extracellular matrices in which they are involved. Biosynthetic and electron microscopic studies of the non-helical structure of Type IV collagen have led to the conclusion that the pro-α-chains are not processed or not *fully* processed by peptidases after being secreted from the cell. The non-processed nature of these terminal regions together with the non-repeating charge profile of the helical portion is believed to be the cause of their inability to form fibres, and thus the reason why they adopt the more amorphous lattice structure that has been proposed for this basement membrane collagen.

Type V collagen too may exist *in vivo* as a partially processed procollagen species. Evidence for this has come from recent work which has shown that the pN-propeptide may be retained following secretion.

REGULATION OF SYNTHESIS AND SECRETION OF PROCOLLAGEN

Because of the unusually complex pathway for synthesis and modification of the procollagen molecule and the large number of enzymes involved in its production, it is not surprising that there are many sites at which regulation and control may be exerted. Control may be exerted at the level of the gene, during transcription, mRNA transport, mRNA turnover, polysome formation, tRNA availability, peptide bond synthesis, entry of polypeptides into the rough endoplasmic reticulum, any of the hydroxylation or glycosylation steps, triple

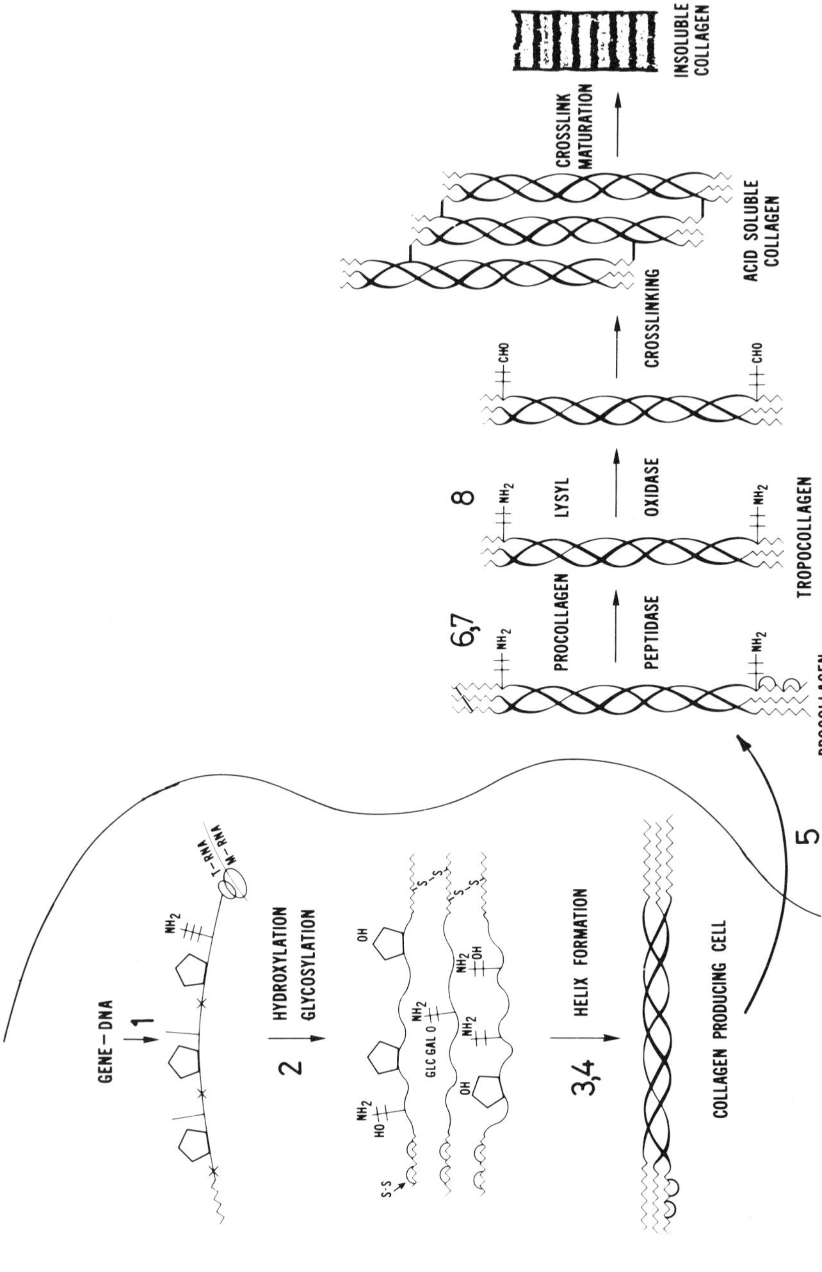

Fig. 3.6 Flow chart showing an overview of collagen synthesis, modification and secretion and the points at which control of the process may be exerted.

helix formation, modification in the Golgi, transport to the cell surface, exocytosis, or during extracellular procollagen processing. In this brief list we have stated fourteen points at which the synthesis of the procollagen molecule may be perturbed, affected or controlled but, in reality, it seems likely that only one or two major steps act as important rate-limiting reactions or control points. These include steps such as the hydroxylation of proline which is crucial for helix stability and the formation of disulphide bonds in associated propeptides to initiate helix formation (Fig. 3.6).

Evidence exists to show that control is also exerted at the level of transcription. In studies of Rous sarcoma virus (RSV)-transformed chick embryo fibroblasts it was shown that lower quantities of Type I procollagen mRNA were present after transformation, indicating control at the level of transcription. Similarly, γ-interferon inhibits RNA synthesis and thereby decreases its levels. A similar situation was illustrated in transformed chick chondroblasts which demonstrated a reduced synthesis of Type II procollagen concomitant with a lowered level of the mRNA for the protein. However, in these cells the mRNA for Type I procollagen was seen to accumulate although Type I procollagen itself was not synthesised. Regulation of gene expression could take place at any one of the splicing steps. There may be two kinds of regulation, through the chromatin structure around the gene or by proteins that react with the promoter regions of the gene.

There is also a large body of evidence, albeit somewhat circumstantial, to suggest that the cleaved pN-propeptide may re-enter the cell and act as a feedback inhibitor of collagen synthesis at the translational level. Another debated mechanism of control is by intracellular degradation of newly synthesised procollagen. Good evidence exists to support the view that a high proportion of procollagen is degraded before it even reaches the cell surface, and some estimates have put this figure as high as 15%. This process may be to remove defective pro-α-chains. It should be stressed that demonstrating these mechanisms unambiguously through experiment has proved extremely difficult and so elucidation of the role of such controls must await more definitive studies.

DEGRADATION

The collagen triple helix is highly resistant to the action of most proteinases, but connective tissue cells have been shown to synthesise and secrete a group of proteinases that act on collagen and function under physiological conditions and are, therefore, almost certainly important in matrix metabolism. These proteases can be divided into four classes: aspartic, cysteine, metallo- and serine active sites.

The best characterised protease is the mammalian collagenase which is a Zn^{2+}-metallo-endopeptidase capable of cleaving the collagen molecule specifically across the three chains to give two triple helical fragments (Fig. 3.7). Fibroblasts, chondrocytes and related cells have been reported to synthesise and secrete collagenase. Following synthesis the leader sequence of the preprocollagenase is processed and the procollagenase secreted, there being little evidence of any storage of procollagenase within the cell. Activation of this latent enzyme is

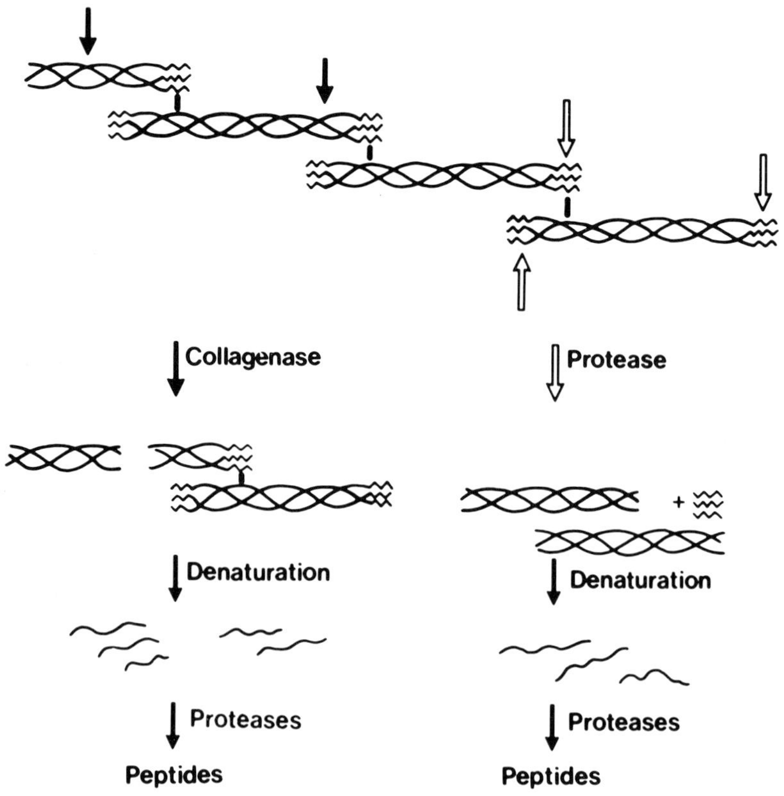

Fig. 3.7 Diagram showing the fragmentation of native collagen molecules by mammalian collagenase and proteases.

achieved by proteolysis of the extension peptide, as occurs with several extracellular enzymes. Many proteases, for example trypsin, plasmin and mast cell proteases, are capable of acting in this way on the enzyme in the extracellular space but the precise mechanism is as yet unknown. Further studies have indicated that the activator of procollagenase is also synthesised as a pro-activator, which in turn requires activation by another enzyme.

Once activated, the collagenase acts on the extracellular collagen fibril by binding to the surface of the collagen fibril in the gap zone. The enzyme is capable of cleaving all the fibrous collagens, although Type II is cleaved at a slower rate than Types I and III, which are cleaved at approximately the same rate. It has been established that the binding of the enzyme is conformationally dependent since it acts on the collagen triple helix at a discrete locus across all three peptide chains, but has little action on the same site on individual α-chains. Cleavage occurs between the Gly–Ile peptide bond at residues 775–776 in the α-1(I)-chain and Gly–Leu in the α-2(I).

cleavage bond
↓
chick α1(I)-Pro–Gly–Pro–Gln–Gly—Ile – Ala–Gly–Ala –
chick α2(I)-Hyp–Gly–Pro–Gln–Gly—Leu–Leu–Gly–Gln–

This locus is three-quarters the length of the molecule from the amino terminal end, thus yielding two triple helical fragments, designated TC^A and TC^B (Fig. 3.7). Both TC^A and TC^B fragments retain their triple helical character but have a denaturation temperature below that of the intact molecules (39°C) and therefore denature at physiological temperatures. The denatured chains (gelatin, see Chapter 10) are readily digested by most proteases, in contrast to the enzymic resistance shown by the collagen triple helix.

The limitation of cleavage to one site and the inability of the enzyme to cleave other Gly–Ile bonds along the molecule indicates a conformation specificity for the enzyme as described earlier and, therefore, a particular orientation of the peptide bond within the helix. The amino acid sequence around the area is low in charged groups and the reduced imino acid content reduces the stability of the helix. In contrast the N-terminal side of the cleavage point is a region of high stability. Presumably the looser helix facilitates cleavage of the Gly–Ile peptide bond. The variation in rates of cleavage of the different fibrous collagens may be due to a difference in the stability of the helix in this region.

It is interesting that the neutral collagenase has no effect on the non-fibrous collagens, Type IV, or on Type V. Type IV has been shown to be cleaved by a different collagenase isolated from a tumour which cleaves in a similar manner to the neutral collagenases across the triple helix but nearer the N-terminus. Similar enzymes have now been isolated from various normal cells, all of which are metalloproteinases that are secreted as proenzymes which can be activated locally by enzymes such as plasmin. An enzyme has also been isolated that specifically cleaves Type V collagen. The minor collagens have not been extensively studied but Type IX collagen has also been shown to be resistant to neutral collagenase. It is probable that these different types of collagen each have a specific collagenase.

The latent collagenase could exist as a true zymogen, that is a proenzyme, or it could exist as an enzyme/inhibitor complex. There has been considerable controversy over these mechanisms but it is now generally agreed that the procollagenase exists as a true zymogen. Collagenases are synthesised by a wide variety of cells in tissue, e.g. fibroblasts and macrophages, but very little can be extracted directly from tissues. The enzyme is therefore usually extracted after synthesis *de novo* in culture. Yields are low but can be increased by the use of specific stimulating factors.

During development and maturation the collagen fibre is increasingly stabilised by covalent intermolecular bonds (cross-links), as discussed in Chapter 4, and therefore degradation of the fibre is more complex. Cross-linking effectively locks the molecules within the fibre by attaching the molecules by both end regions to adjacent molecules. Action of the collagenase in cleaving the three polypeptide chains does not, therefore, release the fragments from the fibre and thereby permit denaturation, since both fragments are still attached to adjacent molecules through their cross-links. Release from the fibre can only be achieved by the action of other enzymes on the non-helical region around the cross-link residue. Degradation of the collagen fibril thus requires the concerted action of at least two enzymes at different locations in the molecules (Fig. 3.8).

Cleavage of the non-helical region around the cross-link site is achieved by the neutral proteases. Cysteine, aspartic and serine proteases act at neutral pH on the non-helical end regions of the molecule between the cross-link and the triple helix, thereby de-

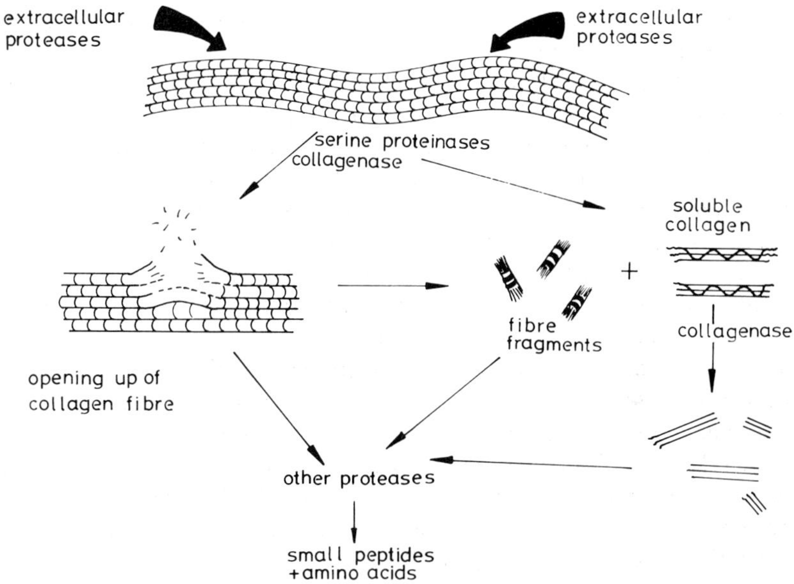

Fig. 3.8 Diagram showing the concerted action of collagenase and proteases on collagen fibres.

polymerising the collagen molecules within the fibre. Two serine proteases secreted by leukocytes, cathepsin G and elastase, also readily degrade collagen. The released collagen molecules are slowly denatured at 37°C and are then rapidly attacked by most proteinases. Prior action of neutral collagenase would release the fragments TC^A and TC^B rather than whole molecules, which would then rapidly denature at 37°C. The relative importance of these two extracellular routes has not yet been established with any certainty, but clearly both mechanisms must be operative in the degradation of a highly cross-linked fibre.

Subsequent digestion of the fragments released by the neutral collagenase and proteinases can occur following endocytosis (Fig. 3.9). In this process the fragments are taken up by cells in vacuoles which are then combined with or incorporated in phagolysosomes. Enzymes such as cathepsins B, L and N in these organelles may then fully hydrolyse the collagen fragments. These lysosomal cysteine proteinases are all capable of degrading collagen. Their optimal activity is around pH 3·5–4·0 and they are therefore expected to confine their

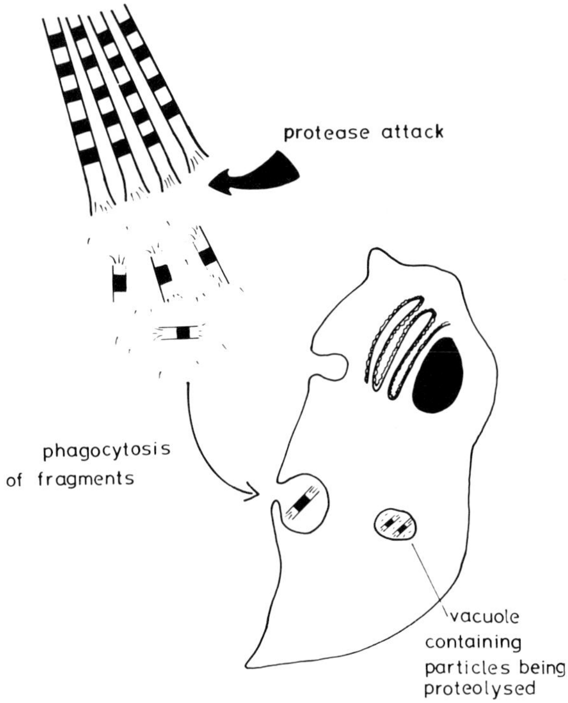

Fig. 3.9 Schematic representation of phagocytosis.

activity to the phagolysosome where conditions of relatively high acidity can be tolerated. Collagen, once released from the fibre, would be unstable at the pH and temperature within the lysosome. On the other hand, there is evidence that the cathepsins can act extracellularly in specific environments, e.g. the ruffled border of osteoclasts. Like the neutral serine proteases, they are capable of cleaving the non-helical regions, thereby depolymerising the collagen fibre.

These enzymes are controlled in the tissue by two types of inhibitors. The plasma inhibitors are $\alpha 2$ macroglobulin and β-anticollagenase, whilst the tissue contains inhibitors of the metallo-proteinases and have been designated TIMPs (tissue inhibitors of metallo-proteinase). The activity of these collagenases *in vivo* will therefore depend on the balance between local concentrations of the activating factors and the tissue inhibitors.

Mechanism of degradation *in vivo*

The physiological mechanism of degradation clearly involves a complex series of enzymes derived from different cell types. These cells may well be tissue specific, for example osteoclasts in bone, or they may be migratory cells such as macrophages. Since it is generally agreed that both the mammalian collagenases and the neutral proteinases must play a role in the degradation of the fibre, we will assume both act at the same time, although the relative effectiveness of each is not yet known. The action of these enzymes results in the release of fragments which may range from TC^A and TC^B to large fragments of cross-linked fibres. The latter may be produced due to lines of weakness in the fibre or a highly localised and continuous action of the collagenase through the fibre. The smaller fragments will denature and be readily catabolised by most proteinases, but the large fibrillar fragments will be removed by phagocytosis (Fig. 3.9). Fibrillar fragments within the cell can readily be observed in tissue undergoing rapid resorption, when examined in the electron microscope. Once inside the cells the collagen-containing phagosome will fuse with the lysosomes, and the lysosomal acid cathepsins will complete the degradation.

Although there are many detailed aspects of collagen degradation that are not understood, particularly with respect to the relative importance of the different enzymes and of their inhibitory factors, the overall scheme appears to be well established. The mechanism of degradation of the Type IV collagen of basement membrane is less clear. Under normal physiological conditions degradation may be achieved by the 'turnover collagenases' described earlier, and supplemented by the neutral protease.

Kinetics of turnover

The overall rates of collagen turnover were initially based on hydroxyproline excretion in urine, or radioactive pulse labelling, and these studies indicated a rapid turnover during growth, but that collagen was virtually inert in the adult animal. However, more recent studies have demonstrated not only a higher rate than previously believed, but a considerable variation in the rates of turnover of different collagen tissues. For example, skin has been reported to have a turnover of 5% per day and lung 10% per day, which are ten times the rates reported earlier. These studies have used new techniques to avoid the problem of the recycling of radioactive labels and of the estimation of the pool size in pulse labelling studies. There are considerable difficulties in determining these rates. It may be that the newly synthesised collagen is degraded at a high rate prior to incorporation into the fibre, so that only a small proportion of the total collagen synthesised is normally incorporated into fibres. The rate of turnover of the mature fibres is difficult to establish from these studies. Again, the early studies indicated that they were degraded but studies on the rapid growth of animals following starvation clearly indicated little, if any, degradation of the pre-existing mature collagen.

The kinetics of turnover were further complicated by the report that a high proportion, 10–30%, of the newly synthesised collagen was degraded *intra*cellularly prior to secretion from the cell. This apparently inefficient collagen production would clearly make a major contribution to urinary hydroxyproline excretion values which would suggest a higher level of turnover of existing collagen than was actually occurring. Establishment of the relative importance of the intra- and extracellular modes of degradation could be further complicated by the different rates in different tissues. Furthermore, the rates of metabolism of the different collagen types introduced another variable into the kinetics. On the basis of current studies the turnover of skin and tendon is small, bone collagen somewhat higher, whilst, at the extreme, the periodontal ligament has a turnover time of a few days. As a generalisation the turnover can be considered to be high during growth, but reaches a low level plateau at maturity.

SUMMARY

Different collagens are synthesised in specialised cells using a range of large genes to provide the genetic code for the twelve or more types identified to date. The genes comprise unusually well conserved exon structures which contain 54 bp, 45 bp (54 − 9 bp), 108 bp (54 × 2 bp) or 99 bp (54 × 2 − 9 bp) and very large introns. The newly synthesised procollagen chains are inserted into the rough endoplasmic reticulum where they are acted upon by enzymes which hydroxylate proline and lysine, and glycosylate selected hydroxylysine residues. The modified pro-α-chains then associate to form trimers in triple helical form with globular terminal domains and are transported to the smooth endoplasmic reticulum and Golgi apparatus before secretion to the extracellular milieu. Here the new molecules are processed by specific proteolytic enzymes which cleave the extra peptides from the triple helix (not in the case of Type IV and possibly Type V collagens) and the molecules then spontaneously form their extracellular macromolecular structures.

The degradation of the supramolecular aggregates of collagen involves a complex series of enzymes acting in concert on both the normally resistant triple helix and on the non-helical terminal regions to depolymerise the fibril. Degradation is then completed within the lysosomal system in the cell. Collagen possesses an extremely long biological half-life compared to most other proteins. Recent results have indicated a higher turnover rate for collagen than previously believed and have also demonstrated a considerable variation in these rates between tissues. Little research has been carried out on the relative rates of turnover for the different collagen types.

FURTHER READING

1. Duance, V. C. & Bailey, A. J., Biosynthesis and degradation of collagen. In *Handbook of Inflammation,* Vol. 3, ed. L. E. Glynn. Elsevier/North Holland Biomedical Press, Amsterdam, 1980, pp. 51–109.
2. Tolstoshev, P. & Solomon, E., Collagen genes. *Nature,* **300,** 1982, 581–582.
3. Bornstein, P. & Sage, H. Structurally distinct collagen types. *Ann. Rev. Biochem.,* **49,** 1980, 957–1003.
4. Kivirikko, K. I. & Myllyla, L. In *The Enzymology of Post-translational Modification of Proteins,* ed. R. B. Friedman & H. C. Hawkins. Academic Press, New York, 1981.
5. Evered, D. & Whelan, J. (eds), Fibrosis. Ciba Foundation Symposium 114. Pitman, London, 1985.
6. Bailey, A. J. & Etherington, D. J., *Metabolism of Collagen and Elastin in Comprehensive Biochemistry,* Vol. 193, ed. M. Florkin & E. M. Stotz. Elsevier, Amsterdam, 1980, pp. 299–460.
7. Woolley, D. E. & Evanson, J. M. (eds), *Collagenase in Normal and Pathological Connective Tissue.* John Wiley & Sons, Chichester, 1980.
8. Barrett, A. J. (ed.), *Proteinases in Mammalian Cells and Tissues.* North Holland, Amsterdam, 1977.

Chapter 4

Fibre Formation and Stabilisation of Collagen

INTRODUCTION

A major biological characteristic of the collagen family of extracellular proteins is their extraordinary strength. Most proteins exist as single molecules or as small oligomers made up of two, four or up to ten polypeptide chains. These are the globular proteins and include all enzymes, most intracellular proteins and most membrane proteins. These proteins and glycoproteins have little or no intrinsic strength and, even as aggregates, do not fulfil roles in life of a supportive or strength-giving nature. Collagen, on the other hand, like the few other fibrous connective tissue macromolecules (e.g. elastin, fibrin, chitin, cellulose), is unusual; it is able to form long, semi-rigid, rod-like molecules, each composed of three polypeptide chains, and these molecules are capable of binding together to form large, insoluble structures outside the cell masses of the body.

Many proteins and other biological macromolecules can form aggregates and insoluble complexes through, for example, heat denaturation (egg white coagulation), acid denaturation (casein aggregation) or through the influences of physico-chemical effects (e.g. freeze drying or gluten formation in dough). However, few of these aggregates have any physical strength, due largely to the random nature of the formation of the supramolecular complexes. In these cases the protein molecules merely unfold into the denatured state and random, non-covalent bonding occurs between 'uncovered' sites which were formerly involved in retention of the native conformations of the proteins.

Collagen, as we have seen in previous chapters, has an unusual structure. The unique primary, secondary and tertiary structure of the collagen molecules is designed to allow highly specific interactions between molecules which lead to the formation of well-defined, ordered and reproducible structures. These can be either fibres, as seen in the major forms Types I, II and III collagens, or structures like the 'chicken-wire' mesh proposed for Type IV basement membrane collagen (see Chapter 2).

Once the specific matrix structure has been laid down and the collagen molecules are aligned and close packed, a second mechanism comes into play. These structures, fibres and networks, have no intrinsic strength as they are retained, largely, by mechanically weak non-covalent bonding between ionic and hydrophobic amino acid side chains. However, to provide tensile strength specific lysine and hydroxylysine residues in the collagen molecules are modified by enzyme action to form aldehydes which can then form intermolecular covalent bonds known as cross-links. It is this family of cross-linking compounds and their derivatives that stabilises the ordered fibres and meshes of matrix collagen and endows these structures with their characteristic strength, rigidity and resistance to proteolytic attack. Furthermore, once these structures are stabilised by lysine-derived cross-links (and, in some collagens, disulphide bonds), non-enzymic reactions occur during physiological ageing of the collagenous tissues which lead to further changes in the cross-linking and, thereby, the stability and strength of the tissues.

This chapter will investigate the processes of matrix formation for both the fibrous and non-fibrous collagens, the mechanisms of cross-linking in the various collagen types and the physical properties of stabilised collagenous tissues. Also considered will be the changes which occur in collagen matrices during physiological ageing and the implications of these changes for the physico-chemical properties of collagen.

FIBRE FORMATION

In the previous chapter we reviewed the processes by which collagen is synthesised and modified before secretion from the cell into the extracellular milieu. We now need to consider the mechanism by which the fibrous collagens (Types I, II and III) form fibres and the nature of the molecular alignments within the fibres. The discussions of fibre formation below will be confined to Types I, II and III collagen only.

As described in Chapter 2, Types I, II, III and possibly V collagens form fibrous structures. As stated above, it is known that Types I, II and III all form fibrils which 'grow' into larger fibres. It is also known that fibre size varies between collagen types and with the physiological age of the animal. Finally, there is also some evidence of the formation of mixed fibres of two different genetic types (e.g. Types I and III and Types I and V collagens). What is still not yet entirely resolved is the mechanism by which:

(a) fibres form,
(b) fibres grow,
(c) fibre growth is controlled,

although, in all of these cases a great deal of evidence now points us to the likely answers to these questions.

The formation of Type I fibrils is a complex process which starts with the procollagen molecule. The principles of fibrillogenesis (the process of fibre formation and growth) have been extensively studied but this research has generally been carried out with acid-solubilised collagen under non-physiological conditions of pH and temperature changes. The growth of the fibrils is characterised by a lag phase, a growth phase and a plateau, and is considered as a nucleation and growth process. These studies have led to conflicting theories of the growth process. In one hypothesis, assembly occurs through a co-operative mechanism with the initial formation of nucleation centres, similar to actin and microtubule formation. In an alternative hypothesis it was proposed that fibril assembly proceeds through a steady coacervation mechanism. More recent studies on the kinetics of fibrillogenesis have demonstrated a critical concentration for the reaction, which supports the nucleation theory. Furthermore, these workers propose that the nucleating unit is the five-stranded microfibril. This unit can grow laterally by addition of other units.

The reprecipitated fibres formed *in vitro* are identical to the fibres found *in vivo,* but the mechanisms controlling fibre growth and diameter cannot be determined by these techniques. The major reason is that the N- and C-propeptides are probably involved in directing the initial stagger and possibly controlling diameter, and these reactions occur only under physiological conditions. Reprecipitation of procollagen in the presence of the N- and C-terminal peptidases would provide a better model but other factors are probably involved during *in vivo* fibrillogenesis.

As stated in Chapter 3, it is possible that procollagen molecules form pre-fibres or pre-fibrils within the 'packages' in which they are transported from the Golgi apparatus to the cell surface for secretion (Fig. 4.1). There are now several reports of the appearance of these 'pre-fibril' packages and at least one report of specific alignment of these structures. It has been proposed that procollagens may form SLS type structures within the secretory vacuoles and that these units may then aggregate to form staggered quasi-hexagonal lattices once they are secreted from the cell. This model obviously requires that groups of pre-aligned molecules, the existence of which is hypothetical, should specifically aggregate into fibrils. However, other researchers suggest that fibrils grow longitudinally through the alignment in quarter-stagger of two molecules, then three and then four, and so on until a long fibril is formed.

Both models suggest that control of fibre growth is through the cleavage of the pN-propeptide from the procollagen. In other words, all molecules at the outer surface of a fibril or growing fibre retain their pN-propeptides and this prevents further accretion of pN-procollagen molecules. Once the pN-regions at the outer surface of the fibre have been cleaved off by the pN-propeptidase more molecules

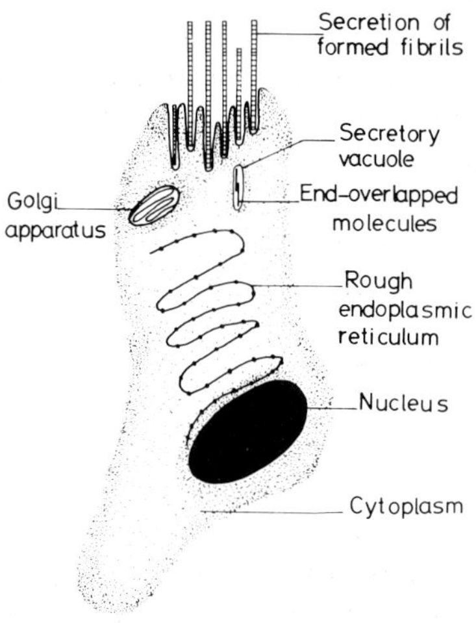

Fig. 4.1 Model of fibrillogenesis and secretion of collagen.

can be added to the fibre and its overall diameter can increase by the sum of two collagen molecule diameters. Thus, control of fibre growth is maintained through the concentration of the pN-propeptidase in the extracellular space. Evidence to support this hypothesis has come from the staining of growing fibres in young tissue with antibodies specific to the pN-propeptide. In this study it was shown that the antibodies did bind to growing fibres, indicating the presence of the pN-propeptide at the outer surface. In contrast, little or no binding was observed on the surface of the large fibres.

CROSS-LINKING OF COLLAGEN MOLECULES AND FIBRES

The high mechanical strength of collagen fibres is brought about through the formation within the fibres of intra- and intermolecular cross-links. These chemical bridges are covalent in nature and result from specific structural features of the collagen α-chains and the way molecules overlap each other in the highly organised manner described in Chapter 2. Without these cross-links the collagen matrix has little or no strength, as illustrated by consideration of diseases such as lathyrism, dermatosparaxis in sheep and cattle or Ehlers–Danlos syndrome in man where cross-linking is inhibited through structural abnormalities of the collagen molecules: the collagen has low mechanical strength and this results in lesions of the connective tissue which can be fatal if occurring in vital organs. Cross-linking, therefore, which represents another step in the post-translational modification of collagen, is a vital stage in the production of fully functioning collagen fibres. Indeed, even in non-fibrous collagens such as Type IV there is a need for cross-linking to produce an extensive stable framework.

Collagen molecules are cross-linked internally and to other collagen molecules by three distinct and well-described mechanisms. Firstly, there are disulphide bonds which are confined to a few types such as Type III and Type IV. Secondly, there are the important but simple divalent bonds linking two collagen chains in the same or different molecules which are formed from lysine or hydroxylysine aldehydes. Because these bonds can be reduced *in vitro* with reagents such as potassium or sodium borohydride these bonds have been called reducible cross-links, and will thus be denoted as such in this text. Finally, there are more complex bonds, joining more than two α-chains together, which are formed from the simpler, reducible divalent bonds described above. Because these cross-links form during the physiological maturation of collagen they have been referred to as mature cross-links and so they will be denoted as such here. Each of these cross-links will be reviewed in turn.

Disulphide bridges

As discussed in earlier chapters, most collagen types do not contain cysteine in the final processed form. Of the common forms only Types III and IV contain this amino acid and therefore are the only types capable of forming intra- or intermolecular disulphide bonds. It may be useful at this stage to define these two latter terms. Intramolecular cross-links are said to be those which form between α-chains within the same molecule. Intermolecular cross-links are those which form between α-chains in different molecules. In the latter case the different molecules may also be in different microfibrils and therefore may be known as interfibrillar cross-links.

The disulphide cross-linking of Type III collagen seems to be of a fairly simple nature. The cysteine residue responsible is located at the C-terminal end of the triple helix in the final triplet before the non-helical C-terminal domain (Chapter 2). Three cysteines are present in each Type III molecule, there being three α-chains in each molecule. Therefore, theoretically, it is possible for two of these residues to react intra-molecularly to form a disulphide bridge. This may be catalysed at the stage of helix formation in the endoplasmic reticulum by the enzyme protein disulphide isomerase. The third free cysteine side chain may then react with other cysteines in adjacent Type III molecules to form intermolecular cross-links. However, it has proved extremely difficult to demonstrate unequivocally the presence of such intermolecular disulphide bonds. Therefore, until further evidence is produced, it must be assumed that both intra- and intermolecular disulphide bonds may exist in Type III collagen.

The case of Type IV collagen is far more clear cut. In this collagen disulphide bonds are known to form between α-chains in the same molecule *and* in different molecules. A two-dimensional model for the quaternary structure of Type IV collagen has been proposed and is currently widely accepted. As described in Chapter 2, the Type IV collagen molecule forms tetramers with molecular weight of about two million. The four molecules are joined at their N-terminal ends by a short 30 nm region known as the 7S domain and extensive disulphide bonding takes place within and between molecules at this point. This forms a four-legged 'spider'. 'Spiders' are then thought to interact through their non-collagenous C-terminal domains (NC1) with the formation of intermolecular disulphide bonds to form a two-dimensional lattice or 'chicken-wire' mesh (Fig. 2.8). Evidence for this latter hypothesis has been produced in the form of the isolation of disulphide-bonded hexamers of the NC1 region from collagenase digests of Type IV collagen.

Thus, in Type IV collagen there is clear evidence for the existence of both intra- and intermolecular disulphide bonds. It should be recognised that under physiological conditions these bonds are as important as any other type of covalent cross-link, particularly the lysine-derived cross-links typical of the fibrous collagens described below.

Lysine-derived reducible cross-links

Two types of lysine-derived cross-link are produced in newly synthesised collagen molecules and fibres. These are intra- and intermolecular cross-links. There is one type of intramolecular cross-link, an aldol, and two types of intermolecular cross-link, both Schiff's bases. All three bonds only form after the modification of specific lysine or hydroxylysine residues in the non-triple helical domains of the collagen molecules by enzyme action.

After the collagen molecule has been secreted from the cell it forms fibres (in the cases of Types I, II and III) or higher macromolecular structures (in the case of Type IV and possibly Type V) as described above. After fibre formation, an extracellular enzyme, lysyl oxidase, binds to the newly formed fibres and oxidatively deaminates the lysine residues in the non-helical N- and C-terminals of each α-chain, converting them to aldehyde groups (Fig. 4.2). This lysine aldehyde is properly called 1-amino-1-carboxy-pentan-5-al (or α-amino-adipic-acid δ-semialdehyde in some earlier texts), but will be referred to as lysine aldehyde for convenience in this book. It should be stressed that the enzyme theoretically acts on all three lysines (one per α-chain) at both the N- and C-terminal ends of each collagen molecule resulting in six aldehyde groups per molecule of collagen. In Type I collagen the C-terminal non-helical peptide domain of the α2(I)-chain does not contain a lysine residue. Two lysine aldehydes at either the N- or C-terminal ends of one molecule could react together to form an intramolecular cross-link, leaving just one unreacted lysine aldehyde free to form intermolecular cross-links. This mechanism seems to take

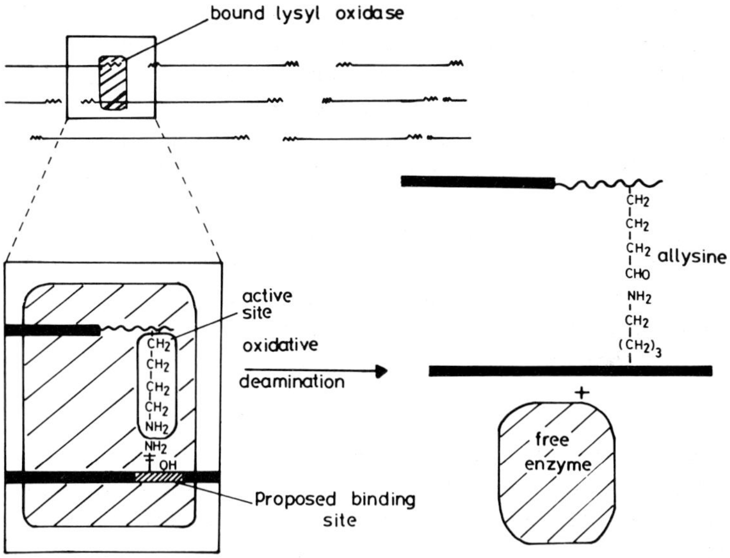

Fig. 4.2 Diagrammatic representation of the formation of lysine aldehyde in collagen by lysyl oxidase.

place at the N-terminus but not at the C-terminus, as will be discussed below.

Lysyl oxidase has been extensively studied and is extremely well characterised. It has a molecular weight of 29 000, contains high proportions of cysteine and tyrosine and is copper dependent, and it may also be pyridoxalphosphate dependent. One of its most important features is that it will only bind to fibrillar collagen; it has no activity against soluble collagen monomers. The sequence of amino acids in the region of the α-chain which overlaps the lysine aldehyde formed by the enzyme is highly conserved in Types I, II and III. This sequence, –Hyl–Gly–His–Arg–, together with the target lysine residue, may represent the binding site for lysyl oxidase (Fig. 4.3). In soluble collagen the target lysine on another molecule is not readily available, but in fibrillar form the two regions (target lysine and Hyl–Gly–His–Arg–) come together in the end-overlapped structure. There is no direct evidence to support this hypothesis, except that cross-links only occur where this sequence is present, including the 7S region of Type IV (here the sequence is slightly modified to –Hyl–Gly–Glu–Arg–).

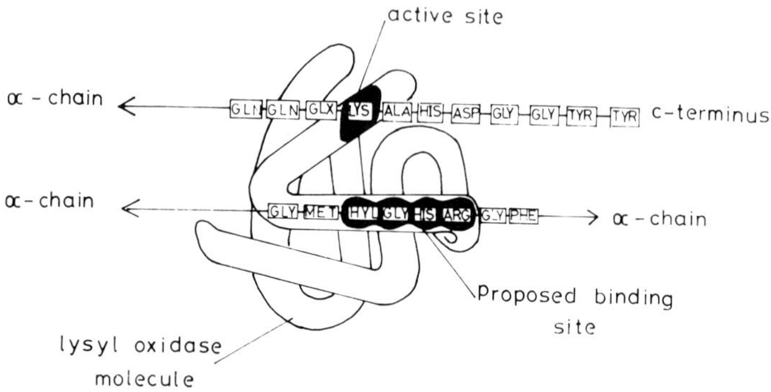

Fig. 4.3 Diagram showing the suggested binding site for lysyl oxidase on the specific Hyl–Gly–His–Arg conserved sequence of collagen.

Intramolecular cross-links—the aldol

The product of reaction of two lysine aldehydes is known as an aldol. As two such adjacent aldehydes can only exist in the terminal non-helical regions within the same collagen molecule (as described above), the bond produced, which links two α-chains, must be intramolecular. This, indeed, readily seems to be the case. For example, when acid-soluble collagen from rat tail tendon is subjected to electrophoresis in order to separate the α-chains dimers, i.e. β-components, are clearly seen. These are the result of two α-chains linked by an intramolecular aldol (Fig. 4.4).

Early research showed that an intramolecular aldol existed at the N-terminus of Type I collagen, thus leaving one lysine aldehyde per molecule at the N-terminus free to form intermolecular cross-links with other molecules. Although much effort has been put into identifying a similar aldol residue at the C-terminus of fibrous-type collagen molecules, no such residue has ever been isolated. It is therefore assumed that aldols do not form at the C-terminus of fibrous collagens, possibly due to steric reasons. It should be remembered that unlike the N-terminal non-helical region of the $\alpha 2$(I)-chain, the C-terminal region of the same chain does not contain lysine. The intramolecular link at the N-terminus could theoretically form between two $\alpha 1$(I)-chains or $\alpha 1$(I) and $\alpha 2$(I), but the major intramolecular cross-link is between $\alpha 1$ and $\alpha 2$, and is referred to as β_{12} (Fig. 4.4).

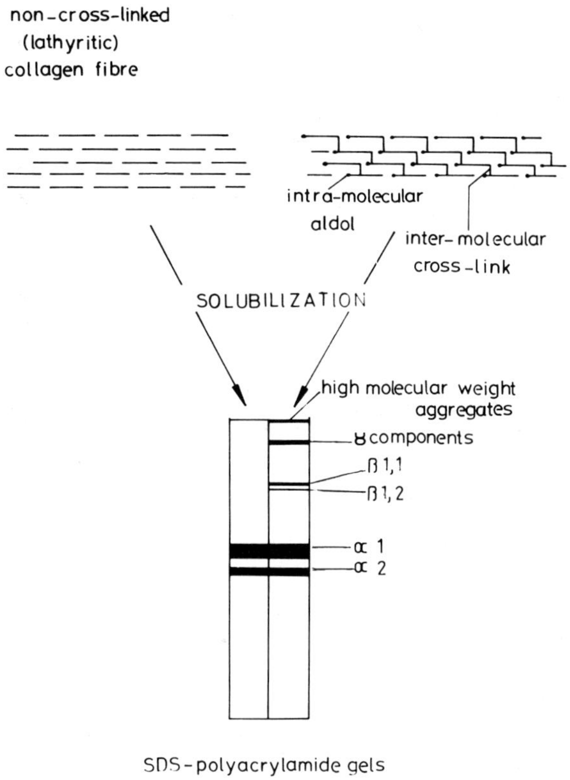

Fig. 4.4 The role of intra- and intermolecular cross-links in stabilising the collagen triple helix and fibril and the characteristic fragments obtained from Type I collagen with and without intra- and intermolecular cross-linking.

It should be emphasised that intramolecular aldols of themselves cannot increase the stability or insolubility of collagen fibres as they are located, by definition, within only one molecule. It needs the presence of intermolecular cross-links to give strength and stability to collagen fibres and matrices.

Intermolecular cross-links—aldimines and oxo-imines

The aldehyde group formed in peptide lysine by the action of lysyl oxidase is located exactly opposite the ε-amino group of the hydroxylysine in the –Hyl–Gly–His–Arg– sequence in the end-overlapped adjacent molecule in the fibril. These two residues can, therefore, undergo reaction to form a Schiff's base cross-link between the two α-chains. This covalent linkage is an *aldimine* called dehydro-hydroxylysinonorleucine (dehydro-HLNL). It contains a double-bonded system formed between the amine nitrogen and the aldehyde carbon (Fig. 4.5) and is completely stable under normal physiological conditions. This bond, formed between a lysine aldehyde and a hydroxylysine, may be chemically reduced with agents such as sodium borohydride (Fig. 4.5). The resulting compound is stable under the conditions of acid hydrolysis needed to break down the protein to a mixture of amino acids and may be isolated by various chromatographic means. In this reduced form it is called hydroxylysino-norleucine (HLNL).

In the case of collagens in which hydroxylation of N- and C-terminal lysines is extensive or complete (e.g. Type I collagen from bone, dentine and, to some extent, tendon; Type II collagen; Type III collagen in embryonic tissues; Type IV collagen), the lysyl oxidase enzyme readily converts the non-helical domain hydroxylysines to aldehydes in the same way as it does lysine. The product of reaction of this hydroxylysine aldehyde with the hydroxylysine in the same helical sequence –Hyl–Gly–His–Arg– of the adjacent molecule is a dihydroxylated Schiff's base. This compound (dehydro-

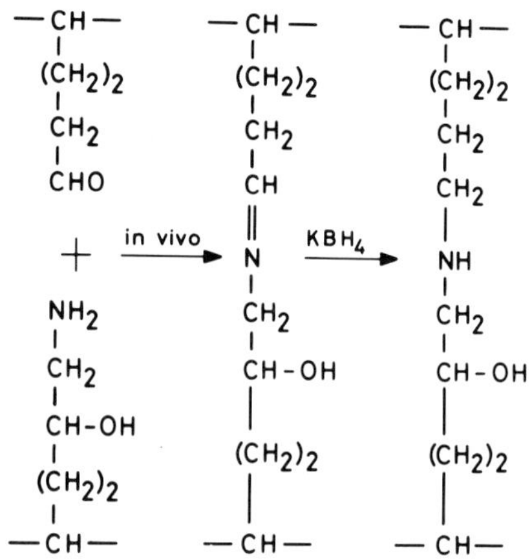

Fig. 4.5 The formation and reduction chemistry of the aldimine reducible cross-links of collagen.

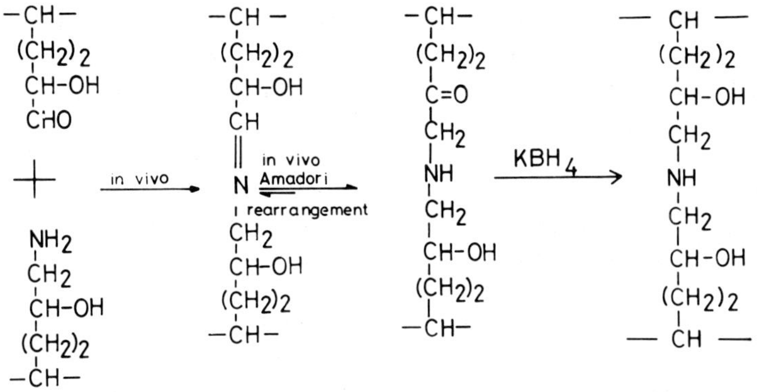

Fig. 4.6 The formation, rearrangement and reduction chemistry of the oxo-imine reducible cross-links of collagen.

dihydroxylysinonorleucine) undergoes spontaneous rearrangement through the Amadori mechanism to form a single bonded system (Fig. 4.6) which is the prevalent form found in highly hydroxylated collagenous tissues. This rearranged component is an *oxo-imine* called hydroxylysino-5-oxo-norleucine (HLONL, note: the 'keto' nomenclature formerly used in place of oxo is now outdated). HLONL can also be reduced *in vitro* with reagents such as sodium borohydride to produce a component stable to acid hydrolysis (Fig. 4.6). This product is called dihydroxylysinonorleucine (DHLNL).

Distribution of aldimine and oxo-imine cross-links

Tissues showing low hydroxylation may be expected to contain high amounts of the aldimine relative to the oxo-imine, whilst tissues having highly hydroxylated lysines will be expected to contain a relatively higher content of the oxo-imine. This is indeed the case, so that skin Type I collagen, which shows very low hydroxylation of lysine residues compared to Type I from other tissues, contains almost no oxo-imine cross-links but only aldimines. Type I collagens found in bone and dentine are highly hydroxylated compared to skin Type I collagen and are cross-linked almost exclusively by the oxo-imine bond. Similarly, Type II and Type IV collagens are both very highly hydroxylated and are found to be cross-linked exclusively by the oxo-imine (Table 4.1).

Table 4.1
Levels of the collagen aldimine and oxo-imine reducible cross-links in various tissues

Tissue	*Percentage aldimine*	*Percentage oxo-imine*	*Age of tissue used*
Tendon, rat tail	100	0	
Tendon, rat achilles	50	50	
Skin, foetal rat	20·0	80·0	Foetal
Skin, adult rat	100·0	0·0	Adult
Bone, rat	0·0	100·0	Newborn
Bone, cow	33·3	66·6	Mature
Cartilage, adult cow	0·0	100·0	Adult

Therefore, the proportion of the two intermolecular reducible cross-links found in any tissue or collagen type is directly related to the extent of hydroxylation of that tissue or collagen type. It should be noted that any one tissue may show different levels of hydroxylation and, therefore, in the proportion of aldimines to oxo-imines at various stages of development. Thus, it has been shown that embryonic skin contains almost exclusively the oxo-imine cross-links, whilst the skin of young post-natal animals contains only the aldimine cross-link. The reason for such a change in cross-linking is not clear but must be related to the level of hydroxylation of the collagens in these tissues. Although this is the general case, it is possible for the N- and C-terminal region lysine to be hydroxylated to a greater extent than expected from the overall hydroxylation of the triple helical lysines. This would suggest that hydroxylation of these telopeptide residues is under the control of another hydroxylase and such a possibility further increases the range of biological diversity of the tissue by selectively varying the type of cross-link.

Location of reducible cross-links

The location of the reducible cross-links within the collagen matrix has remained a central issue over the last twenty years. The reason for this interest is that knowledge of the precise location and mode of cross-linking in collagen fibres or matrices will set limitations on the construction of models of these structures, provide data on mechanisms of enzyme degradation, and account for the variation in mechanical properties, and therefore generally aid in our understanding of these higher macromolecular systems.

Early studies seemed to indicate that collagen could form cross-links between molecules at any point along the length of those molecules. This would mean that, to increase the strength of a collagen fibre, more cross-links could be added at any position. Thus, stability would be proportional to cross-link number. This notion is now known to be completely incorrect. Collagens can only form reducible cross-links, i.e. aldimines or oxo-imines, through the aldehydes produced by lysyl oxidase from the lysine (or hydroxylysine) residues at the N- and C-termini. That is, only a maximum of two aldehydes per chain, each located in the non-helical domains at the termini of the chains.

As already stated above, the C-terminal lysine aldehyde reacts with a hydroxylysine residue in an adjacent α-chain. This latter residue is located in the conserved sequence –Hyl–Gly–His–Arg– and is known to be residue 103 in the Type I α-chain (residue 89 in the triple helix). It is convenient to express this location in terms of the CNBr peptides obtained on digestion of fibrous collagen with this reagent. Residue 103 is the N-terminal amino acid of peptide α1(I)CB5 after CNBr digestion of Type I collagen. The C-terminal lysine aldehyde which reacts with this hydroxylysine is located in peptide α1(I)CB6 after CNBr digestion. Thus, the cross-linked peptide (known as an 'H' peptide), which is liberated from cross-linked insoluble Type I collagen after CNBr digestion, is known as α1(I)CB6–α1(I)CB5 (Fig. 4.7). Similarly, the 'H' peptides released from cross-linked Type II and III collagens involving the homologous residues are α1(II)CB14–α1(II)CB12 and α1(III)CB9–α1(III)CB7 respectively as shown in Fig. 4.7. To date, only the C-terminal 'H' peptides from Types I and III have been unequivocally identified by isolation from the whole insoluble collagens.

The analogous N-terminal 'H' peptide, formed from the lysine aldehyde and a hydroxylysine in the helix of an overlapped adjacent molecule, results from the reaction of residues 9 and 946 (with respect to the Type I collagen sequence). In terms of CNBr peptides this involves a linkage between the N-terminal α1(I)CB1 peptide and α1(I)CB6, resulting in the 'H' peptide designated α1(I)CB1–α1(I)CB6. This peptide has analogous copies in Types II and III collagen which are α1(II)CB4–α1(II)CB9 and α1(III)CB3–α1(III)CB9, respectively (see Fig. 4.7).

The case of Type IV collagen is analogous to that of the fibrous collagens, although the cross-linking appears to be somewhat more specialised. Here, four Type IV collagen molecules align at their

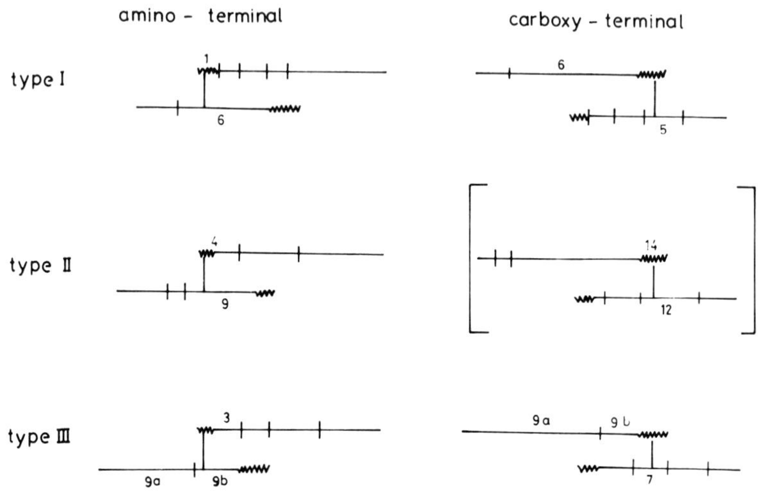

Fig. 4.7 The location of aldimine and/or oxo-imine cross-linked peptides in various collagen types.

N-termini to form the short 30 nm 7S region which is stabilised by disulphide bonds as mentioned earlier (Fig. 4.8). Oxo-imine cross-links also form in this region and their precise location is known in terms of the sequence. Since the tetramers cannot be isolated directly from membranes by disulphide bond cleavage reagents alone, this suggests that oxo-imine cross-links also stabilise the polymerisation of the tetramers into a network. Type IV would then possess cross-link sites towards the C-terminal end of the molecule. The precise location of these cross-links is not yet known in terms of sequence or CNBr peptides, but the discovery of these particular bonds has led to the proposal of an overlapped model for Type IV collagen. Essentially,

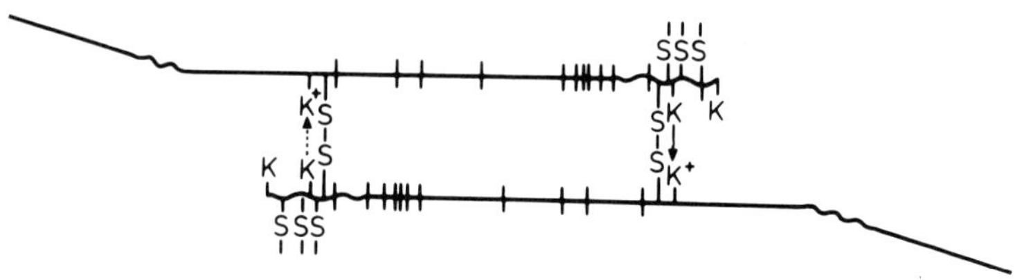

Fig. 4.8 The location of oxo-imine reducible cross-links in Type IV collagen. The 7S region contains intermolecular reducible cross-links as well as disulphide bridges.

this suggests that these intermolecular oxo-imines must form between end-overlapped molecules in the same way as is seen in fibrous collagens. This proposal is supported by the presence of the sequence Hyl–Gly–Glu–Arg– both in the 7S region and along the arms of the Type IV molecule making up the tetramer. Thus, an overlaid system of sheet-like 'chicken-wires' has been proposed in which each sheet is shifted approximately 100 residues with respect to the sheet below it in exactly the same end-overlapped configuration as that adopted by fibrillar Types I, II and III collagens. Confirmation of this model must await further research.

CHANGES IN COLLAGEN MATRICES ON AGEING

As collagen fibres age they become progressively stronger and more rigid. The molecular changes which invoke these alterations in mechanical properties also lead to a decreased susceptibility to attack by proteolytic enzymes, a reduction in the extent of swelling of fibres in acid solution and an increase in actual fibre size. The changes in mechanical properties will be more fully explored in a later section (Chapter 7).

The mechanism of formation of mature cross-links

As they age and mature collagen matrices become progressively stronger and more rigid. This could be caused by an increase in the actual number of reducible aldimines and/or oxo-imine cross-links. However, as also shown above, there is good evidence to indicate that these cross-links can only form from lysine aldehydes located in the non-helical regions of collagen and not within the helix. It also appears that these reducible aldimine and oxo-imine cross-links are laid down by the action of lysyl oxidase on newly formed fibres at maximum concentration and that no more such cross-links are formed during maturation of the tissue. In fact, the number of reducible cross-links present in several different collagenous tissues actually decreases during ageing (Fig. 4.9).

What, then, is the mechanism by which collagens becomes stronger with ageing? Does another type of cross-link exist? It was suggested by our group that cross-links with higher valencies than the simple divalent reducible intermolecular bonds must exist in order to explain these age-related changes. The research work summarised in Fig. 4.9

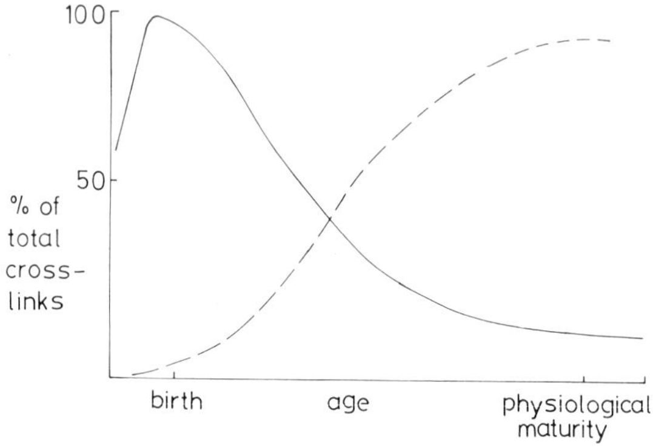

Fig. 4.9 Graph summarising the changes in cross-linking in collagen on ageing. ——, reducible cross-links; – – – – –, formation of mature cross-links.

showed that the apparent number of reducible aldimine and oxo-imine cross-links in several tissues, including intramuscular connective tissue, actually decreased during physiological maturation. It was suggested that these cross-links were not disappearing but were further reacting with other components to become more complex multivalent cross-links that were no longer reducible by borohydride.

This mechanism for the production of multivalent cross-links joining three or more collagen α-chains in three or more molecules has long been accepted as the most favourable to explain the observed increases in strength and rigidity on ageing. However, because there has been little luck to date in preparing a pure sample of such a complex compound, it has proved difficult to categorically identify putative chemicals as true mature cross-links of collagen. Three compounds have been isolated from mature collagens from different sources and variously attributed the name of the mature collagen cross-link. In reality, it would seem most likely that all three may well exist and that each may play a different role in the various tissues in which they are found. These three compounds will be reviewed below.

As stated earlier, CNBr digestion fully solubilises most collagens and gives a simple peptide mixture which may be used to produce a characteristic 'fingerprint' by various peptide mapping techniques. The most popular and useful of these techniques is that of polyacrylamide gel electrophoresis in the presence of the detergent sodium dodecyl sulphate (SDS). When the Type I collagen CNBr peptides derived from young and old tissues are analysed by this method and compared it is easy to see that very few differences are apparent. However, the C-terminal CNBr peptide of the α1-chain, α1(I)CB6, is not present on the gels of digested old material. As this peptide is one of only two on the gels which contain tyrosine, it is possible to label it by iodinating the tyrosine with ^{125}I. When this is done it is again easy to show that the α1(I)CB6 peptide, which apparently has disappeared from the old samples, is actually migrating near the top of the gel, i.e. it is associated with very high molecular weight material. These data, and further work, showed that, at least in Type I collagen, cross-links were forming during the process of physiological ageing in collagen fibres between the C-terminal ends of molecules and other unidentified portions of other molecules. In fact, it seemed that collagen molecules were able to polymerise within the fibre through reaction of multiple C-terminal regions with one another thus forming polymeric structures of variable size. This polymer structure could be solubilised and reduced in size by CNBr treatment and isolated as a polymer of the C-terminal peptide α1(I)CB6. This unit was named poly α1(I)CB6. Polymers probably form between the C-terminal region and α1(I)CB5: good indirect evidence exists to support this supposition. Theoretically, polymers can also form between α1(I)CB6 and the small non-helical N-terminal peptides, but the small size of the α1(I)CB1 and α2(I)CB1 peptides precludes their detection.

Poly α1(I)CB6 has been shown to exist in Type I collagens from many tissue sources, but no analogous polymer has been isolated as yet from Types II, III, IV or V collagen. However, despite this, it seems likely that collagens are stabilised during ageing by a mechanism that involves the partial polymerisation of molecules within the fibre. This means that the cross-links holding such structures together must be at least trivalent in nature, thereby bonding more than two molecules together.

Based on this hypothesis, cross-linking can be considered to take place in two stages. The initial end-to-end overlapped polymers are stabilised by the aldimine or oxo-imine divalent cross-links. This longitudinal cross-linking would give the fibre high tensile strength. Further increases in tensile strength at constant fibre diameter following cessation of growth would then occur by transverse cross-linking between microfibrils in register (Fig. 4.10). This would occur through reaction of the divalent cross-links to form tri- and tetravalent

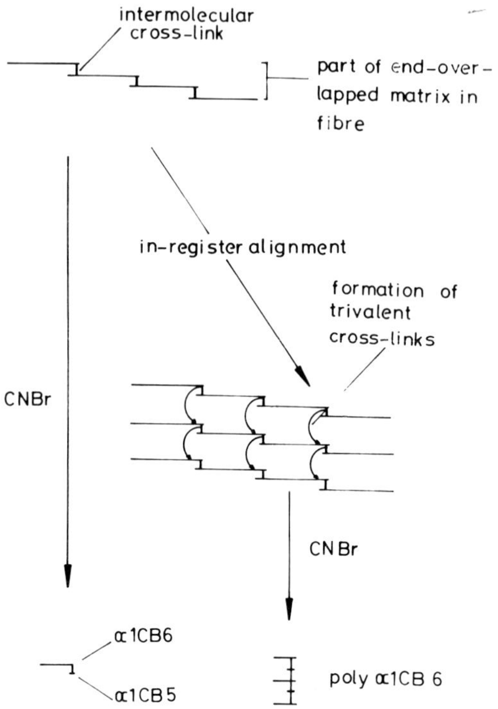

Fig. 4.10 Diagrammatic representation of the postulated pathway for the formation of collagen polymers through C-terminal, in-register cross-linking which would result, after CNBr digestion, in the appearance of the polymeric peptide material, poly α1(I)CB6.

cross-links. In this way a three-dimensional network of cross-links would be built up to produce a fibre of high tensile strength.

Structures of the mature cross-links

In the 1970s several reviews examined the various compounds which had been isolated from mature collagens and denoted as the 'mature' cross-link. Most of these structures have neither been categorically proven to be cross-links of collagen nor denied as such; therefore, they will not be reviewed here. However, three of these compounds currently hold interest in the field of collagen cross-linking and may well be real mature collagen cross-links. The first of these compounds is hydroxyaldolhistidine.

This compound was isolated from mature collagen in the early 1970s and its structure was quickly identified and confirmed by standard physical and chemical methods. It was suggested that the compound formed in collagen during maturation through the reaction of lysine aldehyde, hydroxylysine aldehyde and histidine. Some criticism of the feasibility of this mechanism occurring in collagen has been made as it demands that either hydroxylysine or lysine aldehyde be present in what is considered the receptor sequence Hyl–Glu–His–Arg– in the triple helix. This does not seem likely when the action and mechanism of the enzymes lysyl hydroxylase and lysyl oxidase are considered (see above). Also, the compound was isolated from skin collagen which is predominantly cross-linked by the aldimine bond in young tissue—this means that only lysine aldehydes will be expected to be present in this tissue, thus precluding the formation of hydroxyaldolhistidine by the suggested mechanism. However, recent work has once again implicated this compound as a mature cross-link of collagen and so verification of its role in the physiological ageing process in collagen must await further research. It is of course possible that the compound has been misidentified. Indeed, an alternative structure for the compound has recently been proposed, histidino-hydroxylysinonorleucine.

Pyridinoline is a fluorescent cyclic compound which can be isolated from tissues rich in Type II collagen. Its structure has been well characterised by several groups and different pathways for its presumptive formation have been put forward. The most likely of these seems to be that a hydroxylysine aldehyde reacts across the N–C bond of an oxo-imine cross-link. This results in the cyclic structure of pyridinoline and links three α-chains together (Fig. 4.11). The compound is now well accepted as a cross-link of mature Type II collagen. It is abundant in Type II collagen-containing tissues because this collagen is predominantly stabilised by the oxo-imine in immature tissue. However, its concentration in Type I collagens, such as bone, known to be predominantly cross-linked by the oxo-imine in young tissue, is surprisingly low. It has also been suggested that, in Type I collagen, it may only link two α-chains together (Fig. 4.11). Surprisingly, pyridinoline is not detectable in Type IV collagen which has very high proportions of the oxo-imine in young tissue and matures very quickly. In these tissues another mature cross-link must exist if we are to explain the loss of oxo-imines from basement membranes during ageing. The absence of hydroxylysine aldehyde in skin collagen prevents the formation of pyridinoline in mature fibres, and an

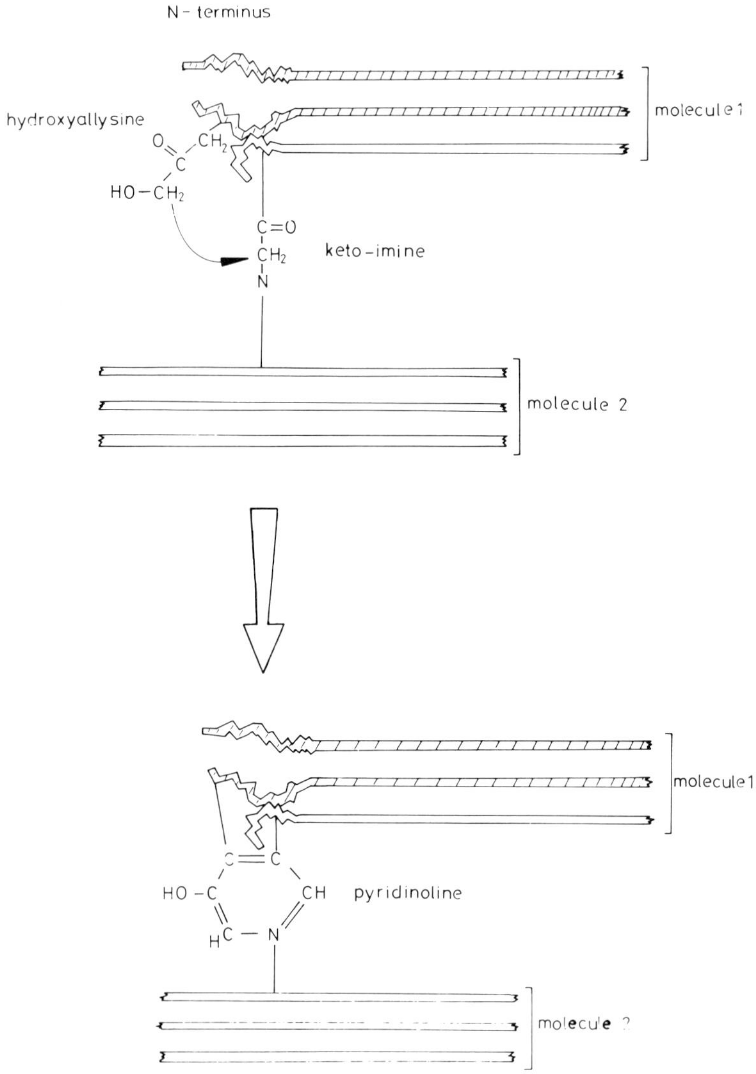

Fig. 4.11 Postulated pathway for the formation of pyridinoline and its possible location linking only two collagen molecules.

additional mechanism must exist for the maturation of skin and similarly cross-linked tissues. For these reasons the actual physiological role of pyridinoline in the various tissues in which it occurs remains unclear.

A compound has recently been isolated from the polymeric material (poly α1(I)CB6) which can be purified from mature Type I collagen. This compound has been shown to be absent from immature and soluble collagen, non-cross-linked CNBr peptides of mature collagen and non-collagenous proteins, and is known to be derived from lysine. As yet a structural analysis of the component is not forthcoming, but it is thought to be a partial breakdown product of a mature cross-link of collagen. It is known to be present in skin, tendon, basement membrane, bone and cartilage collagens and so would appear to be more universal than the other two compounds described above. As yet there is no clue to its mode of formation, but preliminary data suggest that it forms at a reciprocal rate to the disappearance of the reducible cross-links, indicating that it is probably formed from them as was suggested above.

The relative roles of these three compounds in the stabilisation of collagen during maturation and ageing must await considerable further work but, with the knowledge that has been accumulated over the last ten years, it will not be long before we know where, how and why these ubiquitous bonds are formed and how this essential and important mechanism takes place.

SUMMARY

Collagen fibres appear to form from microfibrils which themselves appear to be built from small numbers of molecules which align in specific arrays to form the basic pentafibril unit. It is thought that the fibre is then built up from these five-molecule units. The rate of formation of the microfibrils may be determined by the rate of cleavage of the N-terminal or C-terminal globular domains on their surfaces. It is not yet known whether fibre growth occurs through the longitudinal association of intracellular packages or through the lateral alignment of long thin staggered arrays.

Fibrils and fibres are stabilised by intermolecular lysine-derived cross-links. These lysine-derived cross-links are formed between lysine aldehyde or hydroxylysine aldehyde (produced by the action of the oxidative deaminase, lysyl oxidase) and hydroxylysine to form an aldimine or oxo-imine bond. The former exists as an acid- and heat-labile aldimine intermolecular cross-link and the latter as a stable oxo-imine cross-link. The distribution of these two cross-links depends on the level of hydroxylation of the collagen. Both are replaced during ageing in most collagens by stable multivalent mature cross-links whose structures are only partly known. Histidino-hydroxylysinonor-leucine is a recently identified putative mature cross-link. Pyridinoline is a cyclic fluorescent mature cross-link formed from the oxo-imine, and so is found only in tissues with a predominance of this bond. It is not, however, found in basement membrane collagen. Another mature cross-link has recently been isolated from collagen but its structure is not yet defined. These bonds are multivalent and are believed to link microfibrils, thereby adding a further degree of stabilisation to the matrix.

FURTHER READING

1. Bailey, A. J., Robins, S. P. & Balian, G., Biological significance of the intermolecular cross-links of collagen. *Nature,* **251,** 1974, 105–109.
2. Bailey, A. J., Recent aspects of the organization and stabilisation of the genetically distinct collagen. In *Advances in Molecular Biology of Connective Tissue Fibrous Proteins,* ed. L. Gotte. Scottish Academic Press, Edinburgh, 1987, pp. 1–23.
3. Pearson, A. M., Dutson, T. R. & Bailey, A. J. (eds), *Advances in Meat Research,* Vol. 4: *Collagen as a Food.* Van Nostrand Reinhold, New York, 1987.
4. Hay, E. D. (ed.), *Cell Biology of the Extracellular Matrix,* Plenum Press, New York, 1981.
5. Trelstad, R. L. & Birk, D. E., The fibroblast in morphogenesis and fibrosis. In *Fibrosis,* Ciba Foundation Symposium 114. Pitman, London, 1985, pp. 12–19.

Chapter 5

Elastin, Proteoglycans and Glycoproteins

INTRODUCTION

The roles of elastin in the vascular system and proteoglycans in cartilage have been well researched, but their roles in intramuscular connective tissue have been neglected. This is primarily due to the fact that they are only minor components, but they could play a subtle role in determining meat texture and must therefore be considered.

ELASTIN

The elastic tissues of the body owe their highly elastic properties to the protein elastin. It is a major constituent of blood vessels and certain ligaments (40–60%), and other compliant tissues such as lung and skin (2–5%). In the light microscope the fibres are identified with the histological stains which are specific for elastin, e.g. Van Giesons. In different tissues the elastic fibres are found to exhibit varying suprastructural properties so that in ligaments they are parallel to the axis of the ligament, whilst they form concentric layers in the arteries (Fig. 5.1), and a lace-like network in elastic cartilage. *In vivo* the fibres are in a contracted state, but, in contrast to collagen, are capable of highly reversible extension to about double their length, and their behaviour seems similar in many ways to that of a rubber-like polymer. Both electron microscopy and X-ray diffraction indicate that elastin fibres are basically amorphous. This might suggest a random structure for the elastin matrix but, as we shall see, chemical evidence points to a degree of order.

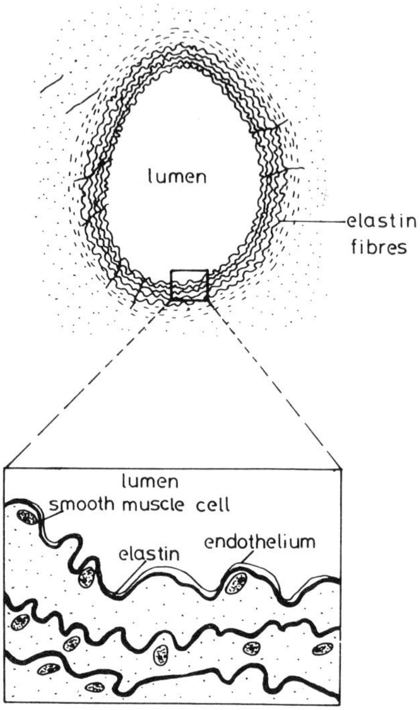

Fig. 5.1 Diagram illustrating the location of elastic fibres in the rat aorta lamellae.

Despite its intractable nature, the biosynthesis, structure and cross-linking of elastin are now generally understood, hence in this chapter we will be concerned with the chemistry and structure of elastin fibres and the relationship between the molecular structure and the elasticity of the fibres.

The elastic fibre, unlike the collagen fibre, comprises two distinct, morphologically different components, representing two separate protein complexes. The presence of these two components was first revealed in the electron microscope. In the mature elastic fibre the major component is elastin. This substance appears to be amorphous; that is no banding or repeat pattern can be discerned. In developing fibres a microfibril 10–12 nm in diameter is readily observed in the electron microscope around the periphery of the amorphous elastin component. This microfibrillar protein is, in contrast to elastin, rich in polar amino acids, and therefore is readily detected by cationic stain.

With increasing maturity amorphous elastin develops around and between the microfibrils, which act as a template during early development. The microfibrils represent only a small proportion of mature elastin and do not play a role in the mechanical properties of the elastin fibre.

Studies of the synthesis of microfibrils in tissue culture reveal the presence of about five different glycoproteins. These glycoproteins may be related to the structural glycoproteins previously isolated from such tissues as aorta. Characterisation of the microfibrils is limited owing to the difficulties of solubilising the glycoprotein complex. Using mild methods of extraction with guanidine hydrochloride and collagenase to remove the collagen, a protein fraction is obtained with a consistent composition. It is highly acidic and contains a high proportion of cysteine, which, through disulphide bonding, almost certainly accounts for its high insolubility.

Isolation of elastin

Elastin is virtually insoluble unless the peptide bonds are cleaved. This has proved, in effect, a ready method of isolation in that elastic tissue can be extracted with 0·1N NaOH for 45 min at 100°C or autoclaved in water, leaving an insoluble residue which is pure elastin. Although very little damage is incurred by the insoluble elastin through peptide bond cleavage with this method, milder techniques have to be employed when the detailed structure is being investigated. These methods have included extraction of other proteins with guanidine hydrochloride and β-mercaptoethanol, and, if necessary, cyanogen bromide and trypsin treatment. Elastin is resistant to all of these reagents and thus remains insoluble (see Chapter 12).

The amino acid composition of elastin is unusual, containing about 40% glycine, plus proline, and 40% hydrophobic amino acids. A comparison of elastin composition throughout the animal kingdom revealed some very similar characteristics, which, as will be discussed later, play an important part in the elasticity of the elastin fibre. In comparative studies elastin was found in all the vertebrates except the primitive cyclostomes, the serpentine-like animals devoid of jaws and paired fins, e.g. lampreys, slime-eels and lung-fish. Elastin is not present in the invertebrates. The appearance of elastin therefore coincides with the presence of a closed circulatory system in which arterial pressure is pulsatile.

Although attempts were made to determine the partial amino acid sequence from cross-linked insoluble elastin, the first real progress was made when soluble elastin was extracted from the tissues of copper-deficient or lathyritic animals. Under these conditions the animals suffer aortic aneurysms since the elastin is not sufficiently cross-linked and, therefore, is more readily extractable. The protein isolated, known as tropoelastin, possesses a similar amino acid composition to insoluble elastin apart from the absence of the cross-links and an increase in the lysine content. Tropoelastin precipitates from solution in a characteristic way (a process known as coacervation) on warming to physiological temperatures, analogous to the precipitation of neutral salt-soluble collagen on heating.

Tropoelastin has a molecular weight of 74 000. Tryptic digestion and sequencing of selected peptides of tropoelastin have allowed some general conclusions to be made about the overall primary structure. The tryptic peptides fall into two groups: those over 20 residues and those of about four residues. The striking features of the larger peptides are that they are rich in hydrophobic residues and that they contain repeating sequences such as Pro–Gly–Val–Gly–Val which imply order, and this should be reflected in the tertiary structure. Indeed, synthetic peptides with this type of sequence all have a preferred secondary structure. The peptide Pro–Gly–Val contains a β-form which produces right-angled bends in the peptide chain, giving support to the idea that elastin is not a random chain. The small peptides are rich in alanine and derived from the cross-link regions. The sequence Ala–Ala–Lys is repeated several times in the molecule.

It is interesting that tropoelastin also contains a repeating sequence of 17 Gly–X–Y triplets, but since the proline is generally in the Y position rather than the X as in collagen, a triple helical structure cannot form. Tropoelastin also contains a small amount of hydroxyproline (1·5%), but its function is unknown. There are surprising differences in the amino acid composition of tropoelastin from various species, for example avian and porcine. This may be because the restraints on the sequence to produce a random coil are less than those required for the highly ordered proteins such as collagen.

Cross-linking of elastin

The cross-linking of elastin and collagen are based on the same fundamental reactions, i.e. Schiff-base reactions between aldehyde and amino groups. The initial oxidative deamination of the ε-amino group of specific lysines to produce lysine aldehyde (see Chapter 4) is mediated by lysyl oxidase, a copper metallo-protein oxidase. The enzyme has now been isolated in many forms from various tissues.

It is interesting to note that just as lysyl oxidase only acts on collagen in its fibrillar forms, it similarly only acts on elastin in its aggregated (coacervated) form. The enzyme must therefore be secreted from the cell and act in the extracellular space. In contrast to collagen, where only two lysines per thousand are involved, with elastin there is a large-scale conversion of lysine residues, which decrease from 40 to 7 residues per thousand. These lysines are exposed on the surface of the protein following coacervation in such a manner as to be accessible to the enzyme. This observation again indicates specific structural areas in elastin in that it must adopt a regular, composition-dependent native conformation such that these target lysines are available to the enzyme.

The major pathway in the formation of the cross-links in elastin is the reaction of two lysine aldehydes to form an aldol condensation product, and reaction of a lysine aldehyde with lysine to form dehydro-lysinonorleucine. These two intermediates then condense to produce the isomers 1:2 dihydro-desmosine and iso-desmosine which are then oxidised to desmosines, either spontaneously or enzymically, although there is no evidence for the latter (Fig. 5.2). Oxidation of the

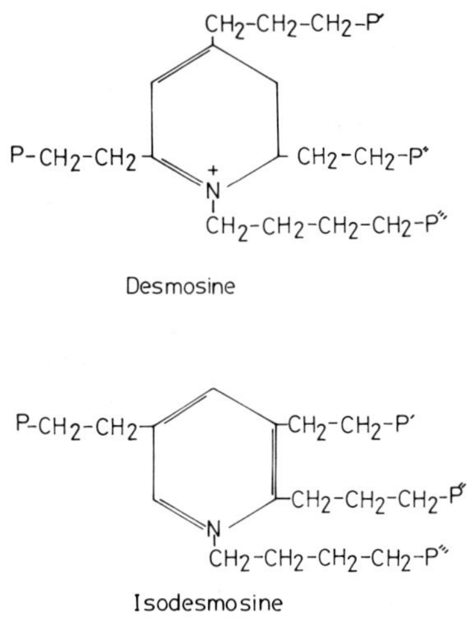

Fig. 5.2 Structure of the isomeric desmosine cross-links in elastin.

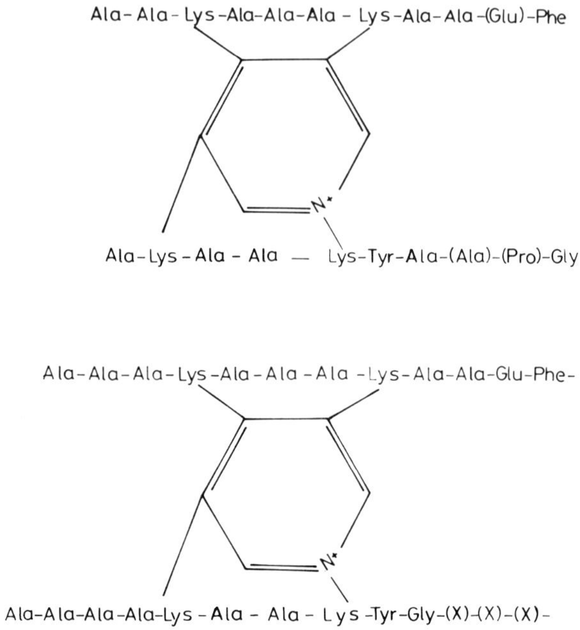

Fig. 5.3 Location of desmosines linking two elastin chains and amino acid sequences in the cross-linking region.

dihydro-desmosine by dehydrolysinorleucine can account for the presence of lysinonorleucine. An alternative mechanism could be the reaction of lysine- aldehyde with dehydro-merodesmosine.

Although the precise mechanism has not been elucidated, the cross-linking of the elastin peptide chains via the tetravalent desmosine is well established. For such a series of reactions and the bonding of four modified lysine residues to take place there must clearly be precisely ordered regions of the elastin molecules in these cross-linking areas. Despite the apparently random structure of elastin, the requirements for lysyl oxidase specificity, its facility of access and the controlled formation of tetravalent desmosines indicate that the cross-links must be precisely located. Examination of the amino acid sequence supports this proposal, since the lysines appear to be in a highly unusual arrangement (Fig. 5.3). They occur as pairs in alanine-rich α-helical regions:

(I) Ala–Ala–<u>Lys</u>–Ala–Ala–<u>Lys</u>–Tyr–Gly–Ala

and

(II) Ala–Ala–<u>Lys</u>–Ala–Ala–Ala–<u>Lys</u>–Ala–Ala

Sequence I could give rise to dehydro-lysinonorleucine and sequence II to the aldol condensation product, as the tyrosine residue in sequence I may protect the adjacent lysine from oxidation by the lysyl oxidase whilst the lysines in sequence II would be unprotected. This would give:

CHO (on the first Lys), NH_2 (on the second Lys)
Ala–Ala–Lys–Ala–Ala–Lys–Tyr–Gly–Ala–

Clearly certain portions of the elastin peptide chains are structured and the cross-links are not produced at random (Fig. 5.4). However, there is still some controversy over the structural integrity of the non-cross-linked regions. To satisfy the theory that elastin acts as a 'rubber', the chains would have to be kinetically independent and not

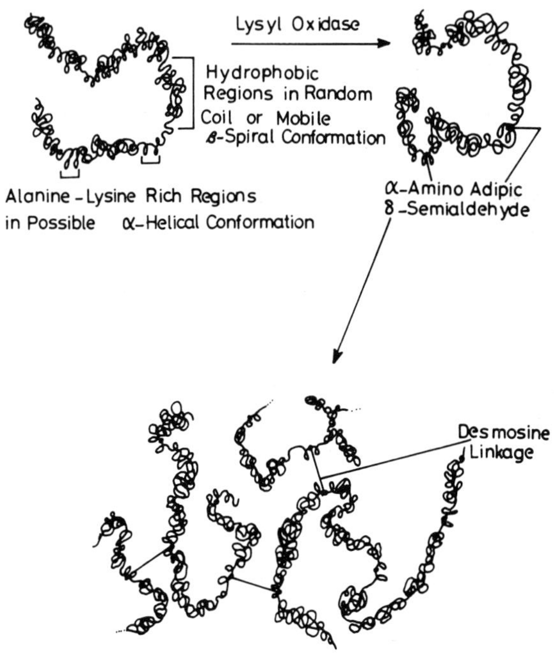

Fig. 5.4 Diagrammatic representation of the process of elastin coacervation and cross-linking.

possess any inhibiting structures. In contrast to this, it has been demonstrated that specific regions in the non-cross-linked domains possess structure, e.g. β-turns, and that these account for the thermophysical and NMR properties of elastin as well as its appearance in the electron microscope. If these structures occur to a significant extent in the polypeptide chains, elastin cannot be considered a true elastomer.

With increasing age elastin becomes less elastic. This may be due to excessive cross-linking, although no change in the quantity of known cross-links has been detected following maturation despite obvious changes in physical properties and an increase in fluorescence and yellowing of the fibres. The latter may be due to a continuing interaction of the cross-link moieties or the interaction with carbohydrate-containing components of the matrix.

Turnover and degradation of elastin

As in the case of collagen, the turnover of elastin in adults is very slow compared to other proteins and is not much higher in developing tissue. Turnover studies have suggested that, once laid down in aorta, the elastin remains virtually intact throughout life. If this is true, then growth must occur by expansion of the existing fibrous network with the newly synthesised elastin being laid down in the fenestration created. In the rat aorta (Fig. 5.1) the number of elastic lamellae is constant after five weeks and subsequent growth occurs through thickening of these lamellae, possibly by this mechanism.

In old age there is histological evidence of deterioration of the fibres, and partially lysed fibres can be observed. The insolubility and hydrophobic properties of elastin essential to its function as an elastic component also render it highly resistant to proteolytic attack. Over the last few years, however, the properties of a tissue protease capable of degrading elastin (elastase) have been established. The enzyme is secreted by granulocytes and macrophages. The enzyme is first synthesised as a pre-elastase which is then activated by a trypsin-like enzyme which is itself secreted by these cells.

At present the nature of the factors or circumstances initiating elastin degradation are unknown. It has not been reported whether phagocytosis of elastin fragments occurs as in the case of collagen. However, some work has been carried out on tissue inhibitors. In the medical field, subjects who are deficient in $\alpha 1$ anti-trypsin are known to be prone to emphysema, where extensive degradation of elastic tissues in the lungs occurs through elastase activity. Presumably a number of other elastase inhibitors are present in normal tissue.

Structure–function relationship of elastin

Physically elastin exhibits properties typical of amorphous polymers, being a soft rubber solid with an amorphous X-ray diffraction pattern. Physical investigations have demonstrated the excellent properties of elastin as a rubber, showing that it is deformed to large extensions by relatively small forces and in a highly reversible manner (see Fig. 7.3 for a typical stress–strain curve for aorta). Over-extension of such tissues is prevented by the highly inelastic collagen fibres always present in elastic connective tissues. The energy needed to restore the shape of a deformed elastomer can be derived from two sources, enthalpy and entropy. The enthalpy term is overwhelmingly the predominant factor in materials like steel and collagen, whilst with rubbers the entropy term is predominant. A rubber network is in a state of maximum entropy, i.e. maximum disorder. Stretching increases the order by orientating chains in the direction of stress, thereby decreasing the entropy. On release the polymer will spontaneously return to its disordered state, seeking a state of maximum entropy.

Two views of the structure–function relationship for elastin have been developed. The physical chemists consider that the protein chains have a preferred conformation to which they always return after stretching. The reasons for the presence of ordered structural determinants which would control such behaviour have been discussed above. In contrast, physicists prefer to consider that the polypeptide chains of elastin are randomly coiled and, theoretically, free as in classical rubbers. According to this theory there is no interaction between chains and the cross-links are distributed at random, elastin being considered as a single phase liquid-like structure.

Microcalorimetry techniques to determine the energy changes on stretching have led to the conclusion that all the elastic energy is stored in the reversible exposure of hydrophobic groups to water during stretching, and that their interactions take place in a totally random and kinetically free polymer network. However, NMR studies have indicated only 80% of the peptide chains of elastin exhibited mobility similar to rubber, and circular dichroic spectra have indicated secondary structures associated with the alanine-rich cross-link regions. Electron microscopy has revealed the presence of thin filaments under certain conditions, which would not be consistent with isotropically mobile chains.

A full discussion of the entropy changes associated with randomly coiled molecules has been presented by several workers, and this analysis has been applied to elastin. A recent analysis of elastin in these terms has been reported on the basis that elastin is a special case of the random network model, since the elastic energy can be stored in the disruption of hydrophobic forces by water during stretching. The results of physical measurements made to test this analysis have given the best fit to the calculated measurements yet achieved. However, as emphasised above, elastin is different from rubber in its ability to coacervate to a fibrous structure, its possession of specific repeating amino acid sequences and its ability to form specifically located cross-links. It cannot, therefore, be a true elastomer, and consequently statistical theories of rubber elasticity are not strictly applicable.

Use of synthetic polypeptides as models has suggested that the structure of elastin is dynamic. There seem to be preferred interchain hydrophobic interactions, the β-turn repeating unit forming β-spirals. How frequently these might occur in the elastin sequence is not yet clear.

Several alternative models have been proposed in which some higher structure is envisaged. The first model proposed a two-phase system with hydrophobic groups stabilising the interior of globules and the few polar groups on the globular surface towards the water phase. In this model overextension would be inhibited by cross-links at specific locations keeping the chains in register, hence, the structure would return to a preferred conformation following extension. Other models involving alternating segments of α-helical cross-link regions and the formation of β-turns and β-spirals containing β-turns have been proposed for the 'structured regions' of the elastin polymer. On this basis elastin would not be a true elastomer since it would possess a significant enthalpy term due to the organised part of the structure.

Despite its hydrophobic nature elastin contains about 55% water, a proportion of which must be bound to the peptide backbone since there are few polar side chains. Forty per cent of the water is loosely bound since it can be readily removed, whereupon the elastin becomes brittle. Reduction in water content would decrease the mobility of the chains and seriously affect elastin's properties as an elastomer.

Antibodies to elastin

Elastin, like collagen, is a poor antigen (Chapter 6), but antibodies can be raised in donor animals by injection of insoluble elastin under suitable conditions. The antibodies obtained cross-react with tropo-elastin. Similarly antibodies to soluble tropoelastin have been obtained. The antibodies obtained by injection of elastins from different tissues do not appear to be tissue specific. However, these antibodies have proved to be valuable in localisation studies in tissues. This is particularly useful since the amorphous structure of elastin does not lend itself to ready identification. The technique is of particular importance in the identification of elastin in tissue culture, but could also be most effectively applied in the study of elastin in the connective tissue tracts of meat.

PROTEOGLYCANS

Although proteoglycans are a minor part of intramuscular connective tissue in terms of their proportionate dry weight in the tissue, studies in tissues where they are present in significant quantities, such as cartilage, have revealed characteristics that might be relevant to the properties of the intramuscular tissues.

The proteoglycans can be defined as macromolecules that possess a protein core to which glycosaminoglycan (GAG) chains are covalently bound by specific carbohydrate sequences of three or four monosaccharides. The number of GAG chains per proteoglycan molecule varies from 1 to 100 and each proteoglycan usually contains one or two different GAGs (Fig. 5.5). The molecular weight therefore varies considerably from about 50 000 to several million due to the size variation of the protein core and of the GAG chains.

The GAG can be divided into four classes:

(i) chondroitin and dermatan sulphates,
(ii) keratan sulphates,
(iii) heparan sulphates and heparin,
(iv) hyaluronic acid.

They are polymers of repeating disaccharide units of a hexosamine and usually a hexuronic acid and are highly negatively charged due to the presence of sulphate esters and the carboxylate group. Self-repulsion of the like-charged GAG chains therefore gives the proteoglycans an

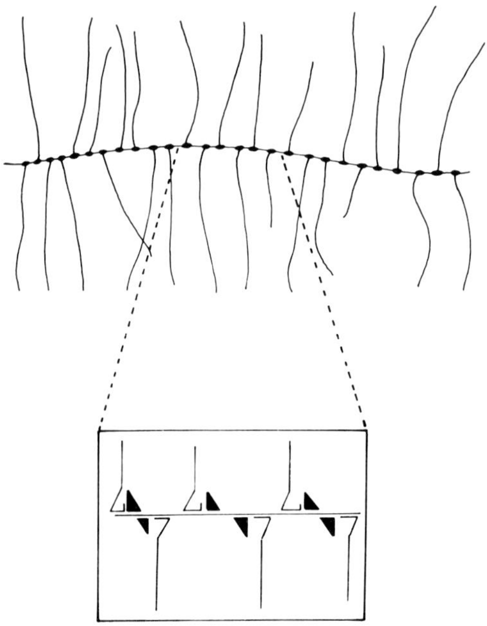

Fig. 5.5 Diagram of a generalised proteoglycan showing the classical 'bottlebrush' structure and, in detail, the nature of link protein (◣) binding of side chains.

extended structure and they occupy a large space in the matrix. The highly charged nature of these large molecules also enables them to bind large quantities of water.

In some tissues, such as cartilage, the proteoglycans aggregate to form supramolecular aggregates in which they are linked through the N-terminus of the protein core to hyaluronic acid. The region of the core protein is specific for hyaluronic acid. Although the binding is quite strong another group of proteins, known as the link proteins, attach to both the core protein and the hyaluronic acid to provide additional stability. The three link proteins identified have molecular weights of about 50 000, and are specific for different proteoglycans. Many proteoglycans, each of molecular weight $1–4 \times 10^6$ can bind to a single strand of hyaluronic acid to form an aggregate with a molecular weight of about 200×10^6. The aggregation of the proteoglycans sterically hinders their movement within the collagen framework, at the same time conferring elasticity to the tissue.

Biosynthesis

The biosynthesis of proteoglycans is a complex process, involving the separate synthesis of both the core protein and a GAG chain. Different types of proteoglycan have different core proteins, so there are probably many different genes coding for these structures. The mRNA for the core protein is translated on polysomes on the rough endoplasmic reticulin membranes, the nascent peptide passing into the cisternum of the reticulum (see Chapter 3). After oligosaccharide addition the core protein then passes through the cisternum of the smooth endoplasmic reticulum to the Golgi stack. The biosynthetic process is completed here by addition and modification of GAG chains, and by substitution with a second class of oligosaccharides. GAG chain synthesis occurs by transfer of sugars, (from their UDP-nucleotide derivatives), to specific amino acids in the core protein (usually serine or threonine). The first sugar of the linkage region is usually xylose. Further sugars are transferred to the terminal sugar on the growing chain. Each transfer is catalysed by a glycosyl transferase enzyme of high specificity. The signal for termination of chain polymerisation is unknown, but different GAGs have characteristic molecular weights ranging from about 4×10^3 to 5×10^4 (5×10^6 for hyaluronate), so a mechanism for limiting length is likely. Many cells synthesise two or more different GAGs, for example chondroitin sulphate and hyaluronic acid.

Polymerisation is followed by structural modifications which probably generate the specific molecular features of particular GAG chains necessary for the function of individual proteoglycans. Three types of modification occur: sulphation (e.g. chondroitin sulphate); epimerisation and sulphation (e.g. dermatan sulphate); and deacetylation, sulphation and epimerisation (e.g. heparan sulphate). These complex biosynthetic and multiple modification processes generate many different types of proteoglycans. Even molecules carrying the same type of GAG chain may vary from one tissue to another, according to the extent of the post-polymerisation modification and the number and length of the chains.

Hyaluronic acid is a polymer of a repeating disaccharide unit of glucuronic acid and *N*-acetyl glucosamine. It is the largest GAG having a molecular weight up to several million. It does not appear to possess a protein core, and is not sulphated. Hyaluronate is found in all connective tissues in small amounts (e.g. about 10% in skin), but is a major component of the vitreous body of the eye and the umbilical cord. It is generally agreed that the role of hyaluronate is to hold water in the tissue.

Hyaluronic acid (n 50–10 000)

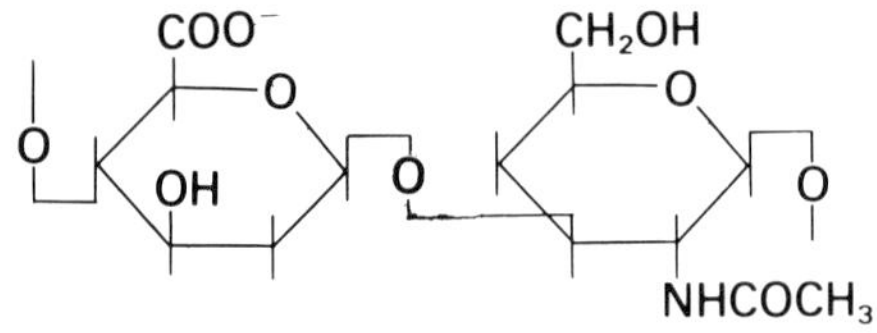

β-D-GlcUA β-D-GlcNAc

Chondroitin sulphate is a polymer of a repeating disaccharide unit of glucuronic acid and *N*-acetyl galactosamine which can be sulphated at the 4 or 6 carbon of the *N*-acetyl galactosamine residue. The number of units varies from 20 to 60 and the sulphate groups average one per unit. Cartilage and intervertebral disc contain a high proportion of chondroitin sulphate (~10%), whilst the fibrous tissues such as skin and tendon contain very little.

Chondroitin-6-sulphate (n 20–60)

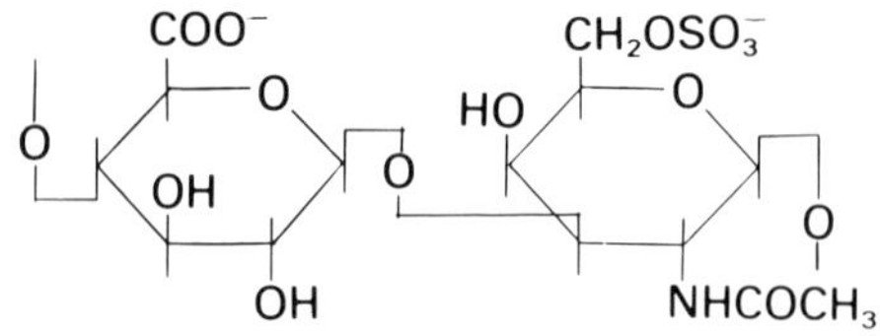

β-D-GlcUA β-D-GalNAc

Dermatan sulphate has a similar structure to that of chondroitin sulphate but it contains, in addition to the repeating disaccharide of chondroitin sulphate, the disaccharide in which iduronic acid is substituted for glucuronic acid. In fact, iduronic acid is a C-5 epimer of glucuronic acid. Dermatan sulphate is present in the fibrous tissues such as skin and tendon, but its function is unclear although it has been reported to be attached to specific locations along the collagen fibril and possibly plays a role in regulating fibre growth and in determining fibre diameter. Sulphate esters are located at C-4 and C-6.

Dermatan sulphate (n 30–80)

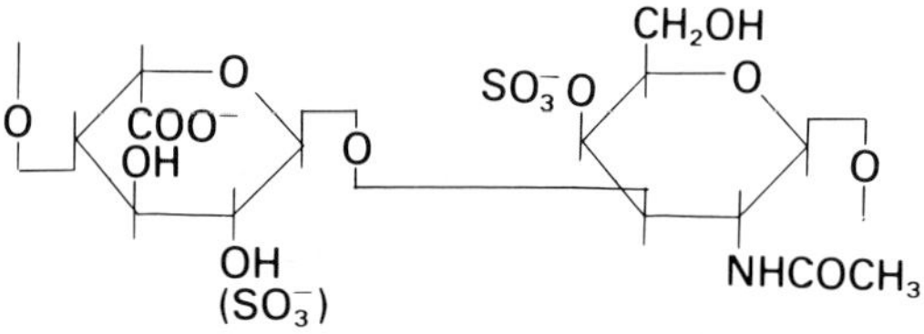

α-L-IdUA β-D-GalNAc

Heparan sulphate and heparin are copolymers of two disaccharide repeating units—glucuronic acid (or L-iduronic acid) and N-acetyl glucosamine. Segments of one repeating unit are interrupted by segments of the other type. Sulphate esters are located at the C-2 and C-6 positions of the glucosamine residue, but in many cases the N-acetyl group is replaced by N-sulphate groups. Heparan sulphate is present in the fibrous tissues but has recently also been shown to be an important component of basement membranes.

Heparan sulphate (heparin)

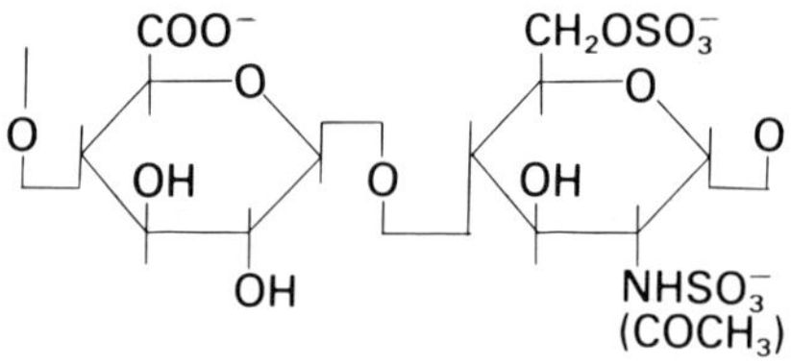

β-D-GlcUA β-DGlcNAc

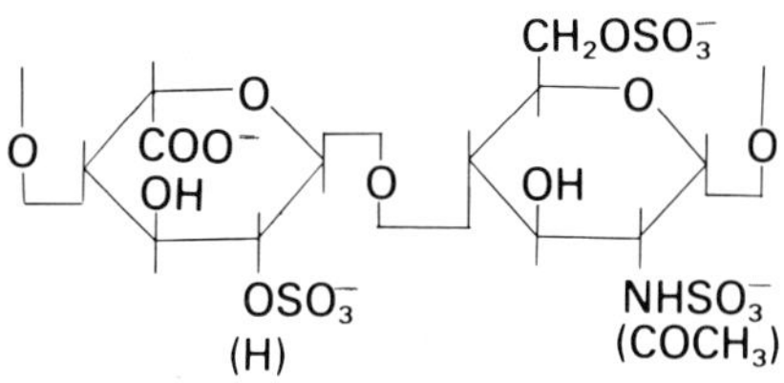

α-L-IdUA α-D-GlcNAc

Keratan sulphate contains the repeating disaccharide galactose linked to N-acetyl glucosamine, the sulphate group being on C-6 of the latter residue. Keratan sulphate is present in cartilage and intervertebral disc but is not present in fibrous tissues. It forms 50% of the glycosaminoglycans in cornea, but in this tissue it has a different linkage to the protein core.

Keratan sulphate (*n* 5–40)

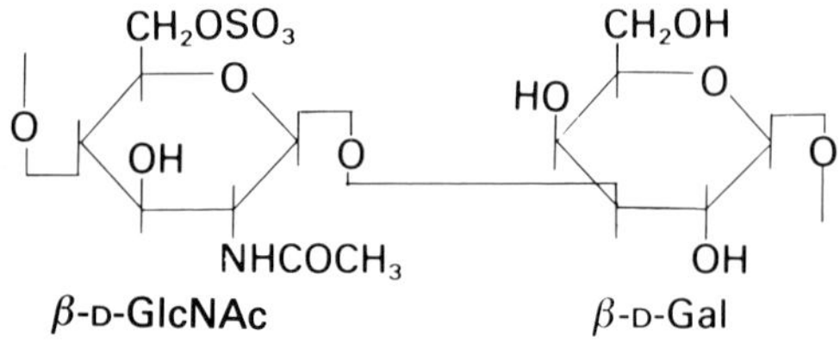

Small proteoglycans

Recently there has been considerable interest in the small proteoglycans, of molecular weight about $50–600 \times 10^3$ (Fig. 5.6). As yet little is known of their function but they are present in most fibrous tissues where tensile force is a major function, e.g. tendon. These low molecular weight proteoglycans can be divided into subclasses based

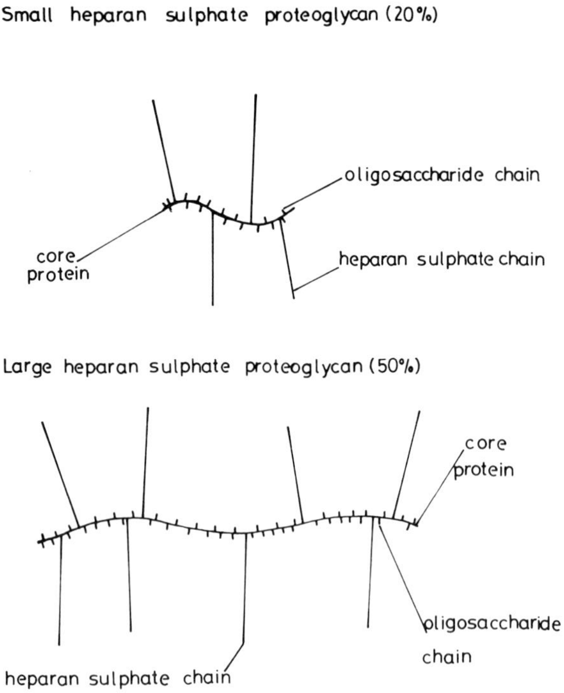

Fig. 5.6 Diagrammatic representation of heparan sulphate molecules.

on the structure of their protein core. They all contain just a few chondroitin or dermatan sulphate side chains. They can be divided into categories of interstitial, cell surface and intracellular proteoglycans. Their precise role remains to be elucidated but they bind to collagen and may have an important function in tissue stability.

Interaction of proteoglycans and collagen

The interactions of proteoglycans and collagen have been much studied but are little understood. Early microscopical studies using ruthenium red stains indicated a specific location of the proteoglycan in relation to the striations of the collagen fibril. More recently, using better stains such as cupromeronic blue, the binding of dermatan sulphate to the collagen fibril has been specifically located (see Chapter 6). The use of antibodies that do not involve dehydration or a change in physico-chemical characteristics of the proteoglycans during the staining process would provide more accurate data on these interactions and, more importantly, include the identification of the particular proteoglycan.

Degradation of proteoglycans

The use of specific degradative enzymes on the proteoglycan complex may provide some understanding of the nature and importance of the interactions between proteoglycans and collagen fibres. The breakdown and degradation of proteoglycans has been extensively studied, primarily in relation to cartilage degradation. Neutral proteases have been identified that act on the protein core, the initial cleavage probably occurring near the hyaluronate binding region of the protein core. Stimulation of enzymes involved in this process can be achieved by the action of chemical factors known as interleukins on active cells such as chondrocytes. As in the case of collagen, degradation *in vivo* is controlled by a balance between proteases and inhibitors. The GAG side chains can be degraded by glycosidases. The sulphate groups may first be removed by a number of sulphatases and the hexoses by specific iduronidases and glucuronidases, thereby aiding the chemical characterisation of these chains.

Function of proteoglycans

The extreme variety of proteoglycans is associated with a wide range of functions, of which many are mediated by interactions of proteoglycans with other molecules. In cartilage their major function appears to be the retention of water in the tissue so that it is stiff enough to resist compression by load. They also act as space filling molecules and limit the diffusion of molecules through the tissue. In tissues like tendon, dermatan sulphate proteoglycans interact with collagen fibrils and may limit radial growth of the fibril. They could also be important in aligning the fibres and thereby controlling rigidity and flexibility of the tissue matrix. Indeed, analysis of tendon collagen has revealed that the shrinkage temperature was linearly dependent on the chondroitin sulphate concentration. *In vitro* studies support the possible role of proteoglycans in stabilising and controlling fibre size. The inclusion of specific pure proteoglycans during the *in vitro* precipitation of collagen fibres results in fibres of different diameters, and of different stabilities as determined by melting temperature (Fig. 5.7). A similar study analysing the amount and type of proteoglycan in different tissues and relating this to the shrinkage temperature revealed a reasonably linear relation between proteoglycan concentration and increasing shrinkage temperatures.

Heparan sulphate proteoglycans, which are located in basement membranes, are associated with cell membrane receptors and may be involved in cell–matrix interactions. Such interactions may be important in cell adhesion. Cell surface heparan sulphate may also be involved in binding fibronectin, another component of the basement membrane. They exist as longer molecules in the glomerular basement membrane and may interact with laminin, a multidomain protein which has receptors on its surface for heparan sulphate. In the renal

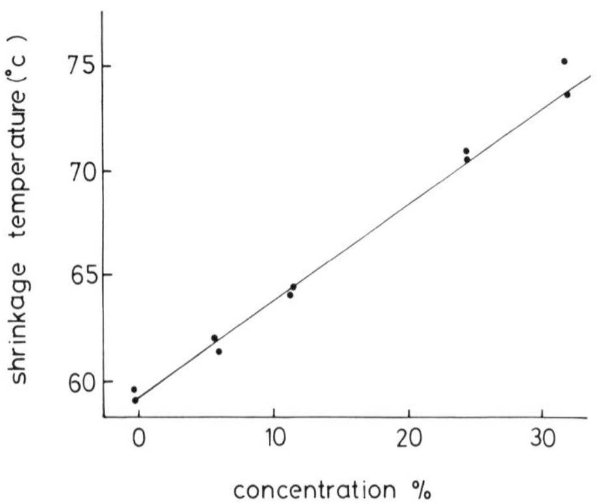

Fig. 5.7 Graph showing the linear relationship between the concentration of GAG in connective tissue and the shrinkage temperature (T_s) of collagen.

basement membrane they are important regulators of the normal filtration process.

In the context of meat and meat products the proteoglycans could be important in stabilising perimysial collagen and perhaps more importantly play a role in the structure and the functioning of the endomysial basement membrane.

GLYCOPROTEINS

The presence of large glycoproteins in the extracellular matrix has been known for a considerable time, but their contribution to the stabilisation of the various components is only just beginning to be unravelled. In multicellular organisms glycoproteins appear to play a role in anchoring cells to basement membranes or the connective tissue and in controlling the highly organised movement of cells through the matrix. A number of these glycoproteins have been isolated and characterised and, in view of their potential role in the matrix, will be considered here.

Fibronectin

Glycoproteins of this class have a multidomain structure and are present in a soluble form in plasma but also exist in an insoluble fibrous form in the extracellular matrix. The differences between the plasma and tissue forms are small and are believed to be primarily due to differences in glycosylation. Recent studies have demonstrated that not all the exons of the fibronectin gene are expressed in each molecule which consequently gives rise to about twelve different molecules. Different types of cells synthesise these different variants.

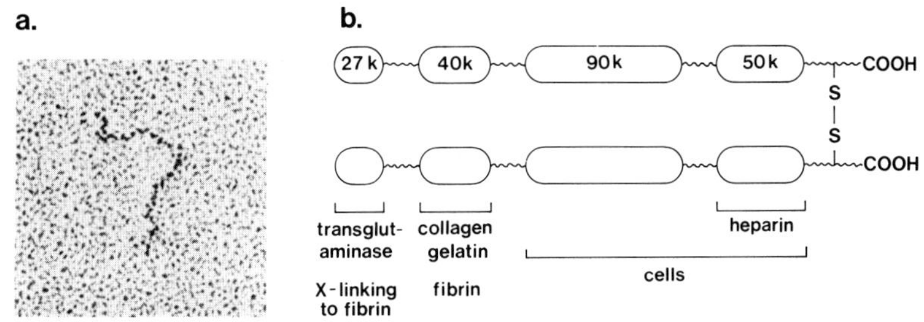

Fig. 5.8 The structure of fibronectin: (a) electron micrograph of rotary shadowed molecule, (b) diagrammatic representation of active domains in the molecule.

The molecular structure has revealed the molecule as a dimer of two identical polypeptide chains of molecular weight 220 000 linked by a disulphide bond close to the C-terminal end of the molecule (Fig. 5.8). The molecular shape has been elucidated by the rotary shadowing technique which, in the electron microscope, reveals a V-shaped configuration, the two flexible arms being about 60 nm long. As expected in a multifunctional protein, the structure possesses different domains each with specific binding sites for collagen, cell surface determinants, the blood protein Factor VIII and GAGs. The fibronectin binding site on the collagen molecule is believed to be close to the collagenase cleavage site. Recent studies have indicated that the cellular adhesion site on the fibronectin molecule is analogous to that found in a number of other cell binding proteins, i.e. Arg–Gly–Asp–Ser (referred to as RGD).

Numerous functions have been suggested for the biological role of fibronectin (e.g. at cell surfaces controlling cohesion and movement, in the matrix binding to proteoglycans and collagen, in wound healing and in homeostasis) and the molecular basis of fibronectin's varied role is now reasonably well understood.

Immunochemical techniques reveal fibronectin under the endothelium of all basement membranes. It does not appear to be an integral component of basement membranes but is always closely associated with them. It may therefore play an important role in the adhesion of the cells to the extracellular matrix.

Laminin

This glycoprotein is an important and major constituent of basement membranes (30% by weight). The protein has a high molecular weight (600 000) and using the rotary shadowing technique the structure appears as a cross-shaped molecule with three short arms and one long arm (Fig. 5.9). Laminin binds specifically to Type IV collagen and to heparan sulphate.

Immunofluorescence studies reveal that laminin is distributed uniformly throughout the kidney tubular basement membrane, but in other membranes may be restricted to the lamina rara. This location is consistent with its interaction with heparan sulphate and with attachment of epithelial and endothelial cells to Type IV collagen. Laminin is clearly important in the optimal functioning of the basement membrane.

Although many roles have been postulated for these glycoproteins in the matrix, in the context of this book we are interested in their

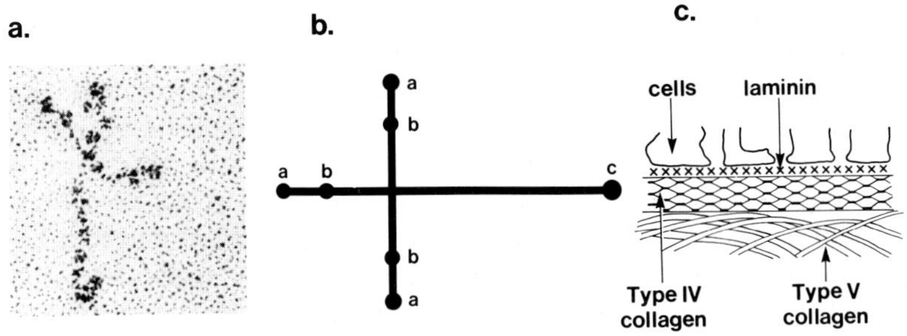

Fig. 5.9 The structure of laminin: (a) electron micrograph of rotary shadowed molecule, (b) diagrammatic representation, (c) proposed location in basement membrane.

possible roles in the stabilisation of specific tissues in meat, in meat conditioning, and following heat denaturation during cooking. The glycoproteins, like the small proteoglycans, have been reported to affect the properties of collagen fibres. Modification of these properties could lead to a change in the shrinkage properties and ultimate strength of the denatured fibres of collagen after cooking, thus affecting the texture of meat. Their presence in the basement membrane of muscle endomysium is known but studies of their role in meat have not been undertaken. Certainly the proteases present in muscle (Chapter 9) are capable of degrading all these components, thereby affecting meat properties.

Although the subject of the non-collagenous connective tissue components in muscle is one which has received singularly little attention in meat research, it is clear that they are involved in the organisation of the fibrous matrix and the functioning of basement membranes. However, their role in determining the mechanical properties of the matrix, particularly following denaturation during cooking, remains to be established.

SUMMARY

Non-collagenous components of connective tissue consist of large glycoproteins like elastin, fibronectin, laminin, and the glycosaminoglycans and proteoglycans.

Elastin is an unusual highly hydrophobic protein which forms highly insoluble fibre matrices with elastic properties. It is found in abundance in elastic tissues such as blood vessel walls and the skin, but is also found in small quantities in intramuscular connective tissue. Large elastin fibres exist in a contracted state and are capable of extension to up to twice their length with full elastic recoil. The elastin monomer, tropoelastin, has a molecular weight of 74 000. Tropoelastin aggregates by a process of coacervation and is cross-linked in the fibre form by tetravalent heterocyclic bonds called desmosines derived from lysine aldehydes. These are formed at precise locations in the structure through the action of the enzyme lysyl oxidase. The specificity of this reaction suggests that the structure of the aggregates is not random. The protein behaves somewhat like a rubber network with a high degree of disorder, a high entropy term and has an amorphous X-ray diffraction pattern but possesses some structural features which prevent it being a true elastomer. Elastin contains about 55% water, 40% of which is easily removed to leave a brittle product. The water is thus necessary for the elastomeric properties of the protein.

Proteoglycans are complexes of proteins (core proteins) and glycosaminoglycan (GAG) side chains. The protein element can vary in size but is normally 50–100 000 in molecular weight and the GAG element can vary considerably both in size and chemical nature. The GAG chains are generally polymers of hexosamine units and can be sulphonated. The latter are exemplified by chondroitin and dermatan sulphate, heparan sulphate and keratan sulphate. The other major GAG is hyaluronic acid. Molecular weight of proteoglycans can vary from 50 000 to several million depending on the size and number of GAG chains attached to a protein core. In some tissues proteoglycans can bind through a link protein to a single strand of hyaluronic acid to give supramolecular aggregates with molecular weights of up to 200×10^6. Proteoglycans form the so-called ground substance of connective tissues and interact with collagen and thereby alter its physical properties.

Glycoproteins are found in the basement membrane; these include laminin (600 000) and fibronectin (440 000), both of which are partially resistant to proteolytic attack. Both have specific regions which are able to bind Type IV collagen and other basement membrane components such as heparan sulphate as well as the cell surface.

FURTHER READING

1. Piez, K. A. & Reddi, A. H. (eds), *Extracellular Matrix Biochemistry*. Elsevier, Amsterdam, 1984.

2. *Advances in the Molecular Biology of Connective Tissue Fibrous Proteins,* ed. L. Gotte. Scottish Academic Press, Edinburgh, 1987.
3. Sandberg, L. B., Gray, W. R. & Franzblau, C. (eds), *Elastin and Elastic Tissue,* Advances in Experimental Medicine and Biology Vol. 79. Plenum Press, New York, 1977.
4. Evered, D. & Whelan, J. (eds), *Function of the Proteoglycans,* Ciba Foundation Symposium No. 124. John Wiley and Sons, Chichester, 1986.

Chapter 6

Structure and Localisation of Connective Tissue Components in Muscle

INTRODUCTION

To discover more about the structure and function of connective tissues in meat, as well as gaining an insight into their role in determining or contributing to the textural properties of cooked meat, there are various questions which need to be addressed. Perhaps the foremost of these are:

(a) what is the supramolecular structure of the connective tissues of meat?
(b) how much collagen is present in different parts of those muscles?
(c) where are the various collagen types located in muscle?
(d) what are the structures of the non-collagenous components of the fibrous meat connective tissue and what is their relationship with collagen?

Early studies of meat connective tissues included some work, using light microscopy, on the structure of collagen matrices as well as the non-collagenous component. This was supported by a small amount of research on the hierarchical structure of the meat connective tissues using electron microscopy. This early work on the distribution of collagen in muscle led to the definition of the three main structural domains, the epimysium, the perimysium and the endomysium. As no reliable method has been produced to separate these three domains physically it was extremely difficult to measure the collagen content of the various domains of meat connective tissue accurately. Early research work was, therefore, confined largely to the qualitative or semi-quantitative description of the three hierarchies of connective tissue in meat, largely through light microscopy, and to gross biochemical measurements such as solubility of intramuscular collagens. However, in the 1970s the fast developing method of immunolocalisation was being applied in collagen biochemistry and was used to probe the ultrastructure of all the connective tissues in meat.

In parallel, new methods for the investigation of non-collagenous connective tissue macromolecules were being developed, but these have not yet been applied to meat. On the other hand, there has been a considerable amount of work on the estimation of the total quantity of collagen in various meat connective tissues and this has led to some disagreement regarding the influence of this parameter on the structure of meat. Therefore, this rather contentious issue will be left for discussion until a later chapter (Chapter 8) where it will be considered with other work on the role of connective tissue in determining textural properties in meat.

The present chapter, then, will discuss the work that has been done to date on the microscopic structure of meat connective tissues and the localisation of the different connective tissue components within muscle.

COLLAGEN IN MEAT CONNECTIVE TISSUE

Microscopic localisation

As discussed earlier (Chapter 1) skeletal muscles have the generalised structure of a cellular mass enclosed in a sheath of connective tissue, the epimysium, and are given cohesion and strength by a labyrinth of intramuscular connective tissue. The muscle fibre bundles are surrounded and held together by the perimysium, and the individual muscle cells by the fine emdomysial sheaths. Standard light microscopy using classical stains such as Van Gieson show that the tendons, epimysium, perimysium and endomysium all contain large amounts of fibrous collagen. In addition, a small number of muscles can be shown to contain variable amounts of elastin by the Verhoeff Van Gieson method. However, these methods only reveal gross structural details and more sensitive techniques like phase contrast microscopy, high power optical microscopy of unstained samples and light treatment with stains such as the Periodic Acid-Foot or, better still, Sirius Red, have been used to elucidate the finer detail.

Sirius Red stains collagen intensely, making it easy to locate, the colour ranging from green to yellow or gold depending on the diameter of the fibre, whilst the denatured fibres stain red. The use of this stain has the added advantage that the dye molecules align along the collagen fibre, thereby greatly enhancing the birefringence. This only occurs with native fibres so that the technique can distinguish between native collagen and heat-denatured collagen (gelatin) in cooked meats.

The general structure of tendons, epimysium and the gross intramuscular connective tissue (perimysium) can be easily assessed by a simple visual examination of dissected muscle. In some instances, such as the large, powerful limb and neck muscles, the tendons are readily identified, being bulky and extremely strong. In other muscles which have a minor role in terms of strength and endurance the tendons are small and may be difficult to identify. However, in most cases the tendons are cylindrical or ellipsoid in cross-section. The perimysium, which is an extension of the tendonous invasions of the muscle ends, is, however, a flat, planar tissue, spreading in sheets throughout and between the muscle fibre bundles. The first question arising, therefore, in terms of the supramolecular structure of these two tissues, is what is the arrangement and size of the collagen fibres in both tendons and perimysium?

The structure of tendons has been well researched and a variety of tissues examined including rat tail tendon (a classic research material), beef, sheep and horse tendons and even turkey and kangaroo tendons. In general the structures of these strong collagenous tissues are all closely related. The tendon is subdivided into axially parallel units (fascicles) surrounded by a fibrous sheath of variable thickness depending on the role of the tendon (Fig. 6.1) and which is often associated with masses of sticky, mucous-like material containing glycoproteins and proteoglycans. Collagen Type I is the major component of tendons, making up more than 90% of the dry weight of the tissue. The collagen fibrils are linearly aggregated into fibres of varying diameter depending on the age and role of the particular tendon, and these fibres are all aligned in parallel flexible bundles to transmit the force of muscle contraction in one direction. The individual collagen fibres of tendon are circular in cross-section. Recent immunological studies have revealed that the tendon is made up of discrete bundles of these fibres, each bundle surrounded by a separate sheath (the endotendinium) consisting of mainly Type III collagen with some Type IV (Fig. 6.2).

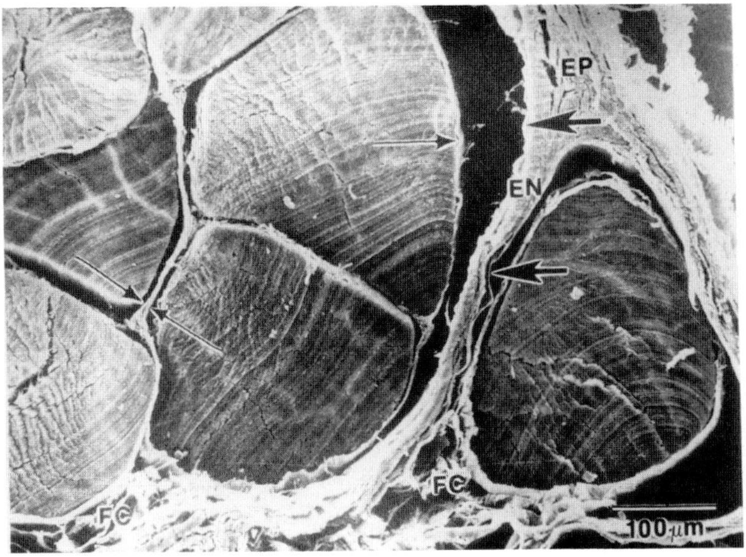

Fig. 6.1 Micrograph of tendon showing the structure of tendon fascicles and the fibrous sheath. (EP, Epitendinium; EN, endotendinium.)

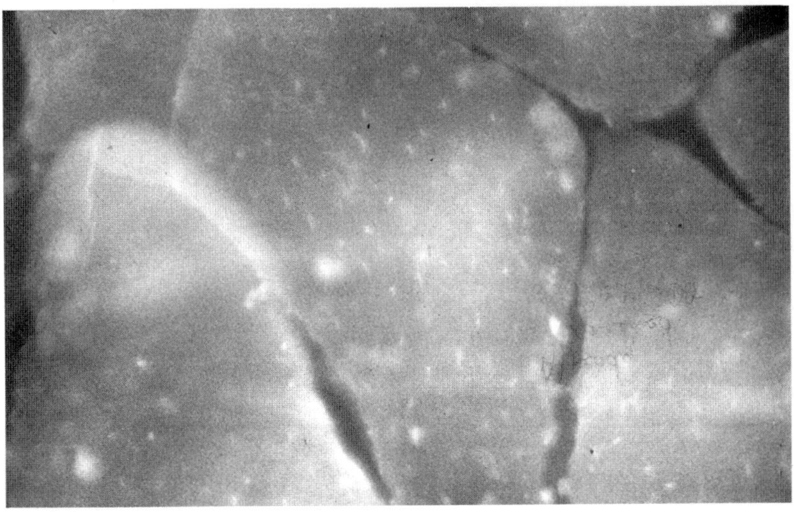

(a)

Fig. 6.2 Immunofluorescence microscopy of tendon showing the location of collagen by staining with antibodies directed against the specific types: (a) Type I collagen.

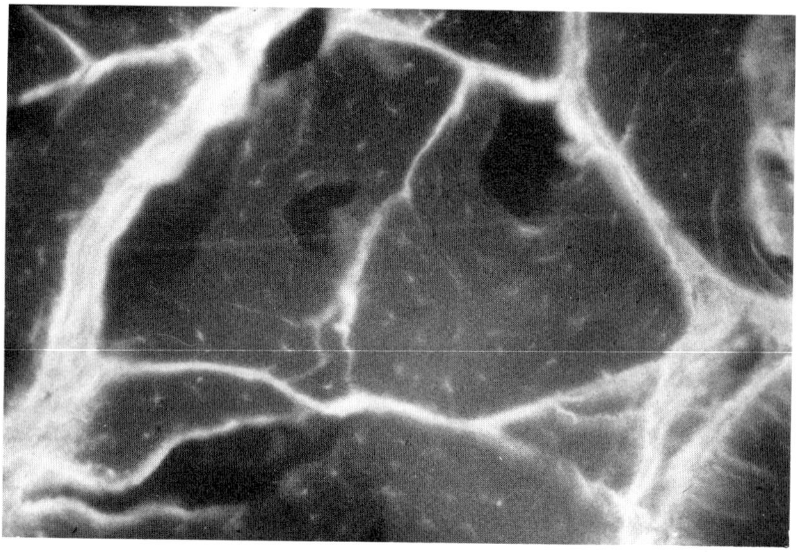

(b)

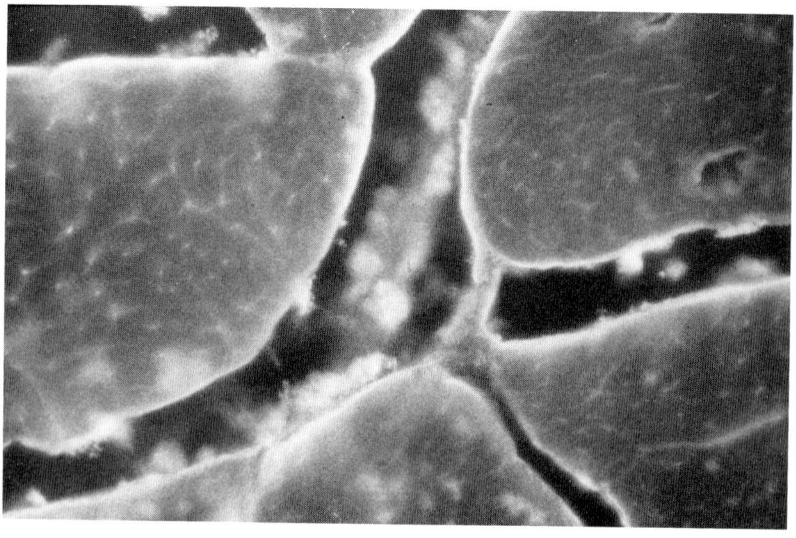

(c)

Fig. 6.2—*contd.* Immunofluorescence microscopy of tendon showing the location of (b) Type III and (c) Type IV collagen.

When samples of tendon are excised from the body and examined by microscopy the bundles of fibrils exhibit a wavy structure. This is planar and appears to be a regular crimping of the fibril bundles. Because of its appearance this supramolecular feature of collagen fibres is known as the crimp. The distance between crimps is around 100 μm, varying somewhat between tendons and with age. Although mostly studied by optical microscopy, electron microscopy has shown that the crimp is sharp, extending over only a few *D* periods of the collagen fibrils thereby giving rise to the 'bend' in the fibril in that region (Fig. 6.3). A considerable amount of work has been carried out on the structure and function of the crimp but, to date, little is known for sure of its role. However, it is thought that the crimp allows collagenous structures to take up the initial shock of movement by 'straightening out' the crimp before any real load is experienced by the tendon. It has also been suggested that crimping may allow fibre bundles to orientate around bends or in a different direction without undue strain on the tissue. As we shall see, there is good evidence for the existence of the crimp structure in meat perimysium.

The tendon passes into the muscle proliferatively although it takes different forms in the various types of muscle. In the fusiform type of muscle the fibres run more or less parallel to the long axis of the muscle, whilst in the pennate muscles the muscle fibres run at an angle to the axis of the muscle. The fibres therefore vary from a few millimetres in length up to several centimetres; fibres 34 cm long have been obtained in adult human sartorius muscle. The muscle fibre diameters in an adult average about 60 μm.

Muscle fibres are classed as red or white according to their pigmentation although there is a gradation of difference, and many muscles contain both types of fibre. White muscles are capable of fast contraction, whereas many red muscles contract slowly. The red fibres tend to be smaller, possess a generous blood supply, thicker Z-bands and thicker sarcolemmas. However, the differences in the connective tissues of the endomysium do not appear to have been properly investigated. An understanding of the differences between muscles is important since the characteristics of the muscle post-mortem are determined by the type of fibres present, for example white fibres are less susceptible to cold-shortening (see Chapter 9) than red fibres.

The attachment of the muscle fibres to the tendon appears to be quite different in fusiform and pennate types of muscle. In the case of the fusiform muscle, in which the fibres run parallel to the long axis of

Fig. 6.3 Electron micrograph of collagen fibres showing the characteristic crimp structure giving rise to a bend in the fibril.

the muscle, the muscle fibres are finely tapered into tendon. The myofibrils appear to be attached by these thin filaments of the terminal sarcomere inserting into the internal surface of the basement membrane at the ends of these insertions. The detailed structure is still not known but actin filaments appear to be attached to these membranes. More recently the presence of other proteins, e.g. talin and integrin, have been identified in these regions and may be involved in these attachments. In contrast, in the pennate muscles in which the fibres lie at an angle to the long axis of the muscle, the fibres have a rounded end butting into the intramuscular tendinous sheets extending from the tendon proper. As in the case of the fusiform muscle, no direct evidence for a coupling of the contractile proteins and the tendinous proteins has been identified.

The perimysium is made up of a series of sheets of collagen fibres arranged in a criss-cross fashion. Thus, fibres in one sheet are organised such that their orientation is planar and in one direction whilst fibres in the next sheet are also organised in a planar fashion but orientated in a different direction to that of the adjacent sheet (Fig. 6.4). There is also good evidence of a regular crimp structure in un-tensioned perimysial samples which is very similar to that seen in tendons as mentioned above. Whether there are links between the different collagen sheets in the perimysium is unknown, but they are embedded in the matrix proteoglycans known as the 'ground substance', which would tend to inhibit any such linking. On the other hand, there is evidence that some proteoglycans bind very strongly to collagen through non-covalent electrostatic bonds, and this bond appears to be site specific along the collagen fibres (Chapters 5 and 8). It is possible, therefore, that these proteoglycans and glycoproteins might bind the sheets of collagen together and act as mediators of the force of contraction. It should be stressed, however, that there is no evidence for such a mechanism and it is well known that proteoglycan and glycosaminoglycan structures have little or no intrinsic mechanical strength. Thus, it is far more likely that the individual collagen sheets in the perimysium act individually but in concert rather than that they are physically linked.

The angle formed by the different fibre directions in adjacent sheets is variable and depends on the state of contraction of the muscle. The angle of the collagen fibre direction to the direction of muscle fibres (i.e. the direction of contraction) is also variable and depends on the state of contraction of the muscle.

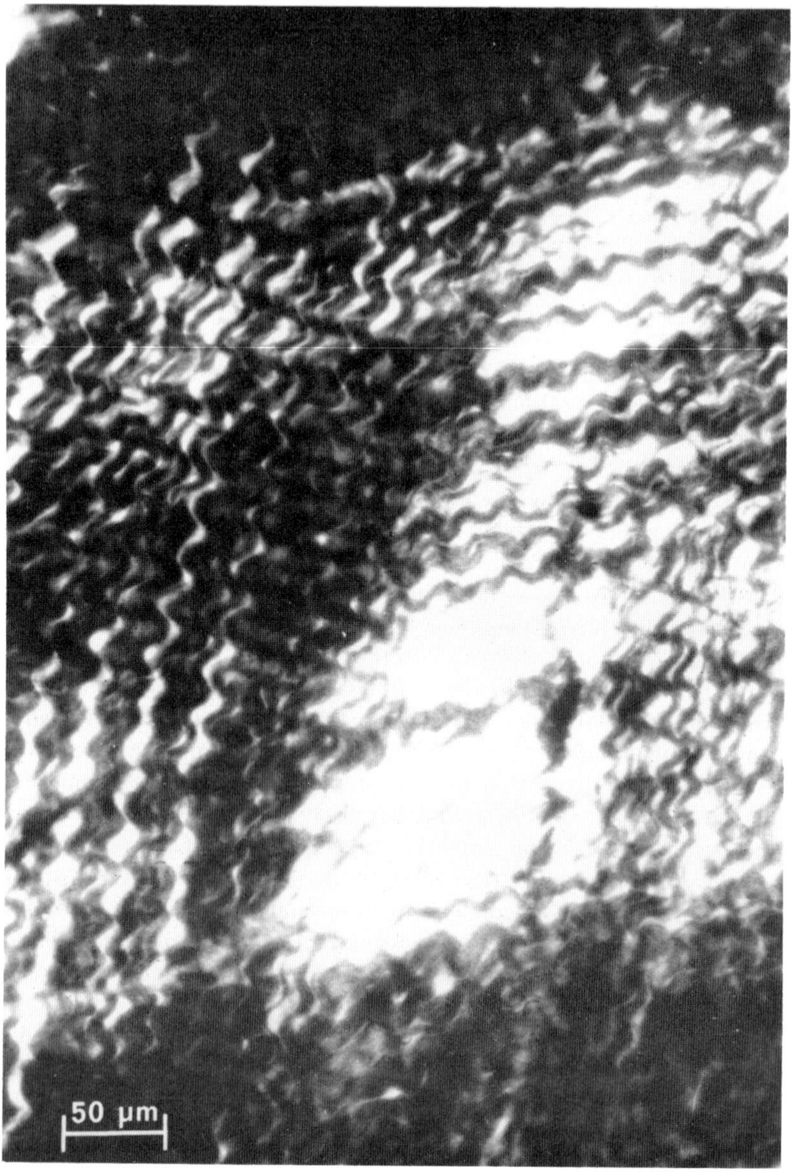

Fig. 6.4 Micrograph showing the sheet-like structure and angular orientation of crimped perimysial collagen fibres from semitendinosus muscle.

In contrast to the fibrous perimysium, the endomysium is a more complex structure consisting of an amorphous basement membrane (as described in Chapters 1 and 8) surrounded by an associated matrix of very fine collagen fibres. Electron microscopy studies have shown that the average diameter of endomysial fibres is 40–50 nm whereas those of the perimysium range from 50–100 nm. Whether the two types of fibre are compartmentally different or there is a gradient in fibre size towards the membrane is not yet clear.

The identification of the different collagen types in intramuscular connective tissue necessitated a simple method to separate epimysium from perimysium and both from endomysium in order to allow reasonably accurate quantitative analysis of the collagens present in each of these connective tissue tracts.

The identification of the different collagen types in intramuscular connective tissue necessitated a simple method to separate epimysium from perimysium and both from endomysium in order to allow reasonably accurate quantitative analysis of the collagens present in each of these connective tissue tracts. This method utilises a primary manual dissection of the epimysium followed by rapid homogenisation of muscle under highly controlled conditions, filtration of bulk perimysium from endomysium and removal of the contractile proteins from the latter to provide endomysial 'ghosts' (see Chapter 12). The washed endomysial 'ghosts' can be observed under the light microscope as open tubes with discrete structures possessing a network of associated so-called reticular fibres on the outer surface of the tubes (Fig. 6.5). The tubes can be of different sizes reflecting the presence in the muscle sample of fast and slow muscle fibres.

Further details of the structure of meat connective tissues have been revealed by the use of these separated and partially refined connective tissue fractions in the electron miscoscope. The scanning electron microscope reveals discrete bundles of fibres in the perimysium, the average bundle diameter being 600–800 nm in diameter, which surround and interconnect muscle fibre bundles. At the surface of the muscle bundles, i.e. close to the endomysium, the collagen fibre bundles are smaller, averaging 100 nm in diameter. Studies of separated perimysial and endomysial collagen fibres at higher magnification in the transmission electron microscope reveal the typical 67 nm axial repeat pattern. It has also been possible to determine the variation in fibre diameter and investigate the intermuscle fibre space. These studies have revealed that the perimysial fibres have an average

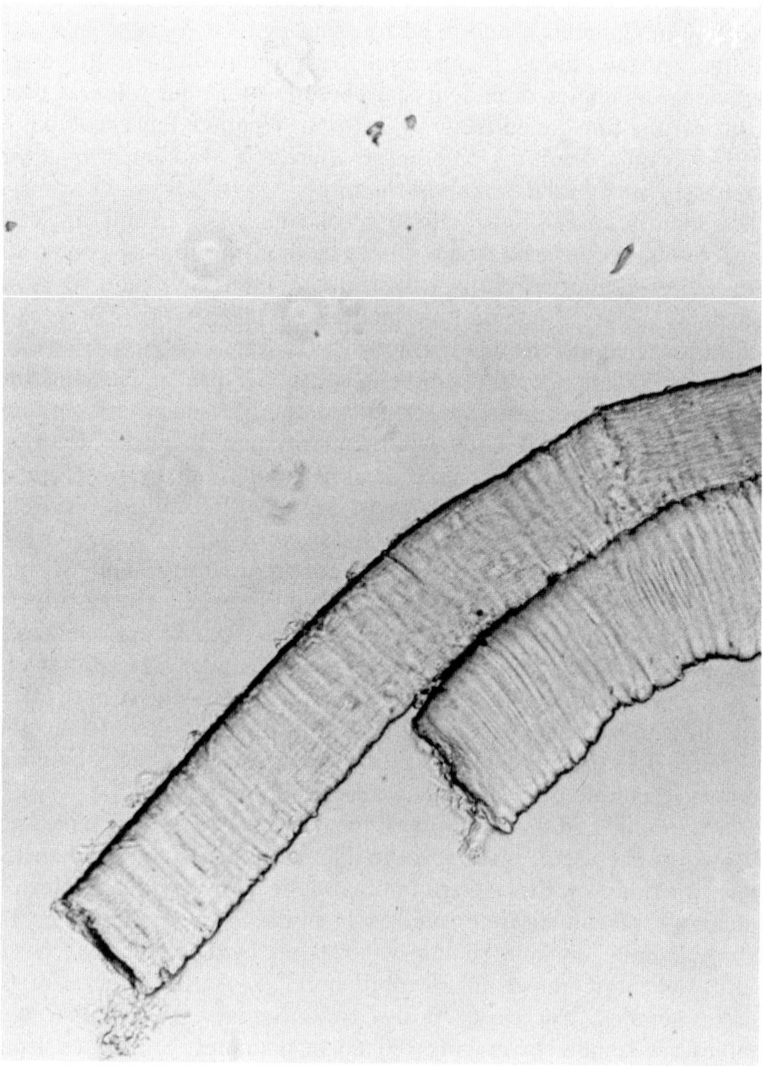

Fig. 6.5 Micrograph of endomysial 'ghosts' (i.e. empty myofibre sheaths).

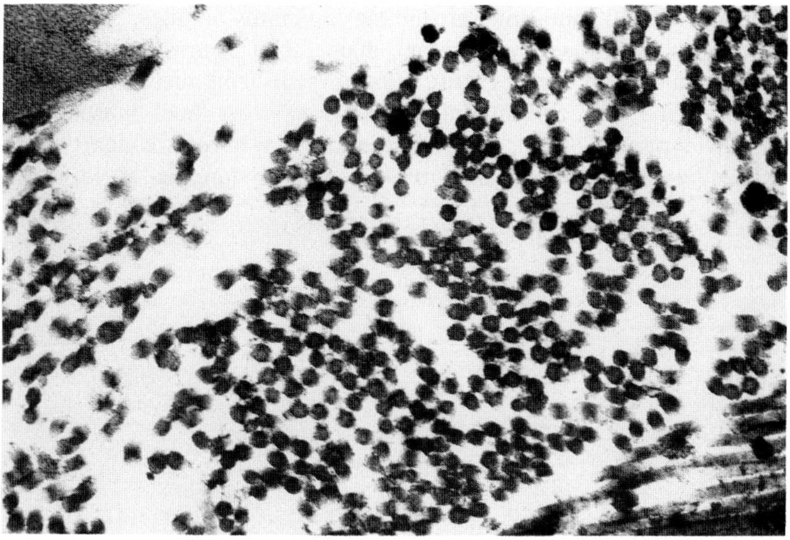

(a)

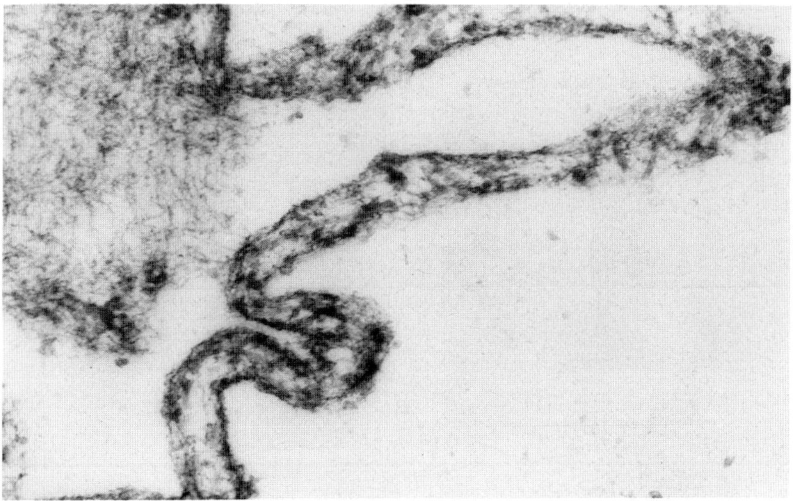

(b)

Fig. 6.6 Electron micrograph of (a) perimysial collagen fibres and (b) an endomysial preparation from beef muscle (longissimus dorsi). [Reproduced with permission from Light, N. D., *Advances in Meat Research,* Vol. 4, ed. A. M. Pearson, T. R. Dutson & A. J. Bailey. Van Nostrand Reinhold, New York, 1987.]

diameter of 100 nm and are made up into bundles, whilst the endomysial fibres, with an average diameter of 50 nm, do not appear to form bundles of fibres. Figure 6.6 shows electron micrographs taken of perimysial and endomysial preparations from beef muscle. The even, well packed, consistently sized perimysial fibres are clearly seen.

The basement membrane of the endomysium is a very thin

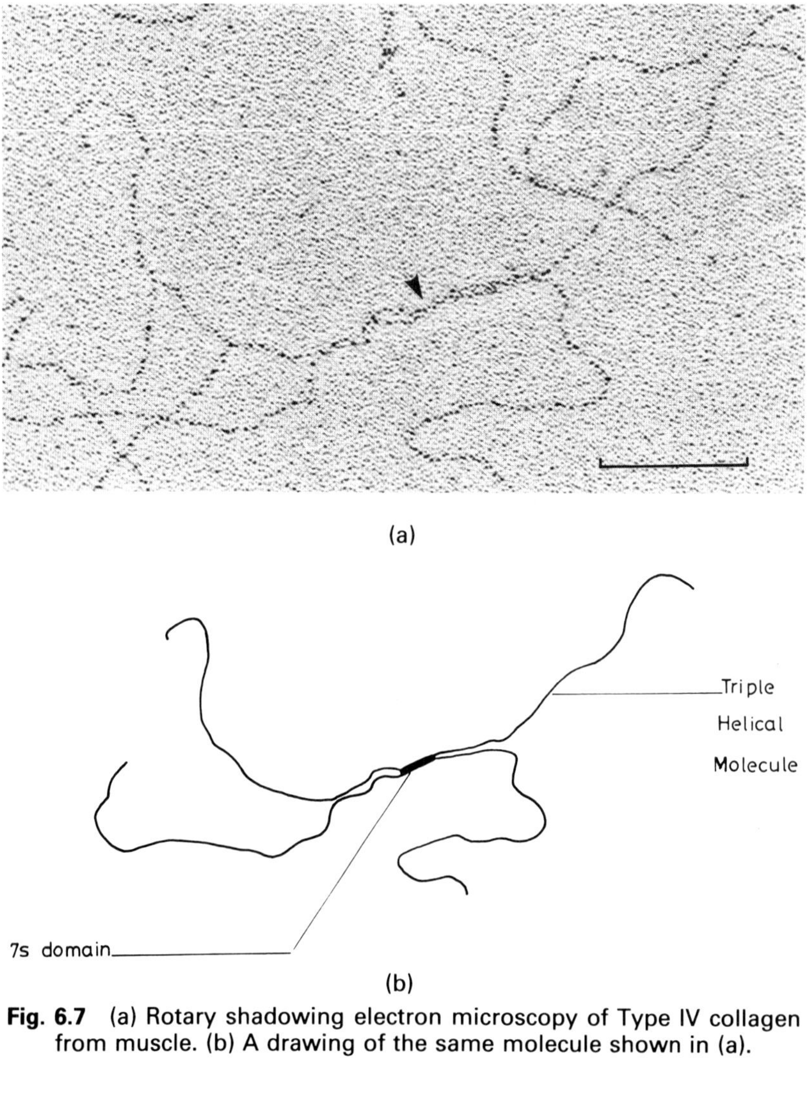

Fig. 6.7 (a) Rotary shadowing electron microscopy of Type IV collagen from muscle. (b) A drawing of the same molecule shown in (a).

membrane about 10–20 nm thick. The amorphous nature of the membrane is well illustrated in the electron microscope, and is analogous to that exhibited by other basement membranes of the body (Chapter 2). Also similar to other basement membranes is the 'spider-like' structure of isolated Type IV collagen molecules obtained from muscle endomysium (Fig. 6.7). The open network structure proposed for the Type IV collagen framework of basement membranes is consistent with its observed amorphous nature. However, some fibrous structure has been observed in the lens capsule and recent X-ray diffraction and electron microscope studies suggest there may be some structural organisation in the form of discrete microfibrils in some basement membranes (Chapter 4).

Immunolocalisation

The use of standard electron microscopy and light microscopy has given a broad and useful insight into the structure of the connective tissue of meat, but cannot indicate the biochemical composition of those hierarchical elements of muscle. This section discusses the use of specific antibodies to different collagen types in order to locate them in different parts of the muscle structure. The technique relies on the use of specific antibodies against physiological antigens in tissue and has been very successfully applied in the investigation of the distribution of the various genetically distinct collagens in a wide number of tissues. Before embarking on a discussion of the findings in tendon and muscle, however, it is necessary to understand something of the method and how it may be applied to tissue localisation of collagen types.

Each collagen type is a distinct gene product (as explained in Chapter 2) and each has its own biochemical identity, i.e. the primary, secondary, tertiary and quaternary structures of genetically distinct collagens are different. Because of these differences, each collagen type has a separate immunological identity. This means that if a purified collagen is injected, under the right conditions, into a laboratory animal, that animal will raise a mixture of antibodies against the invading exogenous collagen which acts as an antigen in the way that any other foreign material would. The methods used to raise and purify specific anti-collagen antibodies are discussed in Chapter 12.

Collagen is a relatively weak antigen but, nonetheless, antibodies will be raised by the injected animal. These antibodies, because of the nature of the immunological response, will be a mixture of various types with a range of avidities and specificities for the injected collagen antigen. However, some of the antibodies produced by the animal in response to the immunological challenge of the injected collagen will react only with that collagen and with no other protein. In other words, when purified they will be specific for that collagen type.

The results obtained with this highly specific and powerful tool are seen in the light microscope with the use of fluorescence optics. The collagens stained with specific antibody fractions and fluorophore-labelled IgG (the sandwich technique—see Chapter 12) show up as bright or diffuse areas depending on their concentration, distribution and, most importantly, the avidity of the particular antibody preparation. This latter point is important because it means that the method can only be used qualitatively. Because each antibody preparation will have different avidities for the tissue collagen it is directed against, and because these avidities, in the actual staining procedure, cannot be measured, it is impossible to quantitate the amount of each collagen type present, for example, by image processing techniques.

In the future, the use of monoclonal antibodies for such staining methods, which are produced in cell culture by mouse hybrid lymphocytes and are specific for only one antigenic site on any one protein (Chapter 12), may allow such accurate quantitation, or at least semi-quantitation, in tissue sections. A few cases have been reported where monoclonal antibodies cross-react with more than one antigen but these are rare and the technique is generally considered to be highly specific and has become a powerful tool to supplement biochemical studies.

The immunofluorescence technique is limited in several ways, most significantly by the non-specificity of the antibodies, some non-specific reactions of fluorescence-labelled second antibodies, the masking of the antigen and autofluorescence. These limitations can be reduced by ensuring that the necessary controls are carried out, and by using filters to compensate for elastin autofluorescence. Exposure of masked collagen may be achieved by hyaluronidase pre-treatment or acid swelling to expose active sites.

For immuno-electron microscopy electron dense stains have been developed by covalently linking electron dense ferritin, a globular protein of molecular weight 400 kD, with an iron core of approximately 1000 hydrated ferrous oxide particles. The ferritin may be coupled to the collagen antibody or anti-γ-globulin Fab fragments.

The iron core provides a resolution of around 5 nm. Unfortunately, ferritin has a tendency to attach non-specifically to other tissue components. Peroxidase-labelled antibodies, although not giving the same resolution, tend to be more specific. More recently the use of gold, sometimes enhanced with silver, as an electron dense stain has been developed and is providing high resolution electron micrographs.

Antibody preparations specific for collagen Types I, III, IV and V have been used to stain tendon and muscle sections and, thus, to locate these different genetic types. In tendon, the bulk of the tissue, as stated above, is Type I collagen and, as anticipated, is uniformly stained throughout the fibre bundles. However, the fibre bundles were found to be surrounded by a thin layer of Type III collagen, which is clearly stained by its antibody, as well as an amorphous layer of Type V collagen. This surrounding sheath is termed the endotendinium. The outer sheath of the tendon consists of Types I and III collagen (Fig. 6.2).

In a wide range of muscles the fibrous Type I and III collagens were found to be distributed at all three levels of connective tissue. The epimysium appeared to be primarily Type I, the perimysium contained significant amounts of Type III and the endomysium stained equally for Types I and III (Fig. 6.8). However, as mentioned above, confirmation of the relative proportions must depend on quantitative chemical techniques.

Type IV is located exclusively in the endomysial basement membrane of muscle (Fig. 6.8), confirming it is a typical basement membrane. Most studies, as discussed in Chapter 1, have shown that this network collagen is located specifically in the lamina densa of the basement membrane. There is no direct evidence as yet to the exact location of the Type IV in the basement membrane of the endomysium, but it seems more than likely that this tissue will fit into the pattern of all other basement membranes.

Type V collagen has been shown immunologically to predominate in the endomysial connective tissues, but in this case the staining with antibody showed a more diffusive and less linear appearance than that demonstrated for Type IV collagen in Fig. 6.8. As stated earlier, Type V is thought not to be a coherent component of the basement membrane itself but, more likely, basement membrane-associated as a supporting mat of fine fibrils. This certainly appears to be the case in muscle but, as can be seen in the figure, Type V also appears to be randomly, if thinly, distributed in the perimysium as well. Little is

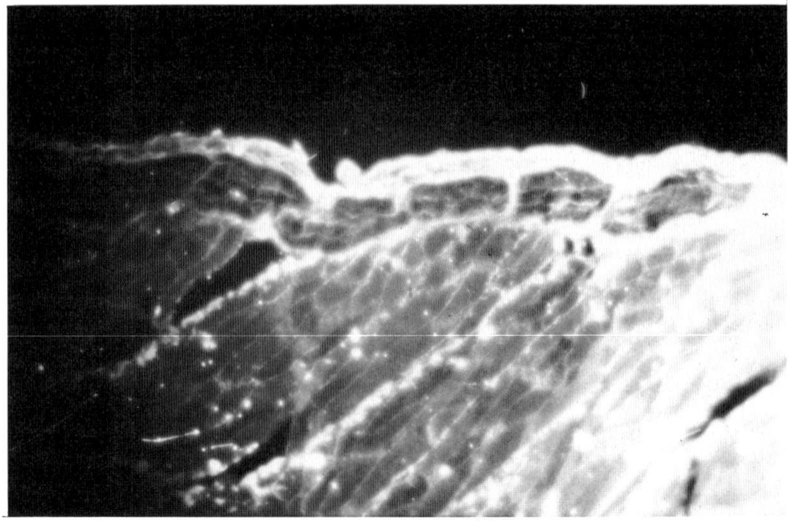

(a)

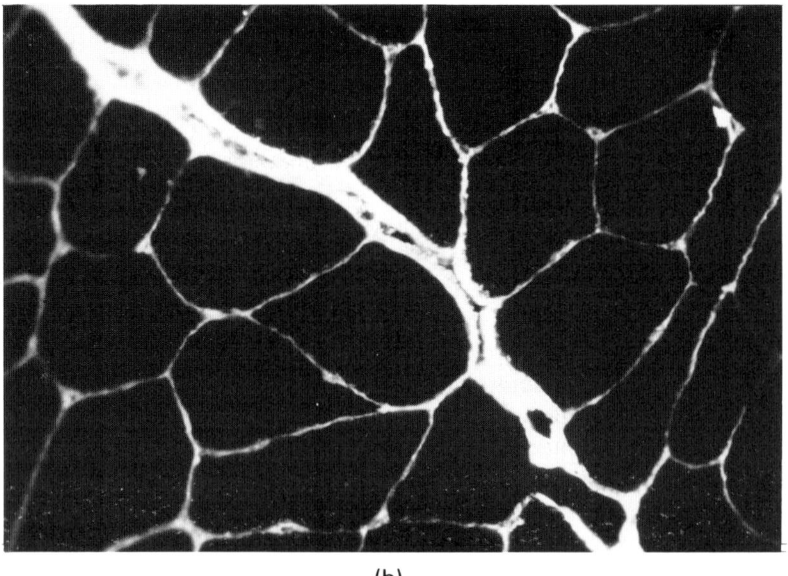

(b)

Fig. 6.8 Immunofluorescence microscopy of muscle showing the location of (a) Type I and (b) Type III collagen by staining with antibodies directed specifically against these types.

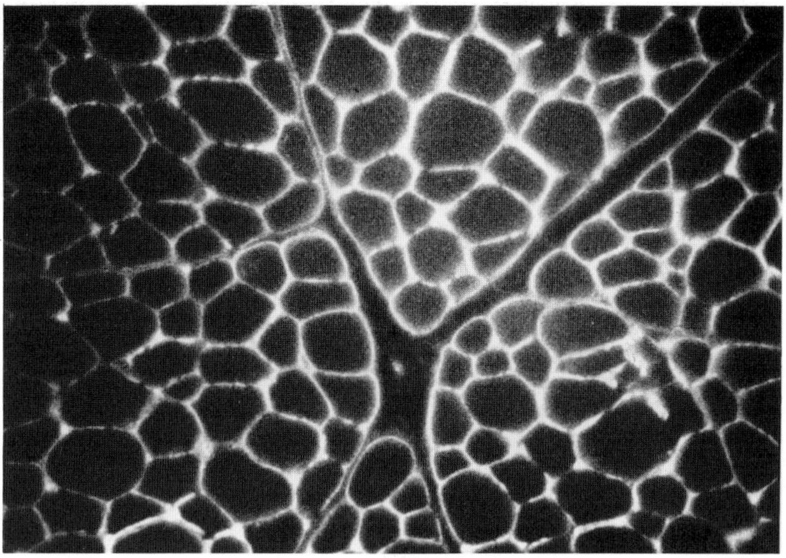

(c)

Fig. 6.8—*contd.* (c) Immunofluorescence microscopy of muscle showing the location of Type IV collagen.

known of the function of this collagen type in most tissues including muscle.

Preliminary studies on Type VI collagen have indicated a general distribution of this filamentous collagen throughout the perimysium by immunofluorescence techniques.

Biochemical localisation

Separation of the epimysium, perimysium and endomysium can be achieved by dissection and filtration and SDS extraction of the non-collagenous components (Chapter 12). The vast proportion of collagen in muscle from slaughter age animals is insoluble in dilute SDS solutions, hence all the collagen types can be investigated biochemically.

The epimysial, perimysial and endomysial fractions can be solubilised with the reagent cyanogen bromide (CNBr) which, as discussed in Chapter 2, brings about the specific cleavage of proteins at methionine residues and thus creates a unique and reproducible soluble peptide mixture from each genetically distinct collagen type. These peptides can then be separated by polyacrylamide gel electrophoresis or a chromatographic method. Marker peptides for each type can then, theoretically, be quantified and used to calculate the total or relative content of each genetic type. The method works extremely well for Types I and III but it is not yet feasible to do this for collagen Types basement membrane in muscle is packed with fine fibrils of collagen

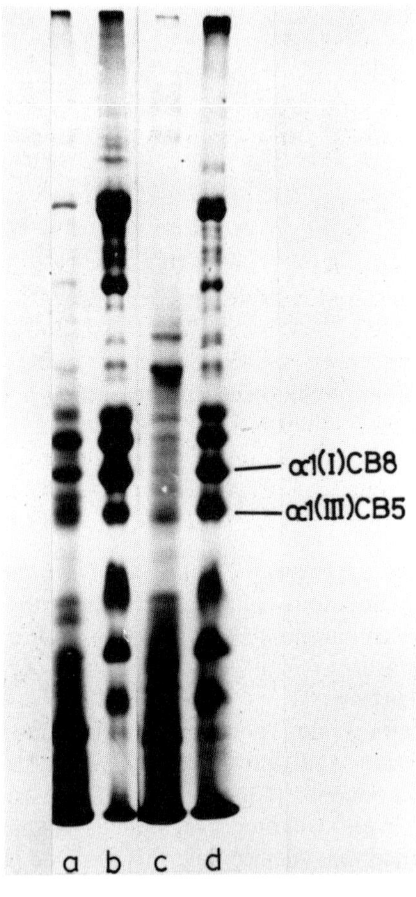

Fig. 6.9 Separation of CNBr peptides of Types I and III collagen by SDS-polyacrylamide gel electrophoresis after preparation of perimysium (a), (b) and endomysium (c), (d) from beef longissimisus dorsi. Samples (a) and (c) were partially purified preparations whilst (b) and (d) were washed with SDS.

IV and V. Figure 6.9 shows a typical result of an experiment in which the collagenous peptide mixture obtained from perimysium and endomysium illustrates how Types I and III collagens are distinguished by polyacrylamide gel electrophoresis. Two peptides are used to quantitate the two collagens as indicated on the figure. In this way the relative content of Type I and Type III collagens in epimysium, perimysium and endomysium has been calculated for a range of muscles from beef (Table 6.1).

These biochemical results support those of immunolocalisation very well. Type I is distributed at all levels in muscle connective tissue. Type III collagen is not present in all epimysia but is found in significant quantities in the perimysium (although Type I predominates here), and at roughly equal levels with Type I collagen in endomysium. As stated above, it has not been possible to apply these techniques to the quantitation of Types IV and V in muscle connective tissues. However, using total pepsin solubilisation of purified epimysium, perimysium and endomysium and subsequent salt fractionation of the soluble proteins obtained, it is possible to demonstrate categorically the presence of Types IV and V collagen in muscle endomysium. Type V collagen has also been shown to be present in very small quantities in perimysium. The total Type IV + Type V collagen in the endomysial fraction only represented about 10% by weight of the total connective tissue in the endomysial fraction. This fits with the knowledge that the pericellular area around the actual

Table 6.1
Type III collagen content of total connective tissue, perimysium and endomysium of six bovine muscles as a percentage of total collagen

Muscle	*Total Type III collagen (% of dry wt)*	*Type III collagen in perimysium (%)*	*Type III collagen in endomysium (%)*
Psoas major	0·6	43·2	55·2
Longissimus dorsi	0·4	24·5	53·3
Pectoralis profundis	1·9	40·7	58·5
Semitendinosus	1·5	41·7	56·4
Gastrocnemius	2·5	36·5	58·6
Sternomandibularis	1·2	40·5	58·1

[Reproduced from Light, N. D., *Advances in Meat Research,* Vol. 4, ed. A. M. Pearson, T. R. Dutson & A. J. Bailey. Van Nostrand Reinhold, New York, 1987.]

which are thought to be mainly Type III. These fibres have traditionally been referred to as reticular fibres or reticulin.

Given the rapid refinement of methods like high performance liquid chromatography for peptide separations, it should be possible in the near future to separate and to quantitate accurately these minor collagens even from relatively intractable tissues such as muscle. These new methods are mentioned in Chapter 12.

ELASTIN IN MEAT CONNECTIVE TISSUES

The belief that muscle elastin was mainly associated with the vasculature and the relatively low content of the protein in muscle has led researchers over the years to the assumption that it plays a negligible role in meat structure and, therefore, detailed studies of its distribution have been limited. In 1967 a study was carried out simply to assess the gross elastin content of a range of bovine muscles (Table 6.2). This was done by removing the total non-elastin material by controlled hydrolysis (see Chapter 12). This study revealed that the content of elastin was highly variable although generally very low, i.e. 5% of the total connective tissue or 0·1–0·5% of the total dry weight of muscle. This would seem to accord with the view that the elastin of muscle is located primarily in the blood vessel walls. However, one or two muscles, notably the semitendinosus, showed significantly higher

Table 6.2
Elastin content of a range of bovine muscles

Muscle	*Elastin as % dry wt*	*Elastin as % of collagen*
Sternomandibularis	0·19	2·5
Trapezium thoracis	0·65	7·4
Biceps brachii	0·11	2·2
Latissimus dorsi	2·0	33·0
Longissimus dorsi	0·07	2·9
Psoas major	0·08	3·8
Semitendinosus	1·82	37
Panniculus	1·13	10·4

Muscles of the hindquarter and loin contain less than 0·2% elastin on a dry weight basis, that is, less than 5% of the collagen. The semitendinosus muscle is the exception, containing 2% of dry weight and about 40% of the collagen.

proportions of the protein, i.e. around 40% of the total connective tissue or approximately 2% of the total dry weight of muscle. Longissimus dorsi was shown to have the lowest elastin content at 0·07% of the total dry weight of muscle or only 3% of the connective tissue.

The finding with regard to semitendinosus is an interesting one in view of the popularity of this muscle with researchers for microscopical and mechanical measurements. The semitendinosus is one of the few skeletal muscles in which the muscle fibre direction is easy to see macroscopically and relatively consistent throughout the muscle structure. It can, therefore, be easily dissected to yield viable sections for microscopy and mechanical tests. However, its unusually high elastin content makes it atypical and it may behave very differently in mechanical terms from the majority of other generally low elastin-content muscles. This factor must be considered when interpreting mechanical and microscopic data.

Detailed studies of the location of elastin in the semitendinosus and longissimus dorsi muscles have been carried out at the light microscope level using the Verhoeff's Van Gieson stain (see Chapter 14). These revealed heavy concentrations of thick elastic fibres (5–19 μm) throughout the epimysium and perimysium which were aligned with the long axis of the muscle fibres (Fig. 6.10). Heavy deposits were also

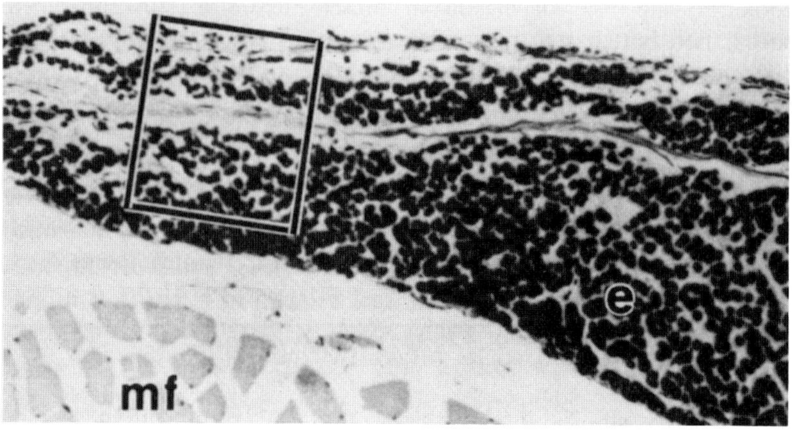

Fig. 6.10 Light micrograph showing the location of elastin in bovine semitendinosus muscle: e, elastin; mf, muscle fibre. Magnification: ×140.

observed at the junctions of perimysial sheets and embedded in the centre of the thicker perimysial tracts of collagen. In addition, finer fibres of elastin (1·2 μm) were observed in both epimysium and perimysium in an orientation parallel to the collagen fibres which were at an angle to the long axis of the muscle fibres. The semitendinosus contained coarse elastin fibres throughout the epimysium and in almost every perimysial sheet studied, and even the thinnest sheet contained fine elastin fibres. Longissimus dorsi, on the other hand, contained less coarse elastin in the epimysium and it was rarely seen in the perimysium as fine fibres of elastin. More detailed scanning electron microscopy studies revealed that the thick elastin fibres in the perimysium of semitendinosus were covered in collagen fibres. The perimysial fibres were in a cross-weave pattern (see Fig. 6.4) and the elastic fibres were longitudinally aligned along the perimysium. The histological staining techniques appear to give a higher elastin content than the previous chemical analyses. It is essential, therefore, that chemical analyses are performed on those muscles demonstrating high levels to confirm the histochemical results.

In view of these findings it would appear that the significance of elastin in meat has been overlooked. We only have data on a small range of muscles but even the muscle with lowest elastin content, the longissimus dorsi, appears to contain organised fibres which may well play a part in determining the textural properties of the tissue after cooking. The role of elastin in muscle tissue is still subject to speculation but, based on its ability in other tissues to provide rapid restoration of the tissue to its original state after deformation (e.g. in skin, artery and ligament), it could act in this way during the contractions of muscle, i.e. as an energy store to facilitate return to original muscle shape after contraction. It may be that this effect is not universal in all muscles but has been developed in certain cases such as the semitendinosus for specific reasons. Further work is obviously necessary to extend our understanding of this protein in muscle.

PROTEOGLYCANS IN MEAT CONNECTIVE TISSUES

In the early studies it was shown that the proteoglycan content (calculated from total hexosamine—Table 6.3) of various beef and sheep muscles closely paralleled that of elastin. The correlation between collagen and proteoglycan content was less close. It was

Table 6.3
Proteoglycan content of a range of muscles

Muscle[a]	*Proteoglycan hexosamine (mg/100 g)*	*Ratio of proteoglycan hexosamine to total N*
Steer PM	14·0	0·35
Steer LD	13·3	0·46
Steer ST	16·0	0·49
Calf PM	16·3	0·52
Calf LD	17·8	0·53
Calf ST	21·2	0·68

[a] PM = psoas major; LD = longissimus dorsi; ST = semitendinosus.

noticed that muscles from young animals contained higher proportions of proteoglycan than those of older animals, as was also the case for elastin and collagen. Later studies showed that chondroitin sulphate was the predominant glycosaminoglycan of skeletal muscle. This GAG was extracted using $CaCl_2$ solvent and was shown to be similar in characteristics to a commercial preparation of cartilage proteoglycan.

In a recent report these early studies have been corroborated by the extraction of a low molecular weight proteoglycan from rabbit skeletal muscle. In this case proteoglycans were extracted from homogenised rabbit skeletal muscle with 4M guanidine hydrochloride and were purified before characterisation. The chondroitin sulphate proteoglycan constituted 44% of the total muscle hexuronic acid content and was, therefore, by no means the sole proteoglycan present. The chondroitin sulphate proteoglycan had a molecular weight of 95 000 and consisted of polypeptide and GAG in a weight ratio of 1·0:0·6 with the protein core having a molecular weight of about 50 000. This small proteoglycan was compared with similar small cartilage proteoglycans and was supposed to be the major muscle proteoglycan.

Light microscopy has generally been employed to determine the location of proteoglycans in connective tissue, although recent refinements of the stains have permitted highly specific localisation studies. Analysis of tendons, usually the model tendons of rat tail or rabbit, have shown that there appears to be a regular orthogonal perifibrillar arrangement of proteoglycan molecules associated with collagen. The proteoglycan can be identified by histological stains such as Ruthenium Red or preferably Cupromeronic complexes. Furthermore, with the advent of monoclonal antibodies it is now possible to identify the individual type of proteoglycan attached to the collagen fibre. Tissues with high hyaluronic acid contents appear to have small collagen fibres whereas tissues with the largest collagen fibres were rich in dermatan sulphate. Unfortunately, no definite data are yet available on the possible roles for proteoglycans.

Detailed electron microscopy showed that the transverse elements of the regularly spaced proteoglycan–collagen aggregates were placed specifically in the gap zone of the collagen fibres. These results are interesting and may be important but, until further supported, should be treated with caution as they can only be obtained with fixed samples and therefore may be the results of fixation artefacts. The use of monoclonal antibodies should provide more definitive data.

In summary, then, it seems likely that proteoglycans are localised along the collagen fibres in a highly specific manner and chondroitin sulphate proteoglycan is an important component of muscle connective tissue, especially the perimysium. Its role may be to control fibre size and fibre organisation, and it may possibly play a role in the mechanical properties of the fibre. As such, it should be studied in more detail as it may have a role in contributing to meat texture through its interactions with other components such as collagen and elastin.

COMPONENTS OF BASEMENT MEMBRANE

In the basement membrane of the endomysium the Type IV collagen only forms about 40% of the total dry weight. The major proteoglycan is heparan sulphate which constitues about 1–5% of the dry weight. The other main constituents are the glycoproteins laminin, nidogen and fibronectin. Thus, it would seem that the chondroitin sulphate proteoglycan in the perimysium (and possibly the epimysium) and the heparan sulphate of the endomysial basement membrane are the two major proteoglycan constituents of muscle connective tissue. However, it must be remembered that the amount of basement membrane is very small compared to the epi- and perimysial tissues.

Table 6.4 shows a compilation of the known major macromolecular constituents of basement membranes. These data have been compiled mainly from the study of model basement membranes such as lens capsule, placenta and kidney glomerulus. Little work has been carried out on muscle basement membranes due to the difficulty in dealing

Table 6.4
Major components of basement membranes

Macromolecule	*Molecular weight*	*Assumed function*
Type IV collagen	550 000–600 000	Formation of network structure; binding to other components (laminin, fibronectin, heparan sulphate)
Heparan sulphate	130 000?	Filtration; binding to cell surface, Type IV collagen, laminin
Laminin	1000 000	Cell attachment to BM[a] and binding to Type IV collagen, heparan sulphate, nidogen and entactin
Fibronectin	450 000	Cell attachment to BM; binding to most other BM components
Nidogen (Entactin)	150 000	Binding to laminin, fibronectin and Type IV collagen

[a] BM = basement membrane.

with this tissue. However, most basement membranes appear to be closely related and so we may draw some conclusions about the endomysium from the study of other homologous tissues.

Studies of the localisation of these proteoglycans and glycoproteins in basement membrane are limited. An early model of basement membrane structure suggested that fibronectin, laminin and heparan sulphate are localised mainly in the lamina rara (otherwise known as the lamina lucida), whereas Type IV collagen is found only in the lamina densa. This model was proposed on the basis of immunolocalisation studies. More recent work using immunoelectron microscopy and rotary shadowing of various mixtures of individual basement membrane macromolecules has suggested a more integrated model. In this model, heparan sulphate and laminin are thought to be embedded in a polymerised matrix of Type IV collagen laid out in the open 'chicken-wire' network discussed in Chapter 2.

However, there must be some structural differentiation between the lamina rara and lamina densa to account for the simple differences in gross appearance of these two layers in standard electron microscopy preparations. Also, there is no evidence that Type IV collagen binds to cell surfaces and so it may not be expected to be found at the very outer surface of the basement membrane next to the cell. The charged macromolecules such as heparan sulphate and the cell attachment protein laminin are thought to be at the surface in the lamina rara.

SUMMARY

Simple light microscopy staining techniques have shown that the main component of meat connective tissue is collagen but that some muscles contain small amounts of elastin. The gross structure of muscle connective tissue is common to all skeletal muscles although variable in composition and properties. The three hierarchical domains of meat connective tissue, endomysium, perimysium and epimysium, are contiguous with the tendon and act to transmit the force of muscle contraction to the bone. The crimp structure observed in tendons is also present in perimysial fibres, which are arranged in sheets, but disappears when the tendon or muscle is under load. The perimysial sheets are arranged at varying angles to one another and these angles change under load. Perimysial fibres are 50–100 nm in diameter and are arranged in bundles of 600–800 nm diameter. These become smaller at the perimysial–endomysial junction (120 nm) where the endomysial fibres can be seen to be finer (40–50 nm) than their perimysial counterparts.

Immunolocalisation techniques have allowed the unequivocal assignment of Types I and III collagen to the epimysium; Types I and III, and to a lesser extent Type V, to the perimysium; and Type IV to the endomysium, together with Types I, III and V forming the underlying fine matrix. Biochemical work has corroborated these findings and shown that the relative amounts of these types in the epimysia, perimysia and endomysia of different muscles is highly variable.

Meat connective tissues also contain small amounts of non-collagenous components which contribute to the structure and function of the tissue in the live animal and the textural properties of the meat. Elastin occurs in all muscles as a component of blood vessel walls but is also found to some extent as coarse fibres in the epimysium. Longissmus dorsi has the smallest quantity of elastin (0·07% of dry weight) and semitendinosus has the highest (2% of dry weight). In the latter the protein is seen as coarse fibres in the epimysium, and also as fine fibres distributed throughout the perimysium. These were also seen to be embedded at the centre of larger perimysial tracts and at the junctions of perimysial sheets.

The major muscle proteoglycan (44% of total muscle hexuronic acid) is a small chondroitin sulphate-bearing molecule with a molecular weight of 95 000. The protein core has a lower weight of 50 000. The molecule probably predominates in the perimysium and decreases in content with the age of the animal. Microscopy has shown that there may be a specific binding of proteoglycans with collagen fibres. These associations are thought to influence collagen properties and parameters such as collagen fibre growth, fibre organisation and mechanical properties.

Basement membranes contain fibronectin, laminin, other minor glycoproteins like nidogen and the major proteoglycan, heparan sulphate. This latter molecule probably accounts for the majority of the residual proteoglycan not accounted for by the small chondroitin-rich molecule mentioned above. The exact structure of the muscle basement membrane is not known but probably resembles other well-described basement membranes. It is not yet clear whether Type IV collagen, fibronectin, laminin and heparan sulphate are distributed through the thickness of the membrane and involved in complex intermolecular interactions, or whether Type IV collagen is restricted to the lamina densa and the non-collagenous components to the lamina rara.

FURTHER READING

1. Rowe, R. W. D., Elastin in bovine semitendinosus and longissimus dorsi muscles. *Meat Sci.*, **17**, 1986, 293–312.
2. Bailey, A. J., Restall, D. J., Sims, T. J. & Duance, V. C., Meat tenderness. Immunofluorescent localisation of the isomorphic forms of collagen in bovine muscles of varying texture. *J. Sci. Food Agric.*, **30**, 1979, 203–210.
3. Duance, V. C. & Bailey, A. J., The nature, structure and function of the vascular basement membrane. In *Biochemical Interactions at the Endothelium*, ed. A. Cryer. Elsevier/North Holland Biomedical Press, Amsterdam, 1983, pp. 41–78.
4. Bailey, A. J., The chemistry of intramuscular collagen. In *Recent Advances in the Chemistry of Meat*, ed. A. J. Bailey. Royal Society of Chemistry, London, 1984, pp. 22–40.
5. Hamm, A. W., *Muscle Tissue in Histology*. J. B. Lippencott, Philadelphia, PA, 1969.
6. Pearse, A. G. E., *Histochemistry*. J. & A. Churchill, London, 1960.

Chapter 7

Mechanical and Thermal Properties of Connective Tissues

INTRODUCTION

As has already been discussed in previous chapters, because of its ability to form coherent and highly ordered cross-linked extracellular matrices (be they fibrous, filamentous or network), collagenous structures have a high degree of inherent strength. All connective tissues have a mechanical function and therefore a requirement for this tensile property. We have seen that the resistance and mechanical integrity of collagen fibres and matrices (networks) are reliant both upon the topography and composition of the macromolecular structure as well as upon the degree and type of cross-linking between molecules, fibrils and, possibly, fibres. In life, it is important to the normal functioning of the individual organism that the connective tissue does not break.

The strength of connective tissues and particularly collagenous structures has been compared by many writers to materials more common to the engineering world than to that of biology. For example, the Achilles tendon of a horse can withstand loads of up to 4000 kg (that is the weight of more than 55 average men) before it breaks. This is expressed, in mechanical terms, as an ultimate tensile strength of 50–100 Newton (N) mm^{-2}. Whilst under this load and before breaking, the average elongation of such tendons is only of the order of 15–30% their original length, illustrating how inextensible such collagenous tissues can be. By comparison, an aluminium alloy might be expected to have an ultimate tensile strength of 210 N mm^{-2} and an ultimate per cent elongation of 12%. However, these impressive comparisons only hold good with collagenous tissues in which the fibres are arranged in ordered parallel arrays like tendons and bone. In connective tissues like skin and intramuscular connective tissue, where the fibre orientation is more random and the content of collagen is lower, the ultimate tensile strengths recorded in mechanical tests are considerably lower. For this reason, and because of the difficulty in interpreting results obtained from a tissue which is not predominantly composed of just one molecular constituent, most studies on connective tissue mechanics have been carried out on tendons. Also, much of the early work was done with rat tail tendons due to the ease of preparation of these tissues and their physiological reproducibility. However, although they have been a good model system for other connective tissues, more recent work by meat scientists has concentrated on examining meat connective tissues directly.

The connective tissues of meat, as noted earlier, help to retain that tissue's mechanical and structural integrity by forming the hierarchy of sheaths and films which bind the contractile elements into a coherent whole. However, it is important to remember that muscle connective tissue must also perform an added and crucial function, that of helping to transmit the force of muscle contraction to the skeleton. Without this facility there would be no vertebrate locomotion. Mechanical strength and reproducibility of response in muscle connective tissue, which is constantly under variable and repetitive load in the normal animal, is therefore of paramount importance.

An additional factor uncommon in the world of biological, biochemical and biomechanical research is that of heat. Because virtually all meat products are eaten cooked, it has been important in meat research to consider the properties of muscle tissue before and after heating. Collagen, with its special properties, has been of particular importance and has therefore been quite extensively studied. It must be remembered that after heating denatured collagen possesses a fraction of its original strength and becomes highly elastic.

This chapter will discuss the basic facts about collagen mechanics, in general as well as in the particular case of intramuscular collagen, and will attempt to put into perspective the diverse and sometimes conflicting evidence about the behaviour of collagens in meat during and after heating. Because collagenous tissues are known to change their biophysical properties on ageing, due largely to changes in their cross-linking, the influence of these changes will also be considered with respect to meat and meat quality. Finally, we will briefly review the use of sensory analysis to determine the toughness of meat, i.e. its mechanical properties, and the contribution of the connective tissue components.

THERMAL PROPERTIES OF COLLAGEN

Research into the thermal properties of collagens has taken two forms. First, and by far the largest proportion of work to date, has been the investigation of the thermal properties of the molecule in solution and this has provided some important insights into the structure of the molecule as described in Chapter 2. Second, there have been the investigations of the properties of collagen as an insoluble matrix component. In many respects the latter, although more complex and harder to carry out and interpret, has proved more informative with regard to the behaviour and role of collagens in meat. The main methods used in this research with dilute solutions of collagen have been viscometry, circular dichroism, birefringence and differential scanning calorimetry (DSC). For insoluble collagen fibres dynamic mechanical analysis, stress–strain experiments with heated samples, isometric tension measurements, microscopical evaluations and DSC analysis have been utilised.

Collagen monomers

When heated in solutions, the rigid triple helical molecules denature over a narrow range of temperatures, the mid-point of which is referred to as the denaturation temperature T_D. The long rod-like helical molecules produce a highly viscous solution possessing a high optical rotation, but at the denaturation temperature all structure is lost and there is a dramatic decrease in viscosity and optical rotation. The denaturation of the triple helix has been shown to be a two-stage process with separation of the polypeptides as a first stage and the denaturation of their helical form as the second. Generally the solution produced, known as gelatin, will contain not only free α-chains but dimer, trimer and higher molecular weight components (Fig. 7.1).

The value of the denaturation temperature is dependent on the genetic type of collagen and the species of origin and can range from 15°C for cod fish skin to 50°C for Ascaris cuticle. This variation is determined primarily by the extent of post-translational hydroxylation

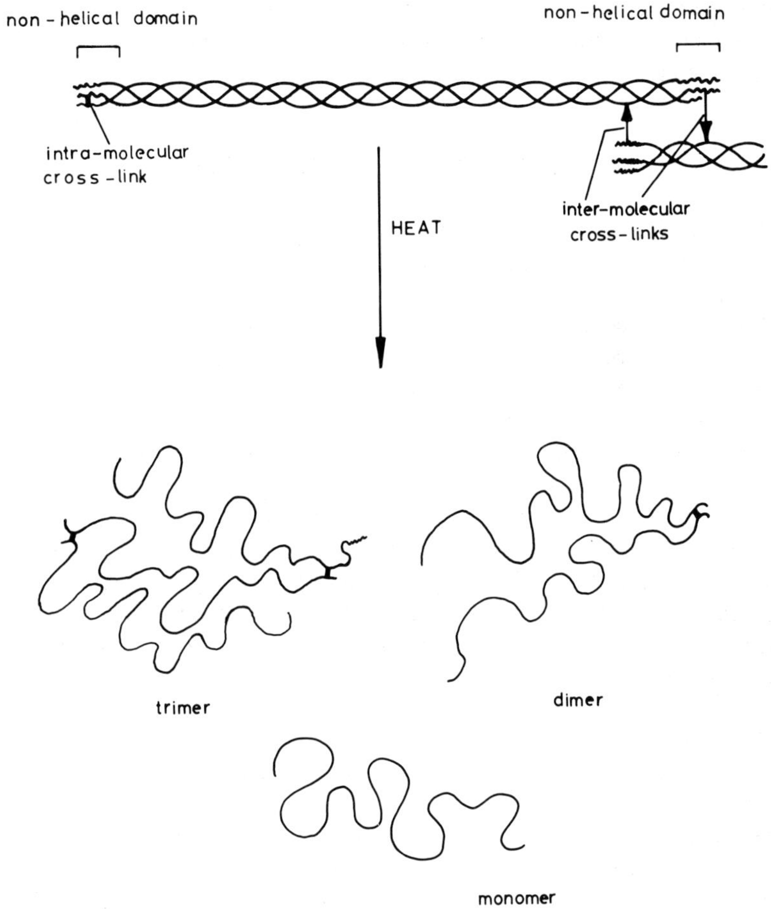

Fig. 7.1 Diagrammatic representation of the denaturation of collagen to gelatin.

of the proline as discussed in Chapter 2. The variation of T_D with species is shown in Fig. 2.4.

Fibrous collagen

The thermal denaturation of the collagen molecules when in the aggregated form of fibres is demonstrated by the collapse of the structure and consequent shrinkage of the fibre. The phenomenon of shrinkage provides a measure of the thermal transition and is

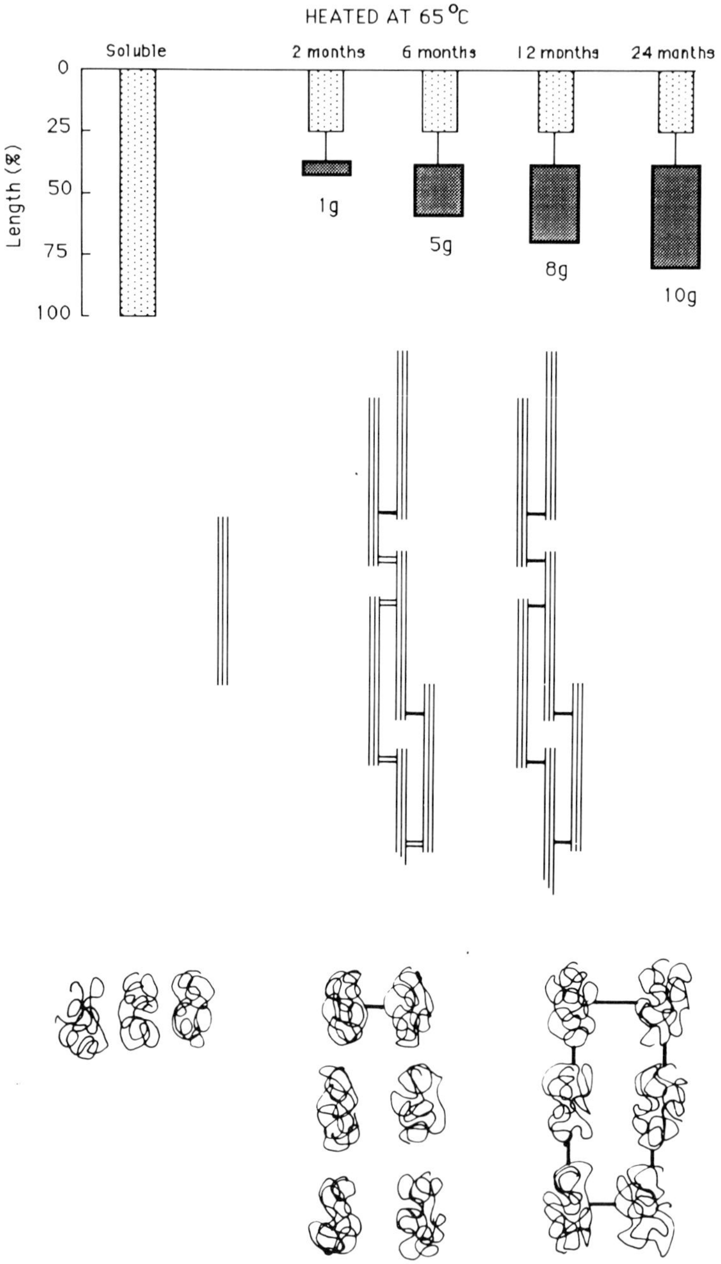

Fig. 7.2 Diagrammatic representation of the effects of heat shrinkage on the collagen fibre.

determined as the temperature at which the fibre, if it is unrestrained, shrinks to one-quarter of its original length (Fig. 7.2). This shrinkage temperature (T_S) is usually about 27°C higher than the T_D of the molecules due to the increased energy of crystallisation derived from the interaction between the close packed molecules in the fibre. The T_S of mammalian collagen is about 65°C compared to the 39°C of the molecule in solution. Similarly the T_S of cod skin is about 47°C compared to that of 19°C for the molecules in solution. The variation of T_S with species naturally follows the hydroxyproline content of the molecule, as shown in Fig. 2.4. The precise temperature of denaturation varies with the environmental conditions at which it is determined.

The exogenous factors which have been shown to affect T_S include pH, heating rate and solute concentration amongst others. Other important factors are the endogenous components such as glycosaminoglycans and proteoglycans. The former, in the shape of chondroitin-6-sulphate, chondroitin-4-sulphate, dermatan sulphate, keratan sulphate and hyaluronic acid have all been shown to have a stabilising effect on T_D, raising it from around 38°C to 46°C, but the effect on the T_S of the native collagen fibre is more difficult to assess.

The properties of the denatured fibre depend on the nature of the polymerising cross-links. In young tissues, with a low content of the heat-stable mature collagen cross-link, the amount of collagen liberated as gelatin by heating is related directly to the proportion of heat-labile aldimine cross-links stabilising the collagen. If this proportion is high then above 70–75°C, at which temperature the aldimine bonds break, most of the collagen will dissolve and form gelatin. This is observed with young rat tail tendon and in the case of cooked veal where there is an obvious high proportion of gelatin production which is exuded from the meat. In young tissues with a relatively higher level of the heat-stable oxo-imine cross-links, there will be little dissolution of the collagen even at very high temperatures (e.g. 100°C). Finally, the collagen in muscles from older animals such as mutton sheep, culled cows and bulls contains a high proportion of the heat-stable mature cross-links. Consequently, these tissues are much less liable to dissolve following thermal denaturation and remain intact and insoluble even at high temperatures (Fig. 7.2). In these cases the insoluble inextensible fibre is converted to an insoluble highly extensive 'rubber-like' material.

A further consequence of the presence of heat-stable cross-links is that as the collagen fibre shrinks to one-quarter of its original length on reaching its melting temperature, the fibre generates tension such that it can lift a weight without breaking the denatured fibre. More importantly the weight a tendon can lift during shrinkage increases dramatically with age of the tissue (Fig. 7.2). Alternatively, if the fibre is restrained at constant length during denaturation, the tension generated can be determined on a strain gauge. This has led to a technique using isometric tension (Chapter 13) to follow these changes and provide a direct relationship between the tension generated and the proportion of heat-stable cross-links in the collagen fibre.

It is worth noting that physical restraint appears to be a factor which influences the shrinkage temperature of collagen fibres. This is important as, in the native state in meat, collagen fibres may be expected to be under some tension and would certainly be restrained against complete shrinkage within the meat matrix. Experiments have shown that T_S increases when fibres are under load and that the stress–strain properties of collagen are also affected if the sample is restrained during heating. This factor, although controllable externally as an exogenous influence in laboratory tests, has to be classed as an endogenous factor under the conditions of meat cooking.

THE MECHANICAL PROPERTIES OF CONNECTIVE TISSUES

Four main methods of measuring the mechanical characteristics of meat have been used. These methods have all aimed to gain some insight into both the individual properties of the meat connective tissues themselves as well as into the part these tissues play in the determination of the overall quality and acceptability of meat. All four methods have been used with both raw and cooked meat or meat connective tissues and have therefore attempted to explain the changes that occur in meat connective tissues on heating in terms of the way the mechanical properties of meat are also affected by cooking. This aspect is obviously very important in view of the fact that, although the tissue is biologically interesting as it exists in the animal or in the laboratory setting, it is in the cooked state that it is eaten and in which condition it will be judged by the consumer.

The four main methods used for the determination of the mechanical properties of collagen in meat research, therefore, are:

(a) stress–strain mechanics,
(b) shear force determinations,
(c) isometric tension measurements,
(d) fracture mechanics.

Stress–strain mechanics

The classical mechanical experiments carried out on connective tissues attempt, in general, to measure stress and strain. The former is best described as the load per unit of cross-sectional area of the connective tissue concerned whilst strain may be conveniently defined as the deformation related to original state of the sample. In the case of strain, because most work has been carried out on parallel fibrous structures such as tendons, i.e. uniaxial structures, this is conveniently expressed as increase in length related to original length (usually as a percentage or unit of the original length). It is the relationship between these two parameters that has been concerned many of the major research groups of the last thirty years. In terms of meat connective tissue much of the work carried out has not considered stress–strain mechanics in isolation, but rather has attempted to correlate the results of such studies in whole meat samples with eating quality as assessed by taste panels, as we shall see later.

Native collagen

The mechanical behaviour of native collagen can be classified as viscoelastic, e.g. tissues such as tendon, or elastic, e.g. tissues such as aorta and cartilage. The stress–strain curve can be analysed from the relaxed state to the point of rupture and is generally divided into three regions, the toe region, the linear region and the relaxation and rupture region (Fig. 7.3). The toe region corresponds to the fibres becoming aligned and the crimp straightening as the stress increases. In the linear region the collagen fibres take the stress and obey Hooke's Law, and in the third region relaxation occurs with slow rupture of the various elements of the fibre. This property allows slippage before final rupture. Further information can be obtained by cyclic deformation of the tissue, or creep tests, to allow for the effect of time. These studies can provide a value for the ultimate tensile strength of the fibre which can be compared with other tissues.

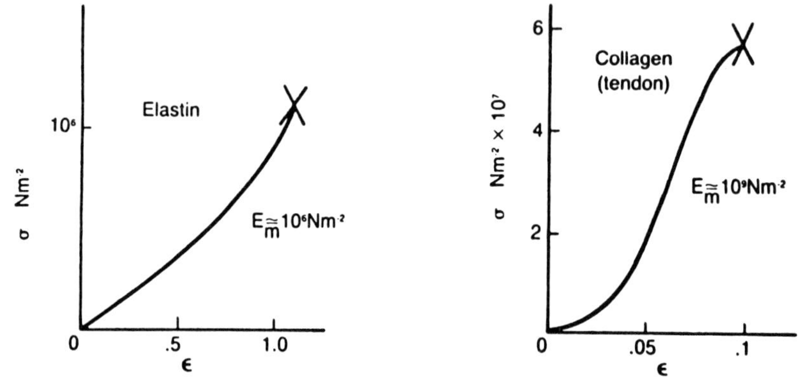

Fig. 7.3 Typical stress–strain curves for collagen and elastin.

Attempts to take into account the various components of the tissue, the inelastic collagen, the elasticity of elastin and the viscous properties of the proteoglycans, generally lead to very complex theoretical models built up of springs and dash pots. The springs represent the collagen fibre rigidity and the dash pots the viscous elements. This has led to useful descriptions of the relationships between the components of such diverse tissues as tendon, skin and aorta.

Figure 7.3 shows a typical stress–strain curve for tendon. As can be seen there is a linear relationship between stress and strain over much of the curve as the tendon is loaded until breaking point (ultimate tensile strength) is reached. Just before this point the curve plateaus, stress falls quite rapidly as strain increases and then rupture occurs. Stress, interestingly, is expressed in newtons per unit collagen as well as in units of $N\,mm^{-2}$. This is because it has proven difficult, if not impossible in some samples, for researchers, to accurately measure or calculate unit cross-sectional area. Because the measurements pertain to the amount of collagen present (especially in samples with very high collagen contents such as tendons), the method of expressing load per unit collagen (and, therefore, assuming a uniform collagen content and distribution in the sample under investigation) has found favour with a number of investigators. Some of these practical problems are discussed in Chapter 13.

A little work has been carried out on the stress–strain mechanics of native meat connective tissues (Fig. 7.4(a)). Most of the research in this field has concentrated on the measurement of yield force or other such parameters in whole meat samples and the correlation of these with sensory analysis of eating quality. In general, the complexity of meat as a tissue prevents the interpretation of such data as is obtained from these experiments in terms of the contribution of the various different elements of muscle.

There are many problems with the type of measurements made with apparatus designed for stress–strain experiments with connective tissues, not least the difficulty of obtaining accurate unit collagen or unit cross-sectional area measurements which can easily be related to the mechanical behaviour. Many experiments have been carried out and much has been written on this subject. Some of the major reviews are referred to in the bibliography to this chapter.

An analysis of the relative contribution of individual components of the tissues is difficult but can be attempted by selective removal of the various components. This can be achieved by the use of specific enzymes to remove proteoglycans and glycoproteins. An alternative approach is to reconstitute purified collagen fibres in the presence or absence of detected proteoglycans. As these new techniques are developed they should provide a better understanding of the interrelationship of the various components.

Thermally denatured collagen

Thermally denatured collagen behaves in a highly elastic manner and stress–strain measurements can be interpreted on the basis of rubber elasticity. It is therefore possible to determine the average molecular weight between cross-links. These measurements are undertaken at about 40°C to prevent any renaturation of the chains to a triple helical structure. Despite high variability of the results, the method does provide a direct measurement of the extent of cross-linking, and is therefore particularly valuable in comparative studies on age changes and effects of enzyme treatment. It should be remembered that the method only gives the cross-linking density for the thermally stable cross-links and cannot provide data on heat-labile cross-links originally present in the native fibre, for example the aldimine cross-link (Chapter 4) and some disulphide bonds.

The method has been used with meat epimysium and Fig. 7.4(b) shows a typical result. The data shows that the apparent number of cross-links present in the epimysial collagen increases in samples from animals aged from 3 months to 10 years, calculated as the average molecular weight between cross-links (M_c). From what is known about

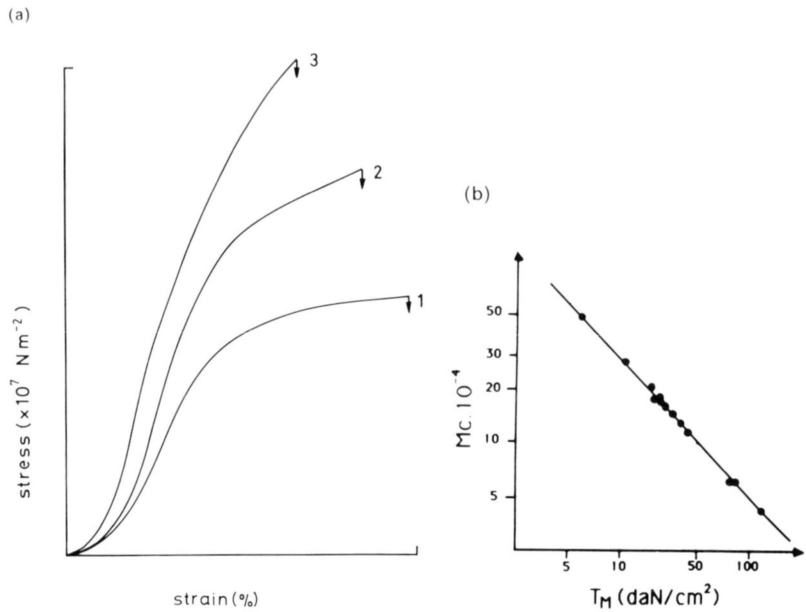

Fig. 7.4 (a) Stress–strain mechanics of epimysial preparations from longissimus dorsi muscle of beef animals with age ranging from 3 months (1) to 10 years (3). (b) Samples were tested after thermal denaturation and the molecular weight between cross-links (M_c) was calculated.

collagen cross-linking and physiological ageing this certainly does not relate to an increase in the number of reducible cross-links, but rather to the gradual conversion of these cross-links to heat-stable mature cross-links (Chapter 4).

Shear force

Another way in which the mechanical properties of meat have been measured is the classical shear experiment. In this, the meat sample is subjected to shearing forces produced by a specially designed knife blade which is pressed through the sample. By measuring the resistance of the meat it is possible to gain some insight into the behaviour of its component parts. As we have already seen (Fig. 1.7), results obtained for the instrumental determination of shear value (Warner–Bratzler or Volodkovitch—Chapter 13) for a range of beef muscles demonstrate distinct differences in shear value between the muscles investigated. This has been interpreted as a measure of their difference in toughness with the tender muscles having low values (e.g. psoas major, psoas minor, longissimus dorsi) and tougher or very tough muscles showing significantly higher shear values (e.g. brachialis, extensor carpi radialis, rhomboideus). In general, the results fit well with the known eating quality of the muscles investigated. Because this type of result allows a rapid, crude measure of quality, the method has been used extensively for quality measurements and inasmuch as it can only give a fairly gross guide to textural properties, has been useful. However, it cannot give a complete picture of the mechanical properties of meat, nor can it even give much information on the relative importance or role in textural properties of the individual muscle components.

Having said this, shear experiments have often been used in an attempt to discover the varying contributions of connective tissue and the contractile apparatus to muscle quality. Figure 7.5 shows some typical shear results obtained with samples from a single muscle cooked at different temperatures. As can be seen, the response curve has two distinct parts. These are thought to correspond to the actomyosin and collagen components and apparently show that the connective tissue is capable of developing higher resistance to shear

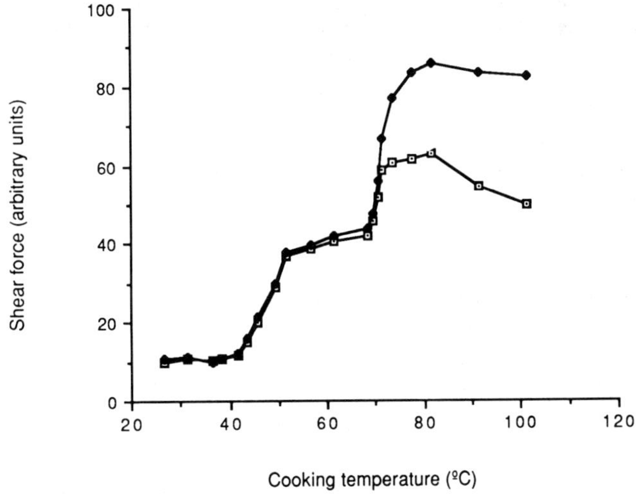

Fig. 7.5 Typical shear force results obtained with samples from a single muscle cooked at different temperatures. □, Calf muscle; ◆, ox muscle.

than the muscle proteins after cooking, and also that this resistance is related to the temperature of cooking. This result, quoted in many pieces of work on muscle toughness and texture, is thought to illustrate the relationship between the denaturation temperatures for the actomyosin and the collagen and their respective role in generating the characteristic texture of meat.

This kind of experiment can be misleading as it attempts to make quite subtle distinctions between the contractile and connective tissue proteins from what is essentially a complex and difficult set of results. The shear experiment is not, by its nature, the best approach to discovering the mechanical properties of elements of muscle. Perhaps a better approach, and one which allows direct measurement of the mechanical properties of connective tissue, in particular at high temperatures, is the technique which measures the isometric tension generated in a sample of connective tissue at a given temperature or over a temperature gradient.

Isometric tension

This method of investigating the mechanical properties of collagenous tissues relies on the fact that collagen shrinks when heated above the denaturation temperatures. When unrestrained by intermolecular cross-links this process is merely a part of the unfolding of the three α-chains which make up each molecule, i.e. the molecule denatures and the chains take on a random conformation in solution. When the collagen molecules are a part of a three-dimensional lattice made insoluble by keto-imine and mature cross-links (see Chapter 4) this shrinkage causes the retraction of the whole connective tissue (see Fig. 7.2). What this means is that, if the ends of a piece of connective tissue can be held at a constant length and the sample heated above the denaturation temperature, a force will be generated in the sample as the molecules try to shrink. By measuring this force (the *isometric tension*) it is possible to relate the behaviour of collagen to its role in the contraction of meat during cooking and, consequently, to the texture of meat.

The original experiments carried out on the isometric mechanical characteristics of collagen were done with rat tail tendon as an ideal tissue and skin as a more complex example (Fig. 7.6). More recently the method has been developed for use with specially prepared muscle epimysium fibres. Figure 7.7 shows typical examples of the kind of result obtained using an epimysial preparation.

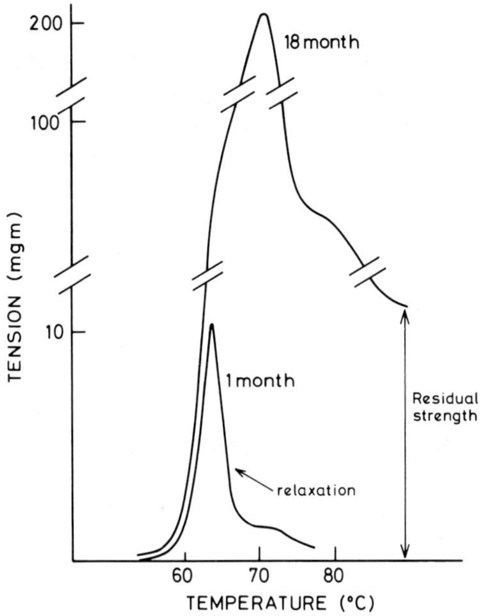

Fig. 7.6 Isometric tension experiment on 1 month and 18 month rat tail tendon showing rapid increase in tension generated with age and the difference in residual strength of the older tendon.

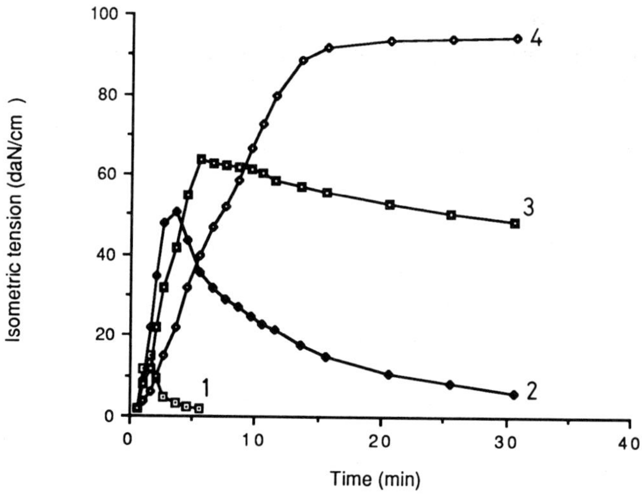

Fig. 7.7 Isometric tension experiment showing contraction and relaxation at constant temperatures (80°C) of bovine epimysium from (1) 5 month old calf; (2) 18 month old bull; (3) 5 year old cow; and (4) 10 year old cow.

The temperature at which shrinkage occurs is roughly constant with little change occurring with increasing age. However, the amplitude of maximum contraction increases dramatically with age. Many different tissues have been examined by this technique including tendon, skin and aorta to provide much valuable data. It is therefore worthwhile analysing the curves for bovine epimysium in Fig. 7.7 in terms of the nature of the cross-links. In young tissue the major cross-link is the heat-labile aldimine, hence following denaturation of the molecule the cross-links are also ruptured reducing the maximum tension generated and resulting in a loss of tension on further heating. Stabilisation of these cross-links with a reducing agent such as sodium borohydride increases the maximum tension generated and prevents the subsequent relaxation. Conversely, mild treatment of the fibre with D-penicillamine, which specifically cleaves the aldimine at physiological pH, results in the generation of a much reduced tension. With increase in age an increasing proportion of thermally stable cross-links are formed, thus increasing the maximum amplitude and decreasing the extent of relaxation on further heating. In aged tendon the tension generated can be twenty times greater, and virtually no relaxation occurs (Fig. 7.7). Therefore in addition to correlation of the maximum amplitude with age, the extent of relaxation also provides a measure of the residual strength of the denatured fibre. This residual strength relative to the maximal tension shows a direct relationship with increasing age of the tissue, again indicating a high proportion of thermally stable cross-links.

Although the perimysial fibres are more important in terms of determining meat texture, it is unfortunately more difficult to obtain uniform fibres than is the case for the epimysium and other model tissues such as rat tail tendon. The intramuscular perimysium exists as a variable network of layered collagen fibres and the tension generated will vary with the preferred direction. Some comparative studies have been carried out and the perimysium clearly follows the age-related changes observed in other tissues.

This understanding of the variation in the extent of thermal contraction, the tension thereby generated and the strength of the residual fibres has provided an excellent basis for developing our theories on the role of collagen in the texture of meat (Chapter 8).

Fracture mechanics

The simple shear force tests discussed earlier have been fairly successful in correlating toughness of meat with taste panel evaluations

(a)
(b)

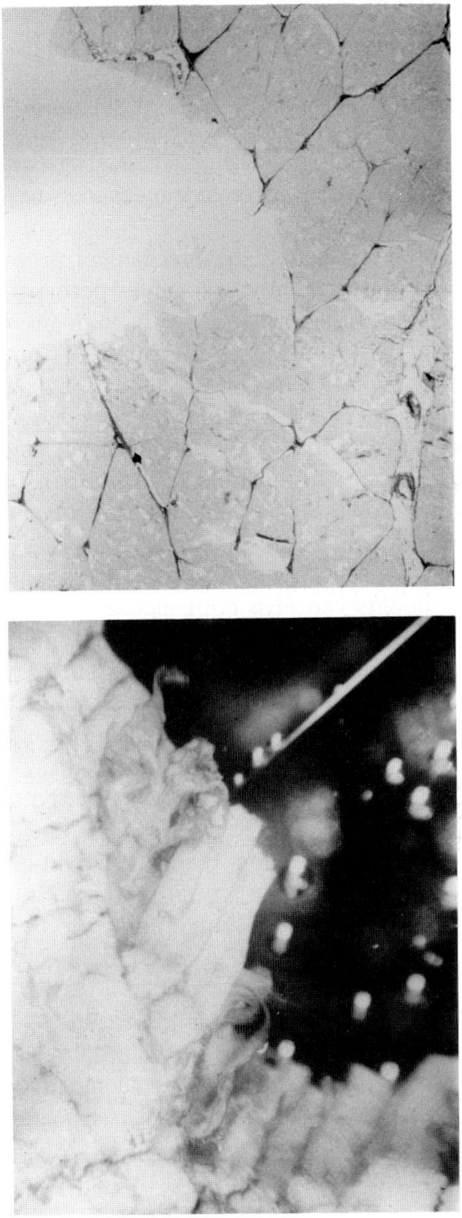

Fig. 7.8 Scanning electron micrographs showing opposing new surfaces created by tearing cooked meat. Note that one surface (a) is almost devoid of perimysial collagen fibres whilst the other (b) is covered in these, implicating the endomysial–perimysial junction in (a) as the major route of fracture in this experiment.

of textural properties. Unfortunately these techniques tell us nothing about the structural events occurring in those tests. We need to know what structures break first and at what stresses. An attempt to achieve this has been the application of fracture mechanics to the properties of cooked meat.

This method of investigating the mechanical properties of muscle has come into use only recently and offers perhaps one of the best opportunities to understand what actually happens in cooked meat tissues when they are subjected to the kinds of forces produced by chewing. The act of mastication brings about breakage or fracture of the elements of a piece of meat. When we consider a piece of meat it seems extremely unlikely that every part of it will be equally resistant or yielding to breakage; it is much more likely that some parts will be more easily broken down than others. It is important to know which parts of the cooked meat resist the fracture process and which parts may act as a route for the fracture.

Fracture experiments can give both mechanical and ultrastructural information as it is possible to carry out the mechanical fracture experiment (i.e. to produce two or more new surfaces in a sample by breaking it), and then view the new surfaces in either the light or electron microscope. Stress and ultimate tensile strength (UTS) can be easily measured using fracture mechanics, but this type of data only describes how 'strong' a material is and can be misleading. For example, glass, in terms of UTS is stronger than wood, and yet because it is so brittle and shatters readily it would not be considered as useful as wood as a strong structural material. In meat mechanics the UTS may not be a very useful parameter to measure. More interesting is the ease with which a material might be torn; or put another way, the force needed to propagate a tear or crack through a material. This type of parameter gives a measure of the resistance to breakage (tearing or cracking) of a material such as meat and therefore some idea as to the parts which may most effectively resist or assist crack propagation—resistance to crack propagation is known as toughness in materials science.

The critical stress intensity factor (CSIF) is the minimum applied stress needed to make an existing defect or crack in a material grow, whilst the critical strain energy release rate (CSERR) is the minimum energy needed to drive a crack through a material. The latter parameter is probably best applied to biological tissues such as meat which behave as non-linear elastic materials. The simplest energy-based measurement is that of work of fracture which may be defined as the absolute minimum strain energy that has to be expended to cleave through unit cross-sectional area of a material. Tough materials have a high work of fracture whilst brittle or frail materials (like glass) have low works of fracture.

To understand the concept of what happens when collagenous tissues fracture it is useful to consider the fact that they do not yield suddenly at UTS but rather show that fibres split into fibrils, then fibrils into sub-fibrils and so on until the tissue finally breaks. This process may be referred to as surface or plane debonding. Equally important is the fact that it is considerably more difficult to tear a tendon across its width, i.e. at 90° to the axis of the tendon fibres, than it is to tear it along its axis. Thus, the orientation of the fibres themselves play a role in determining the force needed to propagate a crack.

Similarly the fracture properties of meat will depend on the direction of crack propagation since the muscle fibres are aligned. The tensile stress needed to break the meat pulling along the muscle fibre direction is over ten times that required in pulling the fibres apart laterally because in the former case both the muscle fibres and the connective tissue have to be broken. However, the initial events are the same in both directions. The meat begins to break by splits opening up along the boundaries between the bundles of muscle fibres. This appears to involve the perimysial collagen layer being pulled away from the endomysium, the muscle fibre bundles remaining intact. Tension applied along the fibre direction also results in fracture of the perimysium, eventually followed by fracture of the muscle fibre. In other words, in cooked meat the weakest component is the intramuscular connective tissue.

Microscopy of cooked meat samples after subjection to the kind of mechanical tests described above shows that some new surfaces created by tearing and crack propagation are devoid of perimysial connective tissue whilst others are not. Surprisingly, the endomysial surfaces on stripped or bare samples remain relatively undamaged (Fig. 7.8). This suggests that the perimysial–endomysial junction is the initial site of fracture. This indeed may be the case as the area has been shown in other electron microscopy studies to be apparently less dense and contain finer fibres than the surrounding perimysial material (Chapter 6). Quantitative measurements for the work of fracture show relatively low results confirming that the perimysium, or perhaps more precisely its junction with the endomysium, is a zone of weakness in meat samples after cooking.

These results shed new light on the issue of the role of collagen in contributing to muscle texture. It seems more likely that the links between the hierarchy of collagenous structures within meat (i.e. between the endomysium and perimysium and between microscopic and macroscopic perimysium) may well provide the major route of cleavage of meat during mastication. This raises interesting issues about the variation in the strength of these hierarchical interfaces in different muscles. Are the perimysial–endomysial junctions in sternomandibularis (a tough muscle) stronger and more inherently mechanically integrated than the same sorts of junctions in a muscle like psoas major (a very tender muscle)?

It is clear that no single mechanical method provides sufficient detail to identify the relative importance of the various components of meat in determining its mechanical properties. The fracture mechanical studies indicate the importance of collagen since the point of cleavage appears to be along the perimysial tracts. Although collagen is the strongest component in fresh meat, its thermal denaturation to gelatin during cooking makes it the weakest component in cooked meat.

SENSORY ANALYSIS

Sensory analysis is an additional method of determining the mechanical property of meat by using human tasters in organised and formalised experiments. Details of the various techniques available for sensory analysis are fully discussed in Chapter 13 and will not, therefore, be dealt with here. Sensory analysis always involves the tasting of whole, cooked meat samples either by trained taste panels using accurate profiling or scoring methods or by consumer panels. These techniques yield data on the general quality aspects of meat treated in different ways, from different sources or cooked in different ways. These quality attributes can then be correlated with data obtained from other methods of physical testing such as stress–strain analysis or with the known biochemical composition of the samples involved. From this type of approach researchers have tried to build up a picture of the relative importance and the role of connective tissues in contributing to meat quality.

Early work with taste panels showed that there is a good negative correlation between tenderness and the amount of alkali-insoluble collagen in meat. In other words, the greater the amount of insoluble collagen that is present the lower the eating quality may be expected to be. This result was subsequently confirmed by later work using both taste panels and mechanical measurements to correlate with collagen solubility data. A later study, along similar lines, suggested that it is both the total content and solubility of collagen and elastin in meat connective tissue that are major factors determining meat texture. The influence of total collagen on eating quality seemed, therefore, well established by sensory analysis.

Another study showed that total collagen content affects subjective tenderness scores whereas collagen solubility affects objective scoring. This result was later extended when it was suggested that differences in the structure of the collagen molecule itself are responsible for perceived differences in texture; i.e. that there is a close relationship between the cross-linking of collagen in meat connective tissue and tenderness scores. Thus, these later studies led to the conclusion that it is the physico-chemical nature of the collagen in meat (i.e. the 'quality' of the collagen) rather than its mere quantity that is responsible for the perceived differences in meat texture between different muscles. This hypothesis will be dealt with in more detail in the next chapter but it is worth noting again that these sensory studies do not provide any primary data on the role of collagen in meat texture. They can only ever provide data about the perceived quality of meat which can then be correlated with other factors. Indeed, when considering the multiplicity of factors which might affect texture it is evident that considerable caution needs to be exercised in making such correlations. However, well conducted sensory experiments are invaluable in providing basic data on the eating quality of meat.

SUMMARY

The most important physical properties of collagen are its unusually high tensile strength, and its highly individual thermal properties. The thermal properties of collagen can be subdivided into those of the molecule in solution and those of the fibre. The denaturation temperature, T_D, of collagen molecules is very sharply defined, usually occurring over an interval of 1–2°C. The temperature at which this denaturation occurs ranges between 5 and 50°C, depending on the species of origin and the hydroxyproline content. Denatured collagen molecules form a mixture of free chains (α, β or γ) in solution called gelatin. The thermal transition of collagen fibres is measured (by direct scanning calorimetry and other methods) as T_S, the shrinkage temperature. This is because the fibres normally do not dissolve, but shrink on heating. $T_S - T_D$ is constant for any given collagen type and is normally 22–27°C. Many factors affect T_S, notably the influence of non-collagenous connective tissue components such as glycosaminoglycans. The denaturation of mammalian collagen fibres occurs at around 65°C and the subsequent extent of solubilisation depends on the proportion of heat-stable cross-links. The fibres shrink to about one-quarter their native length, a factor which leads to the generation of considerable tension within the insoluble matrix.

The mechanical properties of collagenous tissues can be measured by several methods: by stress–strain methods, by shear techniques, by the investigation of the tension developed during thermal denaturation (isometric tension) and by fracture mechanics. Classical stress–strain mechanics involves stretching connective tissue to breaking point and can be used with isolated connective tissues in the raw or cooked state. Collagen shows a three-phase stress–strain curve with a toe region in which the crimp is taken up, a linear region in which Hooke's law is obeyed and a relaxation region in which slippage of molecules and eventual rupture of the fibre occurs. Isometric tension methods can be used to establish the behaviour of various types of collagenous tissues when heated, the extent and type of cross-linking affecting the way such tissues react. Fracture mechanics gives a new approach to the investigation of the physical properties of collagenous tissues. This method allows the combination of controlled damage (fracture) of connective tissues and the concomitant measurement of stress–strain relationships as well as subsequent photographic and microscopic analysis of fracture surfaces. The method has been successfully applied to the study of the fracture of cooked meat.

Sensory analysis using trained taste panels has provided indirect correlative evidence for the role of intramuscular connective tissue, particularly collagen, in determining meat texture. However, sensory methods can only provide an overview of the relative eating quality of meat samples.

FURTHER READING

1. King, N. L., Thermal transition of ovine connective tissues. *Meat Sci.*, **20**, 1987, 25–37.

2. Kopp, J. & Bonnet, M., Stress–strain and isometric tension measurements in collagen. In *Advances in Meat Research,* Vol. 4: *Collagen as a Food,* ed. A. M. Pearson, T. R. Dutson & A. J. Bailey. Van Nostrand Reinhold, New York, 1987, pp. 163–185.
3. Purslow, P. P., The physical basis of meat texture: observations on the fracture behaviour of cooked bovine *M. semitendinosus. Meat Sci.,* **12,** 1985, 39.
4. Viidik, A., Danielsen, C. C. & Oxlund, H., On the fundamental and phenomenological models, structure and mechanical properties of collagen, elastin and glycosaminoglycan complexes. *Biorheology,* **19,** 1982, 437–456.
5. Bailey, A. J. (ed.), *Recent Advances in the Chemistry of Meat.* Royal Society of Chemistry, London, 1984.

Chapter 8

The Role of Connective Tissue in Determining the Textural Quality of Meat

INTRODUCTION

Meat, with one or two notable exceptions (e.g. steak tartare, Japanese raw fish) is always eaten cooked. The process of heating physiological tissues brings about a range of well-described changes in structure and function which, in meat, as well as in all other foods, affect appearance, flavour, odour and, most importantly, texture. Large carbohydrates can undergo structural changes and significant degradation. Proteins also undergo structural changes associated with heat denaturation as well as side chain reactions such as Maillard additions to free ε-amino groups. The components of connective tissue are no exceptions to these general rules. This means that cooked meat, the product that is actually eaten by the consumer, is an entirely different commodity from the raw material. In essence, therefore, when we begin to consider the role of connective tissues in determining the textural properties of meat we have to use our knowledge of both the native, physiological tissue and its denatured form to be able to understand the situation that exists in the cooked product.

Meat has been described as a 'two-component system'—being composed of the complex intracellular contractile apparatus and the compositionally minor extracellular connective tissue. The former was shown by early experimentation to contribute the major element of meat texture through its heat denaturation during the cooking process but connective tissues, particularly collagen, have for a long time been associated with more subtle textural effects. Collagen has been said to contribute the so-called 'background' toughness of meat. If indeed the connective tissue only provided a background contribution to texture, then we would not expect that age of the tissue or limited damage to the connective tissue brought about during conditioning and artificial tenderising processes would bring about significant changes in the overall toughness of the meat. Hence this simplistic view, as we shall see, has now been considerably out-dated by research in which collagen has been implicated as a major contributory factor to the textural properties of cooked meat. The roles of the minor components elastin and the proteoglycans have received little attention, but it is unlikely that they have a dramatic effect except in exceptional circumstances.

The link between intramuscular collagen and meat texture was made as long ago as the beginning of this century and subsequent work between 1940 and 1970 began to reveal a complex picture of the way in which collagen contributed to the overall eating quality of meat. Various research workers suggested different roles for intramuscular collagen in this respect including proposals that the total amount of collagen was directly related to toughness; that the organisational pattern of collagenous fibres within the muscle held the key to textural differences between muscles; and that it was the molecular structure of the collagenous components that was the most important factor in influencing quality.

During research into these areas in the 1970s it became clear, based primarily on the study of the cross-links and hence the factors affecting the shrinkage process, that the 'quality' rather than the quantity of the collagen was important. There are several ways in which the 'quality' of intramuscular collagen could be expected to determine the textural properties of cooked meat. These include the types of collagen present in the various hierarchical domains of meat, the heat stability of the intermolecular cross-links, the overall extent of cross-linking in the collagen matrices of muscle, the size of collagen fibres and fibre bundles, and the damage incurred by collagen during conditioning. In this chapter we will consider each of these parameters in turn and try to obtain an overall view of the way in which connective tissue, specifically its major component collagen, affects meat texture and thereby quality. However, it should be emphasised that research work into the molecular mechanisms of the contribution of collagen and other connective tissue components to meat texture is as yet incomplete.

EFFECTS OF COOKING ON INTRAMUSCULAR COLLAGEN

The thermal properties of isolated collagens and collagen fibres were discussed in detail in Chapter 7. In this section, therefore, we will briefly consider the major changes induced in meat collagen by heating, in order to provide a background for the rest of this chapter.

When meat is cooked collagen molecules within fibres undergo denaturation. Because of the intermolecular interactions within the fibre this leads to shrinkage of the insoluble mass at about 65°C. Although there may be a gradation in shrinkage depending on the cooking temperature and time, in order to develop the hypothesis for the role of collagen we shall make the assumption that the fibre is fully denatured during cooking.

Early microscopical studies showed that at temperatures above 65–70°C fibrous collagen was converted from a fibrillar to a rubber-like amorphous structure, losing its 67 nm periodicity in both perimysial and endomysial fibres at about the same rate. Cooking initially induces thermal shrinkage in perimysial collagen resulting in a loss of the crimp structure. The crossed-lattice structure of the perimysial sheets (see Chapter 7) becomes less distinct as the structural integrity of the matrix breaks down and detail is lost. The denaturation of the collagen fibres can be followed with specific microscopical stains (see Chapters 6 and 12) and application of these techniques to meat has indicated the degree of gelatinisation of collagen fibres in the perimysium and endomysium, although this work was qualitative and the extent of such an effect is hard to judge.

Other studies have shown that meat collagen is further affected by heat in that increased amounts of connective tissue collagen can be released as soluble material by cooking at successively higher temperatures. Several groups have reported the isolation of increased quantities of whole chains and β-components from meat connective tissues after heating, indicating that heat-labile aldimine bonds are broken during the process as would be predicted from the known thermodynamic properties of this bond. However, there has been some misunderstanding in the literature about the extent and importance of this solubilisation. Recent work has indicated that at 90°C, even after extended heating, only about 12–15% of total perimysial collagen is released as soluble material. This means that 85–88% of the perimysial collagen remains as an insoluble shrunken matrix after cooking.

These two effects of cooking, shrinkage and partial solubilisation of the tough collagenous tissues, the mechanisms of which are more complex than previously thought, are crucial to any understanding of the role of collagen in the tenderness of meat.

Influence of the total amount and pattern of intramuscular connective tissue on meat texture

It has long been known that different muscles have entirely different eating qualities. By means of sophisticated taste panel analysis (See Chapter 13) it is possible to 'rank' muscles in terms of their usual tenderness under normal conditions. Thus, psoas major, longissimus dorsi and gluteus medius are normally considered the most tender beef muscles, whilst gastrocnemius, semitendinosus and pectoralis profundis are examples of muscles of intermediate toughness and serratus ventralis, exterior capri radialis and sternomandibularis are generally considered to be tough. All of these muscles have very different structural qualities and different roles in the living animal. As collagen forms strong fibrous structures linking muscle elements together it is not surprising that it has been hypothesised that the total amount of this component must determine meat texture. By comparing the overall connective tissue content of such muscles it is possible to draw some gross conclusions about the role of connective tissue in determining textural quality.

Early research showed that there was some degree of correlation between the total visible connective tissue in meat (as perceived by a subjective visual method) and the eating quality of different muscles. Thus, muscles with high levels of visual connective tissue were generally associated with toughness. However, a number of anomalies arise here which it would be timely to discuss. The proportion of collagen in the muscles of young 'veal' animals is higher than the same muscle in a mature animal, yet the latter is always tougher. Muscles such as gastrocnemius, semitendinosus or sternomandibularis contain large tracts of connective tissue in the form of tendon-like strands. Such structures would obviously contribute to a high overall connective tissue (and, therefore, collagen) content by visual determination or if the whole muscle was sampled and homogenised for biochemical determinations. However, the trimmed areas of such muscles can sometimes be of surprisingly high eating quality. The connective tissue content of the muscle can therefore be misleading unless strictly defined.

An early piece of work suggested that the overall content of collagen in beef muscles varied from 1–9% of the total dry weight of the tissue. Considering its importance in muscle structure and function these are remarkably low levels. Indeed, it is now recognised that only very few muscles have levels of collagen content as high as 9–10% and these tend to be those lower grade muscles which contain major tracts of intramuscular connective tissues which are almost tendinous in nature. The vast majority of muscles appear to contain 1–3% collagen of their total dry weight.

Such quantitations have been correlated with meat quality as assessed by both mechanical and sensory analysis. Some workers have suggested that eating quality shows no correlation with collagen content whilst others have contended the opposite. Today it is generally accepted that there is some correlation of total collagen content with eating quality particularly between animals of the same age. Figure 8.1 shows the correlation obtained between quality, as

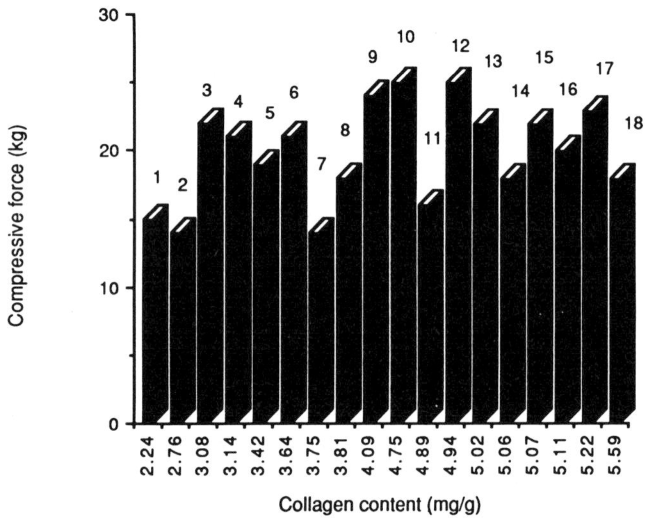

Fig. 8.1 Graph showing the correlation between compressive force as a measure of textural quality and the collagen content of various meats. The bovine muscles used in this study were: (1) psoas major, (2) longissimus dorsi, (3) gastrocnemius, (4) rectus femoris, (5) triceps brachii, (6) gluteus medius, (7) infraspinatus, (8) rectus abdominus, (9) semitendinosus, (10), serratus ventralis, (11) pectoralis profundus, (12) latissimus dorsi, (13) biceps brachii, (14) biceps femoris, (15) supraspinatus, (16) extensor carpi radialis, (17) complexus.

Table 8.1
Collagen content of epimysium, perimysium and endomysium of four bovine muscles as a percentage of the total collagen

Muscle	*Total collagen (% dry wt)*	*Collagen in epimysium (%)*	*Collagen in perimysium (%)*	*Collagen in endomysium (%)*
Psoas major	2·24	15	79	24
Longissimus dorsi	2·76	13	80	34
Pectoralis profundis	4·96	22	98	42
Semitendinosus	4·75	29	54	41

assessed by compressive force measurements, and total collagen content in cooked samples of a variety of beef muscles. It is clear that the correlation is not good but, nonetheless, that total collagen might show some experimental link with quality.

Table 8.1 shows results of the quantitation of collagen content in the epimysium, perimysium and endomysium from four bovine muscles of different eating quality. The collagen content of the dry weight of the three hierarchical domains was found to be 13–35%, 54–96% and 24–42% respectively. The observed variations in collagen content of the epimysium, perimysium and endomysium may represent an expected variability in the purity of each separated connective tissue fraction.

A study of bovine forequarter muscles confirmed that there was an approximate correlation between total collagen content and the mechanical properties of the cooked meat. In general, muscles showing low total work in the mechanical tests had lower collagen contents. However, there were notable exceptions to this observation; for example, infraspinatus, the muscle with the lowest result for the total work done in mechanical tests, had the second highest total collagen content. Also, the mechanical tests ranked gastrocnemius as more tender than gluteus medius, the reverse of the normally accepted sensory ranking of these two muscles.

In summary, it seems likely that the total collagen content of muscles does influence their textural properties but is difficult to correlate directly. A direct correlation of collagen content is obviously simplistic and has led to some considerable confusion. Subtle variations in texture between muscle types are more likely to be influenced by the distribution and quality rather than the mere quantity of collagen present, as we shall see later.

Influence of intramuscular collagen solubility on meat texture

The solubility of collagen in meat is affected by several parameters. For example, in tissues obtained from young animals (suckling pig, veal, etc.) the proportion of heat- and acid-labile aldimine cross-links in the collagen (Chapter 3) may be higher so that acid extraction methods would be expected to show a much higher proportion of soluble collagen in these tissues compared with samples from older animals. Similarly, in young tissue the amount of newly synthesised non-cross-linked collagen will be higher than in older tissue so that the proportion of salt or denaturant extractable collagen will be expected to be relatively high. Finally, as conditioning is known to lead to damage of the collagen in perimysial fractions, conditioned samples would be expected to yield higher quantities of soluble hydroxyproline than meats conditioned for short times at the same temperature. Thus, collagen solubility experiments in meat reflect differences in the age of the animal and conditioning, as well as inter-muscle differences.

Solubility of meat collagen has been assessed by various means using extractants such as acids (acetic acid, acid citrate buffers) and alkalis (sodium hydroxide), salts (neutral salt solution—see Chapter 12), denaturants and chaotropic reagents (urea, calcium chloride) and water (either at low temperature—20°C—or high temperature—85–100°C). Early research showed some positive correlation between the amount of extractable collagen and the tenderness of meat, but later work refuted this conclusion. Indeed, differences in collagen solubility have been found for different regions of the same muscle.

As discussed above and highlighted in other chapters (Chapters 3 and 9), collagen solubility is influenced only by its cross-linking and physical integrity (i.e. whether or not it has been attacked by proteases). Thus, reasons for differences in collagen solubility may be complex and might not necessarily reflect true variations in textural qualitites. It would seem much more relevant to study the cross-link content and proteolytic damage of the collagen fractions of interest directly than to rely on such a derivative factor as solubility.

Influence of pH and water-holding capacity

It is generally agreed that high pH meat is more tender, presumably due to the swelling of the proteins above their isoelectric point. Similarly, at low pH the proteins again swell and there is a progressive increase in tenderness, an effect utilised in marinating. The decrease in shear force values is very rapid between pH 4·5 and 4·0 and then slower to pH 3·5. Clearly the major effect is the water-holding capacity of the myofibrils, but the differences in values of shear force seem to be related to the collagen content of the muscle. Certainly the perimysial sheath can be observed to swell extensively. The swelling of collagen on either side of its isoelectric point has been well researched, and as swelling increases the tensile strength of the fibre decreases, and the collagen is more labile to heat. Although the effect of marinating on collagen has not been investigated, it would appear to play a role in the tenderisation of meat by this process.

Influence of intramuscular collagen fibre size on meat texture

As shown earlier, collagen fibre and fibre bundle size varies from tissue to tissue and with age. Two hypothetical considerations link these parameters to meat texture. First, the larger the collagen fibres and fibre bundles in meat the more likely it is to be tough if the connective tissue provides a barrier to fracture of the meat. Second, the larger the collagen fibres and fibre bundles of the perimysium the less they may be damaged by proteases during normal conditioning (Chapter 9). Although a considerable amount of work has been done on the general morphology of collagen fibres in meat, very little has been done to correlate the parameters of collagen fibre and fibre bundle size with textural qualities.

In a recent study, perimysial fractions were prepared from bovine muscles of different eating quality. Transmission electron microscopy of these samples allowed the measurement of average collagen fibre diameters (Fig. 8.2). It was shown that the collagen fibres of the high quality muscles (psoas major and longissimus dorsi) had a smaller average diameter (51–54 nm) than those of the tougher muscles (gastrocnemius and sternomandibularis), whose diameters ranged from 75–92 nm. Although this appears to be a good correlation it should be pointed out that muscles of intermediate quality (pectoralis profundis and semitendinosus) also had small fibres (49–51 nm) and that the gastrocnemius, which had the largest fibre diameters, is generally ranked as of intermediate quality. No direct correlation between fibre size and texture could be found.

Influence of the different genetic types of collagen on meat texture

We have already seen that biochemical and immunohistological analysis of a number of muscles has allowed the location of various

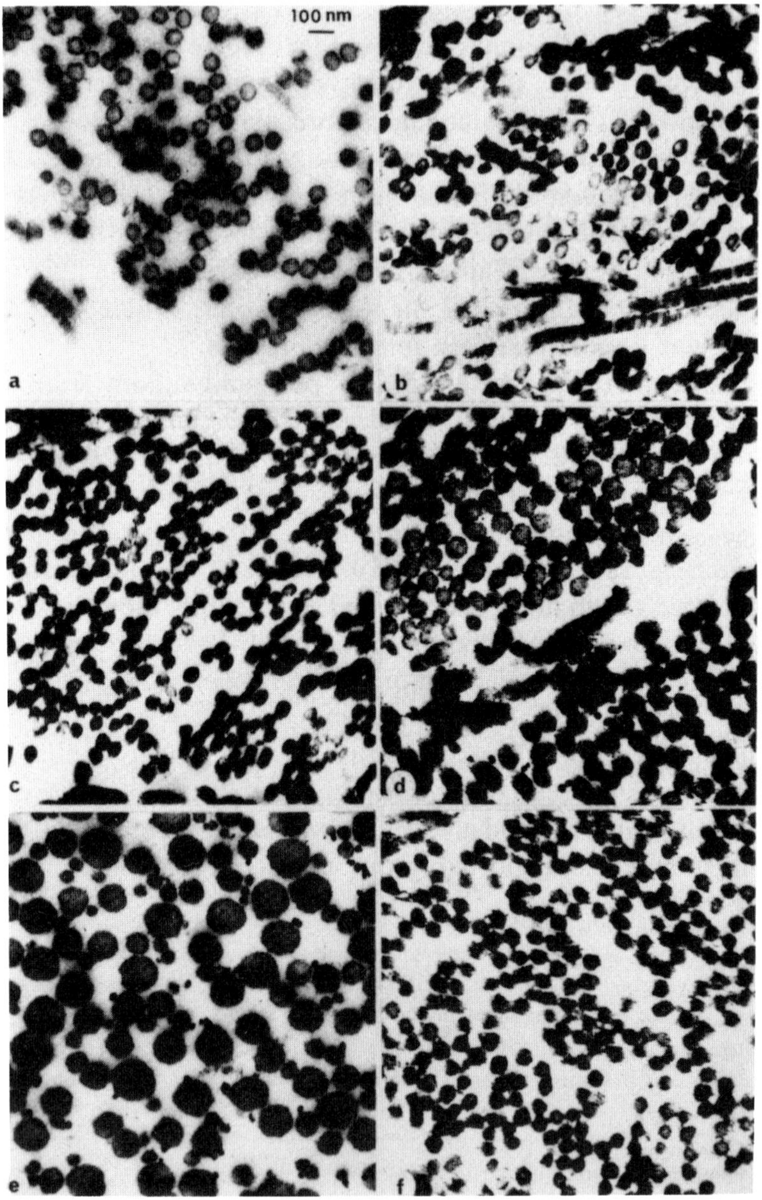

Fig. 8.2 Transmission electron micrographs showing the differences in perimysial fibre diameter in different muscles of varying quality. The bovine muscles used in this study were: (a) psoas major, (b) longissimus dorsi, (c) pectoralis profundus, (d) sternomandibularis, (e) gastrocnemius and (f) semitendinosus. The bar denotes 100 nm.

genetic types of collagen in different parts of the muscle structure (Chapter 6). We have also seen that there are a variety of different types of fibre and matrix structure formed by the different types of collagen. For example, Type I collagen forms large, strong fibres whilst Type III forms thinner fibres often associated with elastic tissues. Type IV collagen does not form fibres at all but rather an amorphous 'chicken-wire' like network and is found uniquely in basement membranes. Type V collagen probably forms very fine fibres but their structure *in vivo* has not yet been demonstrated. It seemed natural, therefore, to postulate the theory that differences in the distribution of the genetic forms of collagen may give rise, in part at least, to the variations in meat texture observed between different muscles.

An early study showed an apparent positive correlation between the amount of Type III collagen present in muscles and increasing toughness. This result seemed to be supported by even earlier observations that reticulin fibres (thought to be Type III collagen) were more resistant to heating than regular collagen fibres and that Type III collagen was more resistant to proteolysis during conditioning than Type I collagen. A later study using prepared perimysial and endomysial fractions from six bovine muscles of varying eating quality confirmed this finding. The results from this latter study which suggest that the overall Type III collagen content is lower in high quality muscles than in intermediate and low quality muscles have already been seen in Table 6.1. Once again, as we saw in the consideration of collagen fibre diameter, there is clear differentiation of high quality muscles but not of the remaining samples. Also, there was no difference in the Type III collagen content of the endomysia of different muscles; this remained very constant at a mean of 56·7% of the dry weight.

It is possible, therefore, that the total Type III collagen content of muscles might play a role in determining texture. This may be due simply to the 'tougher' nature of Type III collagen fibres, but this argument is probably too simplistic. It seems more likely that Type III collagen may be involved in forming boundaries between the perimysial fibres and the endomysial sheath which may be important in determining the textural properties of meat.

The role, if any, of the minor meat collagens, Types IV and V, in determining or contributing to meat texture has yet to be revealed, primarily because of the difficulties in studying such physically minor components *in situ*.

Influence of the number and type of intramuscular collagen cross-links on meat texture

We have discussed in earlier chapters the shrinkage behaviour of collagen molecules when heated (Chapter 7). If the collagen matrix is part of another structure, e.g. meat, then its shrinkage may cause or exacerbate the overall shrinkage of that tissue and result in the loss of fluid and changes in textural properties. Also, if the collagen matrices do not solubilise during heating but remain as intractable masses, then they will form a barrier to the effective breakdown of the tissue during mastication. The factor which governs this heat-insolubility is covalent cross-linking.

The correlation between the increasing toughness of collagen with the age of the animal has been well described (Chapter 4) and clearly depends solely on the cross-linking of the collagen. Consequently, the role of covalent cross-links in collagen in determining textural quality in meat has been postulated. However, these studies have shown only a reasonable correlation between the heat-stable cross-link content of some bovine muscles and their toughness as rated by subjective means. However, there were some significant discrepancies. The best correlation was seen between the perimysial collagen cross-linking and textural quality. The oxo-imine cross-linking in the endomysial fractions was less well correlated with the sternomandibularis and the longissimus dorsi, both of which have approximately the same number of oxo-imine cross-links (Table 8.2). Other studies revealed the lability of aldimine cross-links in meat to the acid conditions existing in post-rigor tissue and proposed a relationship between the number of collagen cross-links and meat toughness. This latter result is possibly

Table 8.2
Proportion of reducible cross-links in epimysium, perimysium and endomysium

Muscle tissue[a]	*Aldimine* (%)	*Oxo-imine* (%)	*Age of animal*
Epimysium (LD)	50·0	50·0	18 month steer
Epimysium (SMD)	21·0	79·0	18 month steer
Perimysium (LD)	42·0	58·0	18 month steer
Perimysium (SMD)	28·0	72·0	18 month steer
Endomysium (LD)	27·0	73·0	18 month steer
Endomysium (SMD)	13·0	87·0	18 month steer

[a] LD = longissimus dorsi; SMD = sternomandibularis.

misleading, however; it is much more likely that it is the ratio of heat-labile (aldimine) to heat-stable (oxo-imine and mature) cross-links that is the more important governing factor in respect of the contribution of collagen to meat texture, based on extent of both shrinkage and solubility. Indeed, the ratio of heat-stable oxo-imines to heat-labile aldimines in the perimysia and endomysia of the muscles did show a good correlation with perceived quality in that the better muscles had ratios less than half those in poor quality meat. A similar correlation was shown when a more rigorous comparison was made between per cent heat-stable oxo-imines and mean compressive force obtained with cooked samples of the muscles (Fig. 8.3).

What these results imply is that it is the *proportion* of heat-stable cross-links in terms of total collagen cross-linking that is the major factor in influencing toughness in meats. This is not surprising in that, as postulated above, it is these cross-links that will remain (holding the insoluble collagen matrices together) after the heating process. In this respect, the influence on meat texture of the mature collagen cross-links, which are known to be heat-stable and to become predominant in mature tissues, cannot be under-rated.

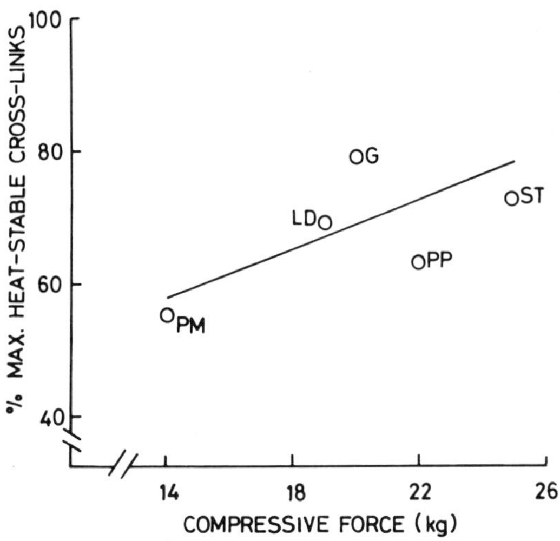

Fig. 8.3 Graph showing the correlation of compressive force as a measure of meat textural quality with the percentage of the total possible oxo-imine cross-links present in the perimysium of a range of bovine muscles. PM = psoas major, LD = longissimus dorsi, PP = pectoralis profundus, G = gastocnemius and ST = semitendinosus.

One of the postulated mature cross-links, pyridinoline (Chapter 4), has been shown to be present in intramuscular connective tissue but very little is known about its role in stabilising intramuscular collagen. Furthermore, it is not known whether the peptide-bound moiety is heat-stable at the pH_u of meat (5·3–5·5). Pyridinoline is not present in basement membrane collagen and, therefore, would not be available to stabilise the endomysial sheath during heating (see below). Further studies are needed to decide if this cross-link has a role in stabilising intramuscular collagen.

The other major postulated mature cross-links (Chapter 4) may be present in meat. At present there is no way of assaying for these compounds accurately but in the future the contribution of these intermolecular bonds must be investigated as it seems likely that they may be present in varying proportions in different muscles. Mature cross-links in collagen matrices must, therefore, play a dominant role in determining the heat-insolubility of those matrices and thus the overall textural properties of the meat. Presumably the best correlations will come from relating the heat-stable divalent plus the heat-stable mature cross-links with the textural quality of the meat.

Influence of the endomysial sheath on meat texture

Earlier it was suggested that, despite the small amount of endomysium in muscle, the collagen content of the endomysial sheath might itself influence the textural properties of meat as it has been found that tougher muscles have a higher collagen content in this connective tissue than more tender muscles. A consideration of the morphology and topography of muscle should allow certain predictions to be made.

The endomysium is composed of the thin basement membrane (20–30 nm) and a network of fine collagen fibres linking the basement membrane to the fibrous matrix proper. These fine fibres are a mixture of Types I, III and V collagen. The basement membrane contains Type IV collagen as the major structural unit. This framework is cross-linked by the heat-stable oxo-imine bonds, giving the membrane mechanical strength and rendering it insoluble. Since the basement membrane is made up of typical collagen molecules it shrinks and generates tension on heating. Studies with lens capsule basement membrane, which does not possess the associated fibres, indicated a lower and broader shrinkage temperature, 55–65°C, than the highly crystalline tendon fibres. Isolated endomysial 'ghosts', whether empty of residual myofibrillar proteins or not, exhibited shrinkage at 65–75°C, the increase probably being due to presence of the associated collagen fibres, which probably shrink at a higher temperature than the basement membrane of the endomysium. The relative importance of this shrinkage compared to that of the associated fibres cannot be distinguished for the endomysium. From these studies it is possible to predict the possible effects on the endomysium during heating.

It is well known that the contractile apparatus shrinks slightly during rigor mortis and then more substantially during the cooking process. Detailed morphological studies have shown that this shrinkage can lead to the production of a significant gap between the endomysium and the insoluble, denatured actomyosin which is filled with cell fluids (Fig. 8.4). On further heating the endomysium will start to shrink. The extent of shrinkage will depend on the nature of the endomysial basement membrane and on the nature of the associated collagen fibres. Extensive shrinkage could result in the loss of the fluid contained between the denatured myofibrillar proteins and the en-

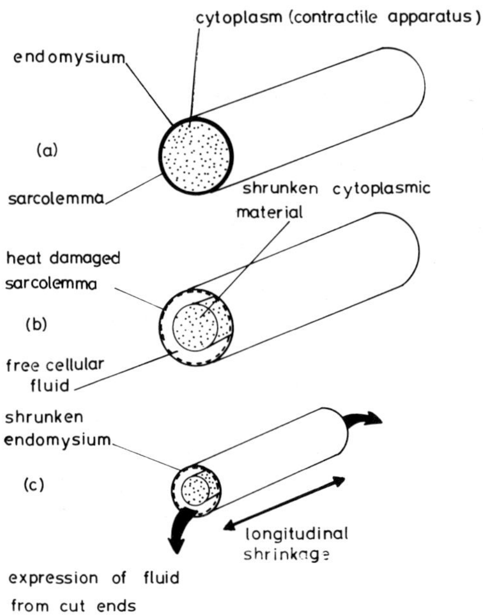

Fig. 8.4 Diagrammatic representation of the postulated process of muscle cell and endomysial shrinkage induced by heating.

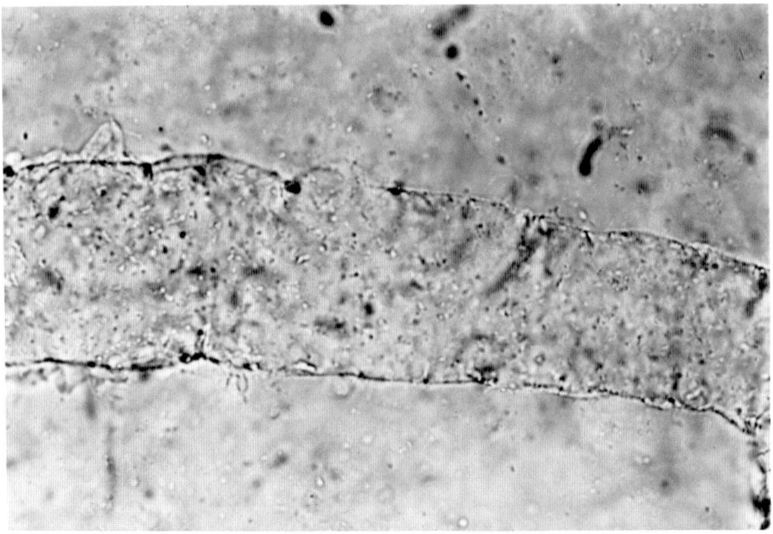

(a)

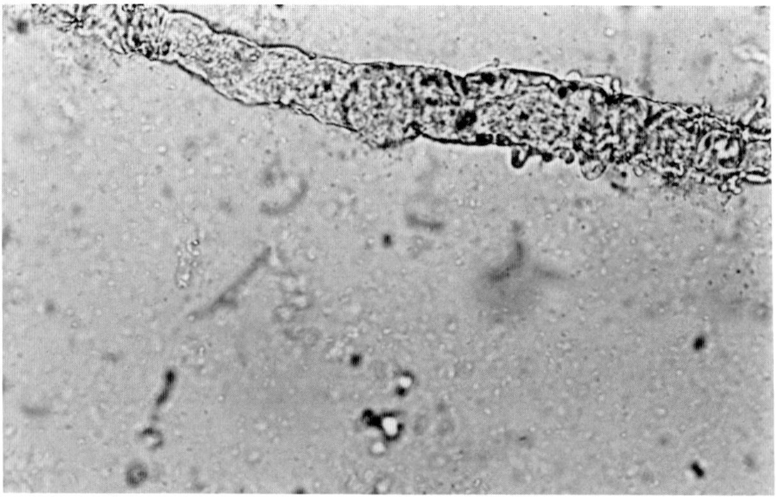

(b)

Fig. 8.5 Light micrographs of an endomysial 'ghost' (i.e. empty endomysial sheath—see Chapter 12) both (a) before and (b) after heat-induced shrinkage.

domysial sheath (Fig. 8.5). Thus endomysial shrinkage could affect the fluid content of the muscle fibres and may play a role in the residual strength of the heated tissue. This assumes that the shrinkage of the endomysium does not lead to fragmentation of the sheath. Further studies on this component are clearly warranted.

The mechanical strength of the endomysium can be demonstrated in swelling studies. The mechanism of swelling involves primarily the swelling of the individual myofibrils, but as discussed earlier the myofibrils are encompassed by the endomysium which would resist excessive swelling just as it resists extensive contraction or extension of the living muscle. The mechanical strength of this collagenous structure in terms of swelling pressure is difficult to determine because of damage during dissection of myofibrils possessing an intact endomysial sheath. The damage could arise by mechanical manipulation but also by enzymic degradation of the collagenous framework prior to dissection. The latter effect would obviously increase the longer the post-rigor period. Indeed, experiments show discontinuous swelling along the length of isolated myofibrils (Fig. 8.6) and this can be

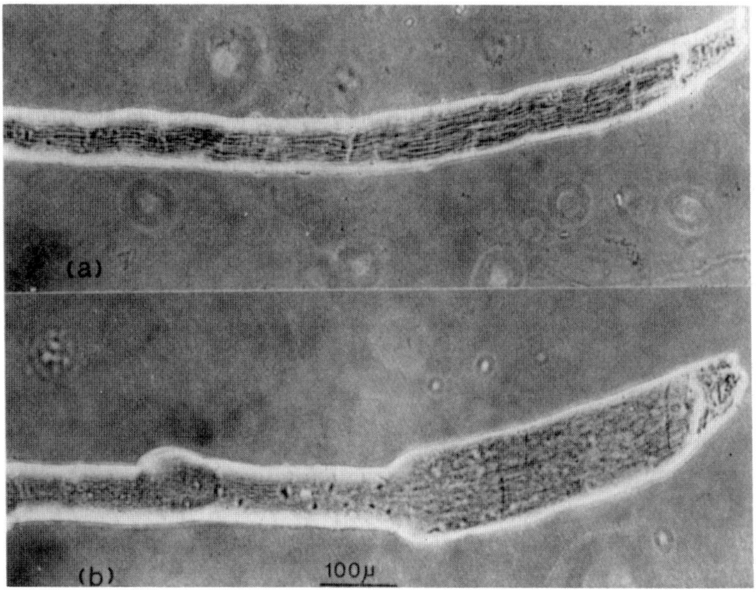

Fig. 8.6 Phase contrast micrographs showing the effect of mechanical damage on subsequent 0·25M KI swelling. (a) A fibre in 0·15M KCl; (b) the same fibre after irrigation with 0·25M KI.

increased by *in vitro* enzyme treatment. Clearly the endomysium acts as a mechanical constraint to swelling, but the tensile properties of this restraining sheath decrease with increased conditioning time. This change in property is clearly very important in the preparation of meat products where swelling by salts is an essential part of the process. Similarly the weaker the endomysium the less the tension generated on shrinkage, which if significant could affect the ultimate texture of the meat. These studies clearly demonstrate the significant strength of the endomysium and it is evident that it must play an important role both in meat swelling and heat-induced shrinkage.

Influence of the perimysial–endomysial junction on meat texture

As mentioned above (Chapter 7), fracture studies and limited biochemical data have suggested that the major route of fracture through cooked meat is in the perimysium and possibly at the perimysial–endomysial junction. This junction must not be regarded as a distinct physical barrier or interface, although it has been discussed as such in this text. It is highly unlikely that such a discrete structural junction exists. Optical microscopy suggests that there is a changeover in fibre size between the intermuscle fibre spaces and the intermuscle fibre bundle spaces (endomysium and perimysium respectively), but little or nothing is known about how the larger perimysial collagen fibres are linked to the smaller endomysial fibres. There may be an area of collagen fibre size reduction at this interface (i.e. a graded decrease in fibre size) or it may be mediated by a discrete collagen type (e.g. Type III or one of the rarer types such as Type VI), but no data exist to support either suggestion.

At present, therefore, understanding of this whole field, although crucial to our understanding of the mechanism for the involvement of collagen at the molecular level in the mechanical breakdown of cooked meat, is incomplete. Only substantial future research can answer the many questions that have arisen recently about this important connective tissue interface.

THE ROLE OF THE NON-COLLAGENOUS COMPONENTS OF CONNECTIVE TISSUE

Elastin

The low elastin content of most muscles has been determined chemically. The belief that elastin is predominantly associated with the blood vessels has led to the assumption that it does not play a significant role in meat texture, and so detailed studies have been limited. However, histological studies (Chapter 6) suggest that its role in meat quality may have been overlooked. Whether elastin does have any significance in contributing to the textural properties of meat is still debatable. Elastin, unlike collagen, is very stable to heat and its properties are generally unaffected by cooking. Hence its significance to texture probably depends directly on the amount of elastin in the perimysium. Elastin could increase the elasticity or the overall residual tensile strength of the denatured perimysium and play a role in maintaining the integrity of the meat through the elastic fibres holding the myofibrils together. In a few muscles, e.g. semitendinosus, elastin constitutes 40% of the connective tissue and in these muscles elastin is likely to play a more significant role. It may contribute to the overall resistance to chewing or resistance to shear. The relative importance of its contribution compared to collagen is likely to be small in other muscles, although it should be investigated.

Proteoglycans

There is no evidence to suggest that the proteoglycans play a role in the texture of meat. These macromolecules are readily denatured on heating and generally do not have a structural role in mechanical terms. It could, however, be speculated that the small proteoglycans which can be observed at specific points along and between collagen fibres may play a role in the thermal stability of that fibre. If this is true then they could affect the way collagen contributes to texture and thereby play an indirect role in texture of meat, but their contribution is likely to be small.

AN HYPOTHESIS FOR THE ROLE OF COLLAGEN IN DETERMINING THE TEXTURE OF MEAT

As described above, attempts have been made to correlate texture with the properties of collagen varying from total collagen, fibre size, molecular composition, solubility, cross-linking, etc., but no single parameter has been able to account for the variations observed. However, the fundamental studies on the structure of collagen have permitted a rational description of its role in the texture of meat. In the following section we discuss our hypothesis, developed over several years, that collagen is not only the major determinant of texture in cooked meat, but that it is the quality of any collagen rather than the total amount that is the predominant factor in the variability of texture.

As the temperature is raised to between 40 and 50°C during the cooking of meat there is an increase in the toughness as determined by the shear value. This change has been interpreted broadly as being due to the denaturation of the myofibrillar proteins, particularly acto-myosin. The increase in shear value is due to the coagulation of the denatured proteins to form a more rigid gel. Microscopic evidence reveals a transverse shrinkage of the myofibrillar proteins within the endomysial sheath, which, being collagen, is unaffected at this temperature (Fig. 8.7). It may, however, have been under tension and shrinkage of the myofibrils might therefore have the effect of releasing that tension by forcing the muscle fluid out of the space created between the endomysium and the denatured myofibrils, thus accounting for the fluid loss at these temperatures.

A second increase in shear value occurs between 60 and 70°C when there is a greater shrinkage of the meat and loss of fluid. This second phase can be interpreted in terms of the intramuscular collagen. As discussed earlier, the collagen fibre denatures at about 65°C and shrinks to one-quarter of its length. The perimysial fibres can be observed to gelatinise, losing their characteristic banding pattern and changing from an opaque fibre to a translucent swollen fibre. However, the extent of the shrinkage and consequent tension generated (Fig. 8.7) depends markedly on the age of the tissue. The endomysium around each myofibril also denatures and shrinks at these temperatures and as there is no evidence that it breaks down under the tension generated, the intact shrunken endomysium should force more water out of the myofibrillar space (Figure 8.7). Detailed studies on the changes in muscle fibre cross-sections indicate smaller spaces between the endomysium and the denatured myofibrils, supporting the proposal that fluid is lost due to contraction of the endomysium. However, although shrinkage could occur, fluid loss would depend on the endomysium remaining intact following shrinkage. Fluid loss also probably occurs between the muscle bundles, that is, in the perimysial channels. Whether the overall force of contraction is great enough to compress the denatured myofibrillar proteins themselves is not clear, but it is certainly great enough to squeeze them together by forcing the fluid between the myofibrils out of the meat. With increasing age a higher proportion of heat-stable cross-links are present in any collagen;

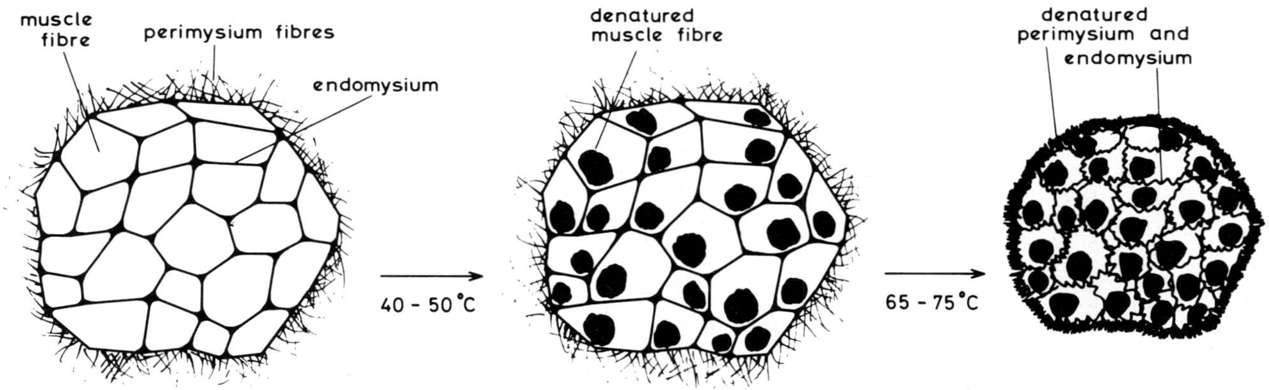

Fig. 8.7 Model of the sequential shrinkage events in meat during cooking, i.e. shrinkage of myofibril protein at 40–50°C and of the perimysium and endomysium between 60 and 75°C.

hence the extent of the tension generated both in the perimysium and the endomysium during shrinkage increases dramatically, resulting in even greater compression and loss of fluid from the meat. The relative importance of the endomysial shrinkage compared to the perimysial is unknown but is likely to be small. The first role for collagen in determining the textural quality of cooked meat, then, is through its shrinkage and the resultant effect on water loss. These changes correlate with the observed increase in the second maximum of the shear value curve and the corresponding increase in toughness.

The slow decrease in shear value on continued heating after the second maximum has been reached is probably due to thermal breakdown of proteins by cleavage of the peptide bonds. Whether this is due to cleavage of peptide bonds in the myofibrillar proteins or the denatured collagen fibres has not been demonstrated chemically. However, based on the proposition that the denatured collagen fibres (gelatin) maintain the integrity of the meat by holding the myofibrils together, it must surely be cleavage of the gelatin chains, and consequent solubilisation, that would have the most dramatic effect on texture under these conditions. Indeed, this is consistent with the observation that prolonged stewing of meat results in easier separation of the myofibrils. Whether the bonds ruptured are the heat-stable cross-links or peptide bonds along the protein chain has not been demonstrated.

An equally important feature of the denatured and shrunken collagen is its residual strength. The older the animal the greater the residual strength of the denatured perimysium holding the myofibrils together, further contributing to the increased toughness of the meat. In meat from young animals the denatured fibres will be weak and tend to dissolve, thus becoming even weaker, whilst those from older animals possess heat-stable cross-links and are therefore stronger and virtually insoluble. Several workers have reported that cooked meat is slightly tougher when measured at 20°C compared to measuring at 70°C. This is probably due to the partial reversion of the denatured collagen to a native collagen structure on cooling. This is the well known Ewald reaction which occurs with collagen extensively cross-linked with heat-stable bonds. However, it must be remembered that the collagen does shrink during cooking, and in view of its extensive cross-linking, generates considerable tension during that time. The partial reformation of the collagen structure increases the 'residual strength' of intermuscle fibre binding.

It could be argued that with increasing temperature in the 70–90°C region further shrinkage and dehydration of the denatured actomyosin occurs. Although this may be partially true, it is unlikely that it could be related to the dramatic changes that occur at 65°C when the collagen shrinks, but would rather be a steady increase in shrinkage. A decrease in sarcomere length could occur, but it is unlikely that either effect could be age-related in the way that the properties of collagen correlate with the extent of the increase in shear value and toughness with age of the animal. In other words, the height of the second maximum of the shear force curve is dependent on the age of the animal, increasing with age. This age-related change could not be caused by the myofibrillar proteins.

The complete reversal in the properties of the myofibrillar proteins, which become stiffer on heating compared to fresh meat, and the collagenous tissue, which becomes elastic and many times weaker on heating, leaves the question of which is now the weakest component, and therefore the major determinant of meat texture. A cursory examination of cooked meat reveals that the adhesion between muscle fibres is readily ruptured; indeed it has long been known from mechanical studies that the tensile properties along the axis of the muscle fibres are an order of magnitude greater than the force required to separate the fibres transversely. Examination of these interfibre strands revealed them to be denatured collagen, and therefore obviously derived from the perimysium. This clearly indicates that the perimysium is now the weakest component. Confirmation of this had been obtained by fracture mechanics. In these tensile tests the initial event opens up splits along the boundaries between the bundles of muscle fibres until the collagen layer is pulled away from the endomysium. Whether the perimysium itself or the perimysial–endomysial bonding is the weakest site is not yet clear. The fine collagen fibres of the endomysium junction suggest that this would be the weakest site, but further studies using this informative technique must be awaited before any conclusions can be drawn.

The role of collagen in conditioning (Chapter 9) can also be explained by this hypothesis. Consideration of the model of collagen fibre structure shows that a very limited number of proteolytic cleavages in the non-helical regions of the molecules could bring about damage to the surface of fibres without inducing a large degree of heat-solubilisation in such an intractable and resistant material. Indeed, research data derived from conditioned meat support this in that it is only possible to detect small changes in the amount of extractable collagen following conditioning. In other words, the attack of proteases on the endomysial structure, endomysial and perimysial collagen fibres and the perimysial–endomysial junction may not lead to extensive solubilisation of the collagen but must lead to a general weakening of these tissues. The cross-linking regions of the collagen molecules within the fibril are susceptible to proteolytic enzymes and the cleavage of a few bonds in these regions would partially depolymerise the collagen fibre, although the effect on tensile strength would be hard to detect. However, on heating conditioned collagen the tension generated would be reduced, resulting in a lower shrinkage of the meat together with a reduced residual strength of the intermyofibrillar perimysial fibres, and hence more tender meat. This feature has also been supported by direct biochemical evidence (Chapter 9) and isometric tension measurements (Chapter 7) where a significant decrease in the tension generated has been observed. Similarly, the integrity of the basement membrane might be disrupted by prolonged exposure to intracellular proteases, thus limiting its ability to shrink on heating and thereby reducing fluid loss. Finally, attack by proteases at the possibly more susceptible perimysial–endomysial junction may lead to a profound weakening of that interface.

Such effects can be readily demonstrated with isolated collagen fibres, but very few data exist to support these contentions in meat. Much sensory and mechanical research has been carried out on whole muscle which shows that conditioning brings about a substantial softening of the muscle tissue and thereby an improvement in textural quality. However, few, if any, of these studies have been able to link these changes directly with proteolytic damage to the collagenous components since the problem is confounded by the simultaneous damage to the myofibrillar components. Indeed, it seems impossible, with present technology, to separate the contribution of proteolysed intracellular or extracellular components as neither can be inhibited during conditioning and the processes cannot be effectively separated and controlled. Resolution of the exact effects on meat texture of the damage incurred by collagen during conditioning must therefore await future research.

Although the basic rationale for the role of collagen in the texture of meat is now clear, the precise details remain to be worked out. The cleavage of the perimysium from the endomysium may involve specific types of collagen fibres. The role of the endomysium itself is not yet clear; it may remain intact following denaturation or it may break down on shrinkage. Preliminary scanning electron micrographs suggest it remains intact but further detailed studies are required. The organisation of the perimysium around the muscle bundles may also affect the extent to which the force generated during shrinkage is exerted on the muscle fibres. For instance, in some muscles the perimysial fibres are laminated, each layer being at a different angle to the next. On denaturation the force exerted on such a muscle would be different from that exerted by fibres aligned along the fibre.

Although differences in the extent of tension generated on shrinkage are related to age, the precise correlation with the nature of the cross-links in the perimysium remains to be established. The perimysium generally possesses a higher proportion of the thermally stable oxo-imine cross-links than the epimysium. Indeed, a rough correlation can be obtained between the content of this thermally stable cross-link and the toughness of the cooked muscle, but further studies need to include the content of the thermally stable mature cross-links (Chapter 4). The ratio of collagen types in the perimysium may also affect the tension generated and further analyses are required to demonstrate whether this is an important factor. The Type III fibres are generally thin fibres, hence one might expect that a high Type III content could reduce the tension generated on shrinkage. On the other hand they are highly cross-linked and would retain considerable strength after cooking. Unfortunately at present fibres composed solely of Type III collagen have not been identified and the problem remains unresolved.

In summary, the sensation of texture is provided by the major proteins in meat, that is the denatured actomyosin, but the collagen, although only constituting 2% of the total protein, plays a crucial role in determining the expression of these denatured myofibrillar proteins as a component of texture.

SUMMARY

Collagen is subject to many changes in conformation and properties when heated. These include denaturation of molecules and shrinkage of fibres, solubilisation of small quantities of material cross-linked by heat-labile bonds and a general loss of structural integrity. Mechanical shear tests show that collagen contributes to the apparent toughening of meat when cooked at temperatures between 65 and 75°C due to shrinkage of the meat.

There is a reasonable correlation between total collagen content and the eating quality of meat in that poorer muscles have higher quantities of collagen. However, this relationship is not strict and there are exceptions. The relative solubility of collagen in meat is not a reliable indicator of the quality of the muscle as some muscles show different collagen solubilities when sampled from different sites, and collagen solubility may vary considerably depending on the length and extent of conditioning. However, there is some correlation of collagen fibre size with quality in that better quality muscles appear to have smaller perimysial collagen fibres. Similarly, there is some evidence to suggest that Type III collagen may have an influence on texture, as poor quality muscles have higher proportions of this component.

The best documented correlation of collagen parameters with quality is that of intermolecular cross-linking. Muscles with a higher proportion of the heat-stable oxo-imine cross-links were the lower quality cuts. There is no direct evidence on the role of the mature heat-stable cross-links in determining texture, but indirect evidence suggests that higher proportions of these bonds in collagenous tissues will severely decrease meat quality. The links between the endomysium and perimysium were also shown to be important by fracture mechanics studies and this interface may provide the major pathway for the breakdown of meat during eating. The influence of conditioning on meat quality cannot be under-rated as this process determines the status of collagenous tissues in terms of the damage incurred through proteolytic attack, and consequently its ability to affect texture during thermal denaturation.

The general hypothesis is therefore that collagen is the major determinant of the texture of meat and that the subtle variations in texture are dependent on the quality rather than the quantity of collagen.

FURTHER READING

1. Bailey, A. J. (ed.), *Recent Advances in the Chemistry of Meat.* Royal Society of Chemistry, London, 1984.
2. Pearson, A. M., Dutson, T. R. & Bailey, A. J. (eds), *Advances in Meat Research,* Vol. 4: *Collagen as a Food.* Van Nostrand Reinhold, New York, 1987.
3. Rowe, R. W. D., Collagen fibre arrangement in intra-muscular connective tissue. Changes associated with muscle shortening and their possible relevance to raw meat toughness measurements. *J. Food Technol.,* **9,** 1974, 501–508.

Chapter 9

The Effects of Ante- and Post-Mortem Factors on Meat Connective Tissue: Growth Rate, Handling and Conditioning

INTRODUCTION

We have seen that connective tissue is the component of muscle which binds it together and gives it structural integrity and strength. These characteristics, and in particular the properties of the collagens, relate to a large degree to the living tissue. However, meat is a different commodity from fresh muscle and cooked meat is entirely different from both.

In the previous chapter we saw how connective tissue forms a variable but very small proportion of the total dry weight of meat (1–9%) and yet there is clear evidence for a major role of collagen in determining meat texture. The amount and quality in terms of fibre size and type of cross-linking in collagen may all play a part in influencing textural properties which lead to the differences in eating quality observed between muscles. Consideration, therefore, of the ways in which these factors influence texture and the use of our knowledge of this ubiquitous tissue, could lead to means of manipulating meat quality at some stage during growth of the animal or during the preparation and handling of meat.

The factors which may be expected to influence the amount and quality of the connective tissue in meat include the breed, age and sex of the animal from which the meat is obtained, the muscle of origin of the meat, the growth rate of the animal and the effect of exogenously administered drugs such as steroids, the length of time and temperature of conditioning, exogenously administered proteolytic tenderisers and the actual cooking process itself. These factors may be subdivided into two categories. These are: (1) the influences on meat connective tissue in the live animal, and (2) the influences on meat connective tissue occurring after slaughter. Only by understanding these influences and the differences in connective tissues in meats of different origin can we begin to form a complete picture of the ways in which this important component influences the textural properties of meat.

This chapter, therefore, will consider the ways in which the amount and quality of connective tissue in muscle is influenced by parameters affecting the growing animal, and then the ways in which the connective tissues are affected after slaughter, particularly during conditioning. A description of the turnover of collagen is essential to any understanding of development, growth and physiological remodelling of connective tissues and for this reason has been the focus of considerable research activity. We will, therefore, begin this chapter with a brief discussion of the development of connective tissue in muscle and the effect of turnover on its role in texture.

MUSCLE CONNECTIVE TISSUE CHANGES DURING ANIMAL GROWTH

Since the first revelations of Mendel on the mechanisms of heredity, man has experimented with the genetics of animals including those of the domestic breeds but, as yet, little is known of the genetic control of connective tissue components (see Chapter 3). In theory, once the genetics of collagen and other connective tissue macromolecules is understood there may be an opportunity for the use of genetic engineering techniques to produce animals with low intramuscular connective tissue content and concomitantly tender meat. However, such an eventuality seems far away. In the meantime, man has investigated the use of breeding techniques to select for animals with the best muscle characteristics. Much has been written on this subject and we do not intend to carry out a general review here. However, it would be useful to examine the changes in connective tissue in the growing muscle in order to be able to understand the effects of endogenous and exogenous factors on these tissues and of the connective tissues on muscle structure.

The growth and development of animals may be divided into three major stages, embryonic growth within the uterus of the mother, early post-natal growth and growth to maturity. In the domestic breeds the last stage is generally irrelevant in terms of today's farming practices as most animals are brought to slaughter well before maturity. Thus, we will consider only the first two stages.

In the early developing foetus precursor cells, known as somites, migrate from the spinal column to positions in the embryo where body muscles and the new limb buds will form. Somites can be grown in culture and have been shown to produce muscle fibres under test tube conditions. Connective tissue is formed in the developing limb from a different set of precursor cells, the somatopleural mesoderma cells. It has been shown that foetuses in which the somites are killed by irradiation will develop limbs with normal nerves and connective tissue but without muscle fibres. These limbs have normal bones and fibrous connective tissue, although patterns of structure similar to the epimysium, perimysium and endomysium are obviously not identifiable. These studies show that the muscle cells and intramuscular connective tissue are produced by different sets of primogenitor cells and produce their muscle and connective tissue proteins independently of each other.

The first stage of myogenesis is the appearance of muscle cell masses around the cartilaginous elements that will later form the limb bones (Fig. 9.1). The pattern of the muscles and muscle shape develops through the controlled repetitive division of these muscle masses. The mechanism of control by which these muscle masses aggregate into the discrete and unique shapes of the limb and body muscles is not known. This genetic information for muscle shape must be contained in all cells of the body but some must be capable of helping express the information. Suggestions of the involvement of nerves in the process have been proved incorrect and the theory that tension in the growing limb exerts an exogenous mechanical control has also been disproved. It is possible that interaction of the connective tissue cells and myoblasts in the developing muscle may play a role in determining muscle structure or that small amounts of connective tissue elements may be important in directing muscle cell migration and pattern development. Some evidence exists for a possible role for glycosaminoglycans and fibronectin in this process, but immunolocation studies with anti-collagen antibodies at the light microscope level have failed to reveal any collagen in the early developing muscle masses. However only small amounts of collagen may be involved at this stage and confirmation at the electron microscope level is clearly needed. There

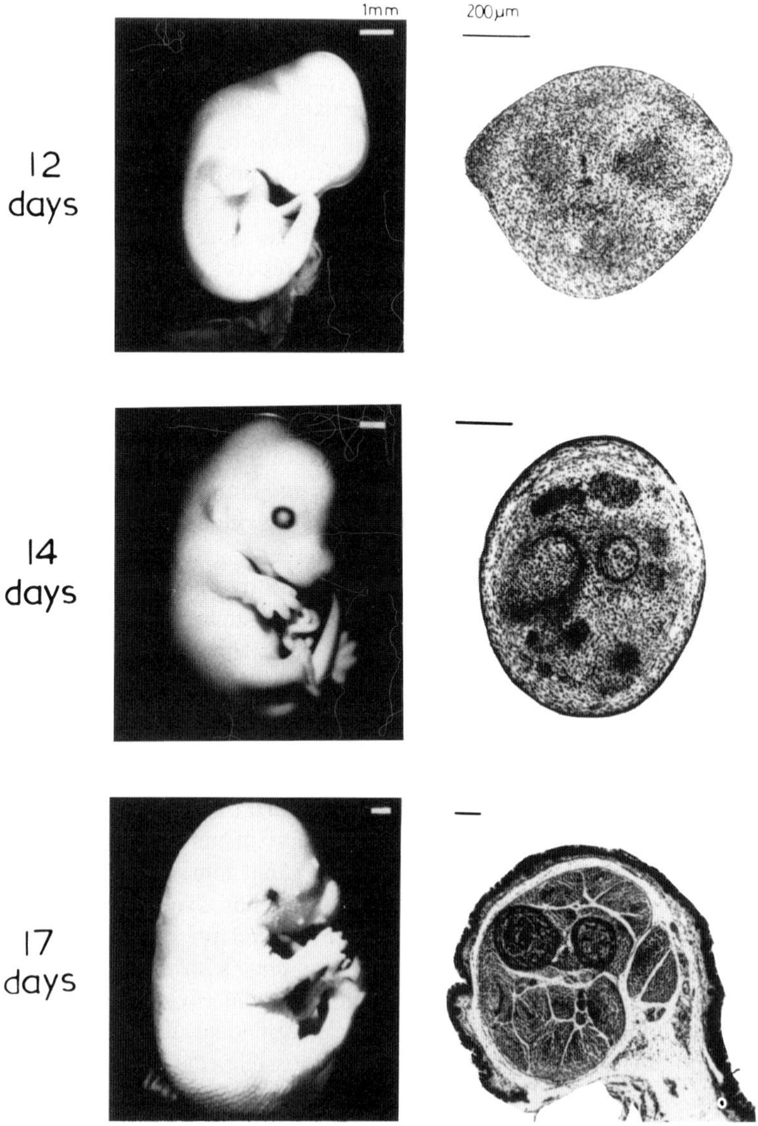

Fig. 9.1 Progressive formation of muscle pattern through the mid-forearm of mouse embryos at various periods of gestation.

is, therefore, only circumstantial evidence for the involvement of intramuscular connective tissue in the development of muscle pattern formation.

Little is known about the changes in non-collagenous connective tissue components during early post-natal growth. There is better evidence for what happens to collagen in some connective tissues. In skin, for example, the ratio of Type I to Type III collagen changes from 1:2 to 5:1 soon after birth and this is matched by a concomitant change from predominantly oxo-imine cross-links to exclusively aldimine bonds. Work with tendon has shown that the average collagen fibre diameter increases with age from birth, as discussed in Chapter 4. There are, therefore, likely to be major and subtle changes in intra-muscular collagen during the growth and development of animal muscles.

Collagen turnover

As described in Chapter 3, the metabolism of collagen has been shown to consist of a series of exquisitely controlled post-translational modifications during synthesis and a complex series of enzymic actions during degradation. These modifications lead to significant differences in properties between and within tissues, including changes with respect to the state of maturity. For example, the extent of the biosynthetic modifications and age-related changes have an effect on the physical properties of the tissue, and hence on its subsequent utilisation as fresh and processed meat.

The early studies on collagen metabolism indicated a rapid turnover during growth, but demonstrated fairly clearly that collagen was virtually inert in the adult animal. Studies over the last decade have challenged this traditional view and recent evidence indicates that tissues metabolise collagen at a significant rate even in mature animals. However, these studies may relate to the rapid metabolism of newly synthesised collagens; it is important to establish whether mature cross-linked collagen is also degraded. Again, early studies indicated degradation of mature collagen fibres, whilst more recent studies on 'catch-up' growth following starvation indicated little, if any, degradation of the pre-existing mature collagen. The details of collagen turnover clearly have not yet been established but, as a generalisation, there is a high rate of turnover during growth which reaches a low level plateau at maturity. The basal level for this plateau varies considerably between tissues. In most adult tissue, e.g. skin and tendon, the turnover is small but now appears to be significant, whilst in contrast bone undergoes a slow continuous process of remodelling throughout life, and at the other extreme the periodontal ligament has a turnover time of a few days and consequently the tissue never matures. The rates of metabolism of different collagen types introduces another variable into the kinetics of turnover. Type III precursors are known to turn over at a slower rate than Type I, but little is known of the rate of turnover of the mature fibrous form.

The rate of turnover and hence the ability of the collagen to mature will affect its physical properties (Chapter 4), in particular the effect of thermal denaturation. This would be important to the role of the perimysium in determining texture but to date no convincing turnover figures for intramuscular collagen have been reported. A rapid turnover rate would result in a high proportion of newly synthesised collagen with few heat-stable mature cross-links and should therefore produce more tender meat.

Examples of the effect of rapid growth rate on the connective tissue can be seen with rapidly growing pigs. One result is the collapse of the tenon and joints of the legs rendering the animals lame. The effect can be reversed by reducing the growth rate. A second example in rapidly growing lean pigs is the separation of the lean from the fat and, in extreme cases, the production of 'lacy fat' when the meat is processed to bacon (Fig. 9.2). The fat is usually soft due to a low saturated stearic acid content but the separation of the fat appears to be due to the immature nature of the adipose connective tissue.

Finally, it is worth considering the condition in meat animals known as double muscling or doppelender ('culard'). This hypertrophy of the musculature, a genetic effect, results in a higher muscle to bone ratio and, thus, a higher dressing-out percentage, relatively less fat and offal and apparently better quality meat. The effect appears to be transmitted through generations via a single gene but the various ways in which transmission can occur has resulted in a variety of hypotheses about the genetic mechanism and a lack of ability to control it. Although these animals are extremely useful, they cannot be bred easily or with any reproducibility.

Various reports have indicated changes in the myofibrillar diameters, fat content and proportion of connective tissue. A definitive study showed that this latter finding was correct in that a selection of muscles from double-muscled Charolais steers contained lower amounts of collagen, finer myofibres and less fat than normal animals and were concomitantly more tender. The perimysia of these muscles

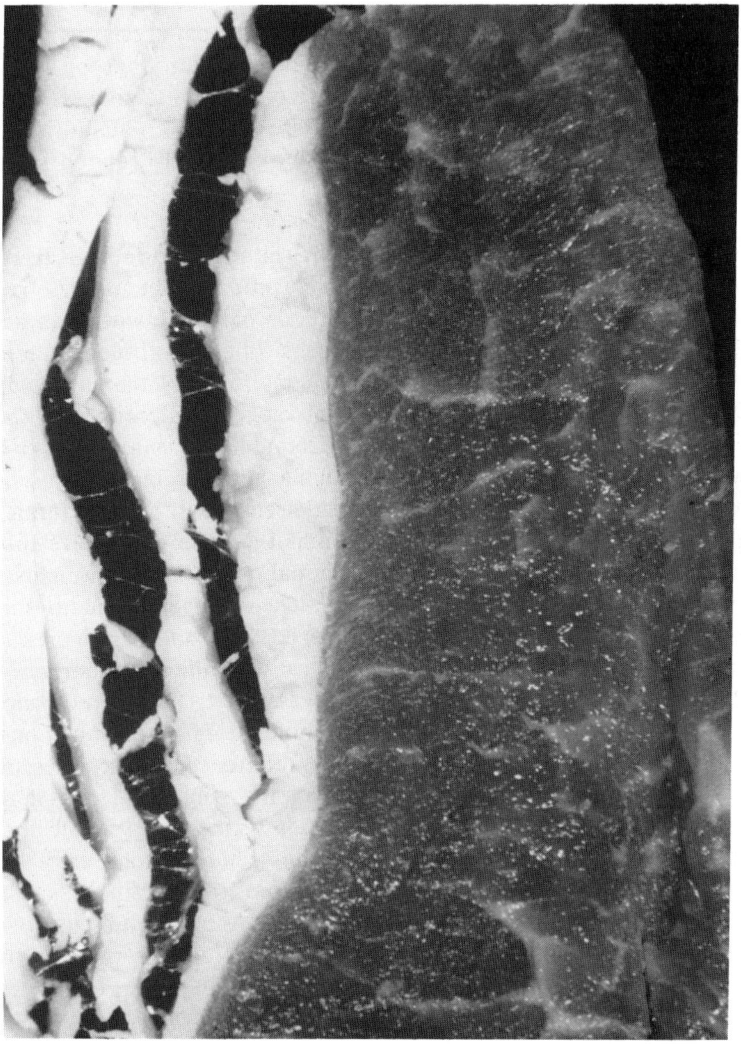

Fig. 9.2 Lacy fat from rapidly grown lean pigs showing the paucity of structural integrity.

contained finer connective tissue fibres than the normal muscles, although the ratio of Type I to Type III collagen was normal. However, the collagen appeared less mature, containing relatively higher proportions of reducible cross-links. Taste panel tests confirmed the double-muscled meat to be more tender than that of normal animals but this could not be attributed to one single factor. However, the major differences in collagen, especially the lower quantities and the immaturity of the cross-linking, may be strongly related to this increased tenderness (see Chapter 8).

Effects of breed and sex of animal on muscle connective tissue

There is a large body of research that has compared the eating quality of meat from a number of different breeds but little work has been carried out to date on the possible differences between breeds in terms of intramuscular connective tissue. It seems unlikely that any major differences would be expected as even interspecies differences amongst the domestic animals are few. For example, the major structural characteristics are similar in bovine, ovine and porcine muscles. All show the typical morphological features discussed in general terms in earlier chapters and all contain collagen as the main connective tissue component. Some breeds of meat animal yield higher weights of muscle per carcass but there is little evidence for better quality meat being produced from one breed compared with another. Certainly, sensory analysis has shown little statistically significant difference in textural quality between breeds as assessed by taste panels. Differences in connective tissue components—for example morphological, quantitative or topographical differences in intramuscular collagen between breeds—have not been the subject of any major study, although many attempts have been made on a small scale. Limited studies on the differences in gross collagen content of four breeds of cattle, Hereford, Friesian, Aberdeen Angus and Charolais steers, have been reported. Fore- and hindquarter muscles were analysed for total collagen content and, as expected, the value for the forequarter was significantly higher than that for hindquarter muscles but there were no significant differences between the four breeds in any of the muscles examined. This result exemplifies the expectations outlined above, although it should be remembered from earlier discussions that breed differences may exist in more subtle connective tissue characteristics such as collagen cross-linking, collagen fibre size, type distribution of collagen and differences in non-collagenous connective tissue components.

The recent increased interest in greater efficiency of bulls as lean meat producers has stimulated comparative studies on textural differences, and consequently on differences in the collagen. Comparison of the same muscles from animals of the same breed but of different sex has shown that those from female are significantly smaller than those of males. Similarly, castrates fail to gain weight during growth in the same way that intact animals do. No major study has been carried out on the sex differences of the gross intramuscular connective tissue but some research has revealed interesting differences in beef longissimus dorsi and semimembranosus from intact and castrated males. It was shown that, although not statistically significant, samples of these muscles from intact animals contained more collagen than those from castrates. This difference correlated with an observed higher toughness of the muscles from intact animals when compared, by taste panel analysis, with those from the castrates. This latter result was also not statistically significant but has been observed by several research groups. The characteristics of collagen suggest that it could play a role in accounting for these differences. Not only would the total amount of collagen be important, but also the physiological age. The reduced solubility and increased denaturation temperatures reported for bull intramuscular collagen suggest a faster maturation rate. The consequent increase in tension generated during shrinkage, together with the increased amount of collagen, probably accounts for the increased toughness.

Other studies have considered the gross differences in quality of meat from animals of different sex although many of these have also concentrated on the differences between intact males and castrates. In one study of beef, for example, it was shown that the general characteristics of quality in longissimus dorsi steaks, e.g. marbling, texture, colour, firmness and tenderness, were better in castrates, although juiciness was higher in the same muscle from intact animals. On the other hand an extensive investigation of the bacon from boars, gilts and castrates failed to show any real differences in the characteristics of the meat. The question of sex differences in terms of meat quality, therefore, seems to be variable with species.

There may be a case for a more detailed study of the impact of animal sex on the quality of meat and especially meat connective tissue. It might be expected that the different rates of growth of males and females would result in differences in the maturity of the intramuscular collagen, with consequent effects on the texture of the meat.

Effect of nutritional regime on intramuscular connective tissue components

Meat animals can be raised on a variety of high energy or low energy feeds or, traditionally, on pasture. It has been shown in a study of longissimus dorsi and semimembranosus in cattle that meat obtained from animals fed on a low energy diet was significantly more tender than the same muscles obtained from animals of the same breed and age fed on a high energy diet. The muscles from low energy diet fed animals also had a lower perceived connective tissue content (from sensory analysis) and a lower overall collagen content, although these results were not statistically significant.

In another similar study cross-bred steers, randomly assigned to ten different groups, were fed on diets including a control and groups fed concentrate, concentrate-maintenance and roughage diets for various lengths of time. Unfortunately there appeared to be little in the way of consistent differences in collagen content or solubility between feed regimes. Those differences that were observed also tended to be rather small.

Undernutrition generally results in an increase in intramuscular collagen, with a decrease in the amount of salt and acid-soluble collagen, indicating that under these conditions there is less newly synthesised collagen. This increase in collagen probably reflects an increased ratio to muscle protein rather than an increase in real terms. The decreased solubility of the collagen is due to its increased maturation by stable cross-links and should, therefore, result in increased toughness.

It is clearly of interest to establish the nature of the collagen in the extreme case of subsequent 'catch-up' growth after undernutrition, or, in the similar but less extreme case, the effect of 'finishing' on the nature of the collagen. Indeed, it is well recognised that the best slaughter time for animals is after a period of rapid growth. During this time newly synthesised collagen will be laid down which is, by definition, less cross-linked. In the experiments described above, fast growing cattle showed a higher proportion of newly synthesised heat-labile collagen than slow growing animals. Similarly, moving the animals to a higher plane of nutrition increased the proportion of heat-labile collagen which was concomitant with an improvement in growth. This dilution effect of the new collagen reduces the overall background toughness due to collagen. On the other hand slow growing animals will produce tougher meat as the collagen has time to mature.

The question arises, what happens to the mature collagen during rapid growth; is it degraded and replaced by new collagen or does the latter dilute out the effect of the mature collagen? It has been generally assumed that there must be degradation of the fibrous network to allow growth to proceed or the mature matrix would be expected to oppose expansion of the tissue. However, most detailed studies suggest that the bulk of degradation is confined to the newly synthesised collagen (see Chapter 3). Furthermore, under rather extreme cases of 'catch-up' growth following severely arrested growth, little if any degradation of pre-existing mature matrix could be detected. Thus, the degradation of older mature collagen fibres does not appear to be a prior condition for growth. This suggests that the matrix consisting of older collagen fibres, though individually highly cross-linked, can expand by fibres slipping past each other and allowing expansion to occur by insertion of newly synthesised fibres. Expansion of the matrix could occur by degradation of the proteoglycans, which have a higher metabolic rate, assuming that these compounds play a significant role in stabilising the fibrous matrix (Fig. 9.3).

The quality of the connective tissue can therefore be modified ante-mortem, in the direction of a higher proportion of new collagen and hence more tender meat by rapid growth in young animals and by rapid growth immediately prior to slaughter of older animals.

It seems, therefore, that collagen in intramuscular connective tissue may be affected in quantity and quality by the nutritional regime of growing animals, but the evidence to date for such variations is sparse and needs to be corroborated and clarified by further study. No studies have yet been conducted on the influence of diet on other intramuscular connective tissue components.

Effect of the age of animals on muscle connective tissue

Animals are taken to slaughter at a relatively early age. For example, cattle are slaughtered in the U.K. at about 18 months and in the U.S.A. and Sweden at about 12 months; sheep are slaughtered as lambs at about 22–30 weeks and pigs as young as 20–22 weeks. In every case, the meat is harvested long before the animal reaches physiological maturity. There are good reasons in terms of economy and quality for this specific husbandry. The younger an animal is when slaughtered the less time the farmer has to handle that animal and the quicker his turn around time. Perhaps more importantly from the point of view of the consumer, the older the animal gets the tougher

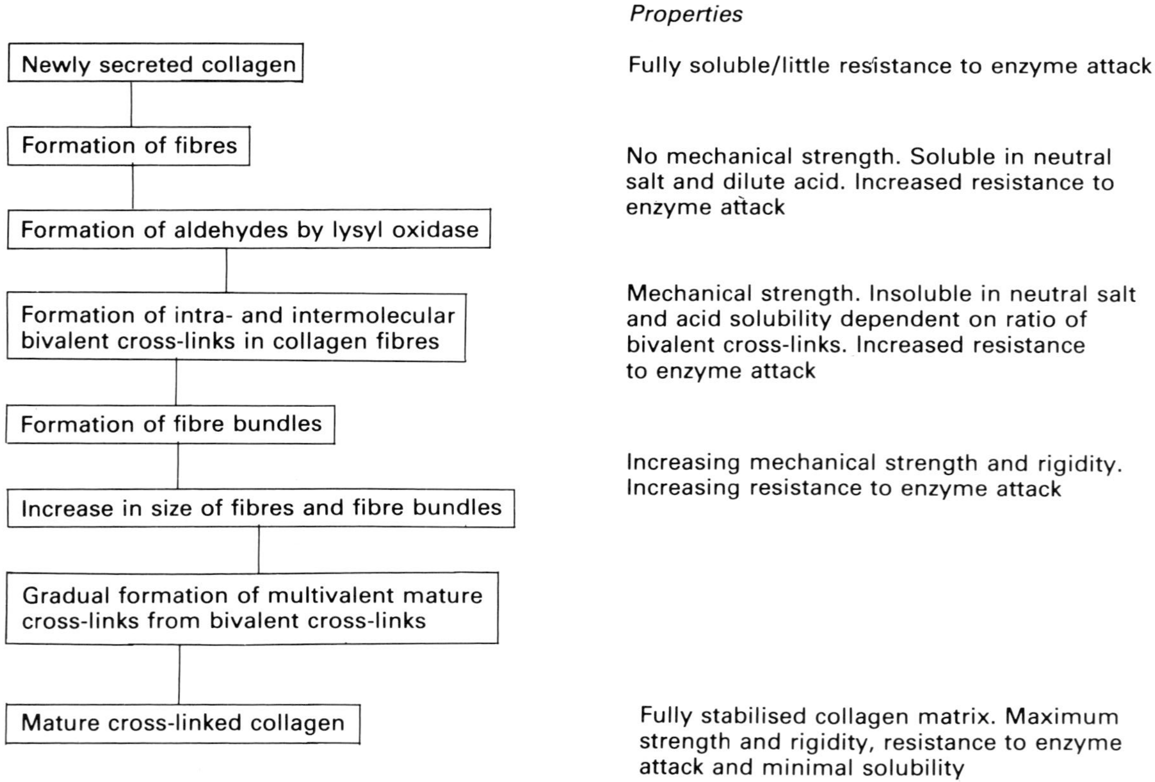

Fig. 9.3 Flow diagram summarising the changes in collagen on ageing and in its properties.

its meat becomes. This is a well proven fact, observed equally often by the lay consumer as by the meat scientist. The reasons for this decrease in textural quality of the meat as the animal ages are two-fold. First, the average muscle fibre size increases and secondly, and more importantly from the point of view of this book, the collagen in the muscle connective tissue becomes progressively tougher, more rigid and resistant and less easily denatured.

In earlier chapters (Chapters 1, 3 and 4) we discussed the changes in collagen with aging with respect to the associated changes in physico-chemical properties of the matrix. Although there is no evidence to show that the total amount of collagen in meat increases with age, it is a well established fact that collagen fibres increase in size as the animal grows (Chapter 1), that fibre bundles become larger, and that the cross-linking of the collagen becomes more complex (Chapter 4). There is also a possibility of an increase in interfibrillar as well as simple intermolecular cross-linking with ageing. All of these changes lead to a general strengthening of the collagen matrix, explaining the well-documented changes in physico-chemical properties such as increased resistance to proteolytic attack, increased rigidity and decreased swelling in acids (Fig. 9.3).

As the collagen of the intramuscular milieu becomes steadily more cross-linked by stable, multivalent intermolecular bonds, collagen fibre and fibre bundle size increase so that there is an increased resistance to enzymic attack by the intracellular proteases in the natural post-rigor process of conditioning. It stands to reason that a large fibre or fibre bundle presenting a low surface area to attack will be less easily degraded by large proteolytic enzymes than smaller fibres and fibre bundles which present relatively much larger surface areas. Similarly, the larger fibres will be more tightly knit and may be expected to have a higher degree of structural integrity due to the presence of maximal amounts of multivalent intermolecular cross-links.

This effect of cross-linking is perhaps the crucial characteristic in terms of the overall strength and resistance of normal collagen fibres. It is, therefore, important to consider all forms of cross-linking which might occur during the ageing process. There is no evidence to date for the formation of the normal lysine-derived intermolecular cross-links at any other site in the collagen fibre than between the N- and C-terminal regions of the molecule and two specific sites in the helix of adjacent molecules (Chapter 4). In other words, the possibility of the fibres becoming stronger through the age-related accumulation of these cross-links in helix–helix sites is not likely. However, related cross-links may indeed form in helix–helix sites through a random, non-physiological process brought about by the non-enzymic glycosylation of the collagen molecule. However, this affect is unlikely to be significant at slaughter age. The enzymes effective during conditioning are therefore confined to depolymerisation of the fibre by cleavage at the non-helical cross-link site.

Many taste panel and shear test surveys have been carried out on meat samples obtained from old animals and the toughness of such samples is well known. A hypothesis linking the age-related changes in the properties of intramuscular collagen with increasing toughness of the meat has been presented in Chapter 8.

Effects of growth-promoting substances

There is an extensive literature on the application of growth-promoting substances to animal production. These reports have naturally concentrated on the rate of growth and carcass composition, but within the context of this book we are only interested in meat quality, particularly relating to the role of connective tissue.

Anabolic agents

The anabolic agents used can be classified as to whether they are oestrogenic, androgenic or progestrogenic in action, and also whether they are endogenous to farm animals or exogenous, steroid or non-steroidal components. The anabolic agents currently permitted by the U.S. Food and Drug Administration are those based on the naturally occurring hormones oestradiol 17β, testosterone and progesterone, and the two exogenous substances zeranol and trembolone acetate. The European Community has imposed a total ban on all anabolic agents. The stilbenes and thyrostats are banned throughout the world as growth promoters.

Anabolic agents have the ability to increase the muscle-to-fat ratio in female and castrate animals, but there is a decrease in this ratio with intact males. In contrast to these generally well established changes, few really consistent effects have been reported in the eating quality, i.e. texture, flavour, juiciness, of the cooked meat. However, texture is probably the most important aspect of eating quality as far as the consumer is concerned and quality studies have been carried out on tenderness. The majority of these studies have failed to report any significant change in texture judged either by shear value or sensory panel with meat from bulls and steers.

There is no doubt that the use of anabolic agents results in improved growth rates, with lesser effects on carcass composition towards leaner meat from hormone treatment of castrated males and females. Anabolic agents tend to reduce degradation of protein, thus increasing overall growth. Increased growth may also be reflected in the rate of turnover of collagen which would suggest that hormone-treated animals would possess more mature collagen, and consequently decreased tenderness. However, the difference, if any, would appear to be small in view of the absence of significant changes in the measurement of texture.

Growth hormone

Long-term administration of growth hormone to lambs causes an increase in muscle growth. Protein turnover increases to a greater extent in red than in white muscle, and collagen synthesis responds in a similar manner. Most other anabolic agents decrease protein degradation rather than increase rates of synthesis. It would be interesting to see if this difference in synthesis rates had a significant effect on the texture of different muscles.

β-agonists

The recent use of synthetic substances that have the ability to repartition nutrients, that is, divert fatty acid into an energy source for the production of muscle, has been demonstrated to limit fat deposition and increase lean tissue in cattle, sheep, pigs and poultry. These repartitioning agents, generally referred to as β-agonists, are substituted catecholamines used in human and veterinary medicine for the control of bronchial asthma and smooth muscle relaxants in control of parturition, but have now found use as lipolytic agents in farm animals. They mimic the physiological actions of the catecholamine hormones epinephrine and norepinephrine and act through the β-adrenergic receptors present in many tissue cell membranes to stimulate cyclic AMP production within the cell. This results in a decrease in turnover of the muscle protein, presumably by limiting muscle protein degradation rather than synthesis, but the specific mechanism is not yet known.

Growth rate in cattle and sheep has been reported to be increased by as much as 15% but the effect is smaller in pigs, although the fat deposited is reduced by 10%. A substantial increase in the cross-sectional area of muscle has been reported. The slow twitch oxidative fibres found in the red muscles are more sensitive to β-agonists than the fast twitch fibres characteristic of white muscle. Much of the published work has concentrated on aspects of growth rate and carcass composition, but relatively little on meat quality. However, there does appear to be a significant increase in toughness of the meat. On the basis of the previous hypothesis of the role of collagen in texture, the faster growth rate might indicate less mature collagen and an increased tenderness. On the other hand if the rate of protein turnover is decreased, then the opposite is true; that is, the slow turnover would allow the intramuscular collagen to form heat-stable mature cross-links, leading to increased toughness.

The mechanism of β-agonists in affecting the texture of meat is an interesting but as yet unexplored area of research, yet would be an excellent model system.

EFFECTS OF ENDOGENOUS AND EXOGENOUS FACTORS ON COLLAGEN

We have seen that the biosynthesis, intracellular transport, secretion, processing and fibre formation of collagen is a long and multistep process. The same is true of the processes involved in the production of proteoglycans, glycosaminoglycans and matrix glycoproteins. Anywhere in these processes chemical perturbations can affect the final product of synthesis. Thus, in theory, it should be possible to affect the extent or quality of connective tissue component production with exogenous chemical agents. Taking collagen as an example, this type of effect can be readily studied in terms of vitamins, minerals, hormones and other factors such as exogenously administered agents.

Effects of vitamins on collagen metabolism

The role of vitamin C (ascorbic acid) in collagen production is well established (Chapter 3), together with the effect of its deficiency in man (scurvy) but, unlike man, meat animals synthesise their own vitamin C and do not become deficient.

Other vitamins, such as certain B-complex compounds and vitamins A and D, have been noted as having some effects on collagen and connective tissue metabolism but extremes of application or deprivation of these compounds have to be used in order to elicit these effects. Such considerations, in terms of the production of healthy meat animals on balanced diets, therefore, seem irrelevant as vitamin deficiencies have only minor secondary effects on connective tissue components compared with the more important effects on the general health of the animal.

Effects of minerals on collagen metabolism

Similarly studies on dietary mineral deficiency have, apart from copper and iron, been both limited and conflicting. As described in Chapter 3, the lysyl oxidase enzyme involved in cross-linking is a copper dependent enzyme, hence a copper deficient diet results in a deficiency in cross-linking of the collagen and a consequent fragility of the tissue. Ferrous iron is a co-factor for the hydroxylation of proline and lysine and a deficiency of Fe^{2+}, like vitamin C, causes reduced hydroxylation resulting in an unstable triple helix. The glycosylation of the hydroxylated collagen propolypeptides involves hexosyl transferases and these enzymes depend on the presence of Mn^{2+}. Both the N- and C-terminal proteases responsible for cleavage of the propeptides from procollagen require the presence of calcium. A deficiency in these minerals, therefore, certainly affects the metabolism and nature of the collagen, but only under extreme dietary conditions and is not, therefore, a problem or a potential control mechanism in normal meat production systems.

Effects of steroids on collagen metabolism

Adrenal steroids, often given to man and animals as anti-inflammatory agents, (e.g. cortisol and the glucocorticoids), are known to have an effect on collagen metabolism, generally inhibiting the synthesis of new collagen and possibly promoting the catabolism of existing collagen. Much work has been carried out on the changes in skin collagen brought about by the application of topical steroids or injection of anti-inflammatory drugs such as corticosteroids and on changes in the tensile strength and biochemistry of tendon collagen in animals fed on similar steroids. In general the long-term effects on collagen are deleterious.

Proline hydroxylase, lysine hydroxylase and the glucosyl and galactosyl transferases are all inhibited by steroids such as hydrocortisone and cortisol. In part, this is due to a general inhibition of protein synthesis. The effects of such agents can also be specific as in the case of dermal connective tissue where the Type I to Type III collagen ratio has been shown to be influenced by the application of such drugs. Another study showed that the short-term effect of systemic application of cortisol on rat skin was to increase collagen cross-linking, although in the long term this reverted to normal and the overall collagen content decreased, resulting in thinning of the skin and loss of mechanical strength. In this study, however, no difference could be detected in the ratio of Type I to Type III collagen.

Effects of exogenous drugs on collagen: lathyrogens

Lathyrism is an induced condition, first observed in animals fed on the seeds of the sweet pea, that has long been known to be due to β-amino proprionitrile, a component of the sweet pea seed, which binds irreversibly to the cross-linking enzyme lysyl oxidase and inhibits its action. The collagen is therefore devoid of or reduced in cross-links and is consequently extremely fragile. β-Amino proprionitrile is known as a lathyrogen. Its action is confined to newly synthesised collagen and β-amino proprionitrile does not effect collagen which is already cross-linked. Administration of lathyrogens to young animals would certainly affect the extent of the cross-linking; however cross-linking of the collagenous tissues is crucial to the physiological well being of the animal. A significant reduction in cross-linking leads to lameness and aortic aneurysm. In addition, the lathyrogens have effects on the nervous system at lower doses than those affecting the cross-links. The use of lathyrogen as a meat tenderiser is, therefore, not a viable proposition.

The administration of D-penicillamine (β-dimethyl cysteine) causes a decrease in the tensile properties of skin. This drug cleaves the aldimine cross-link by shifting the equilibrium in favour of reactions of the lysyl aldehyde with D-penicillamine rather than the lysine of another collagen molecule. The effect is confined to the aldimine, having no effect on the oxo-imine cross-link nor the stable mature cross-link. Administration of the drug has all the disadvantages of a lathyrogen plus the fact that it has little effect on intramuscular collagen which contains the stable oxo-imine cross-links.

EFFECTS OF PRE-SLAUGHTER PROTEASE INJECTION ON CONNECTIVE TISSUE

Animals may be injected intravenously from 1 to 30 min before slaughter with a solution containing a proteolytic enzyme. This method, patented in the late 1950s, is an adaptation of the long known use of enzymes, or natural juices rich in enzymes, to tenderise meat directly. Pre-slaughter injection was shown to give the most efficient penetration of the enzymes to the most remote parts of the carcass. Several sources have been used for such proteases including bacterial and fungal enzymes (protease 15, fungal amylase, hydralase D) and plant enzymes (ficin from the fig, papain from the pawpaw and the pineapple enzyme bromelain). The bacterial and fungal enzymes listed above have little or no activity against collagen and elastin, although there may be some activity in respect of proteoglycans and mucopolysaccharides. The clostridial collagenase, which in its highly purified form possesses a high degree of specificity for collagen (Chapter 12), has not been used for pre-slaughter injection.

All of the plant proteases mentioned above have collagenolytic and elastolytic activities. Ficin and bromelain have high efficiencies in terms of collagen degradation and ficin is equally active against elastin. Papain shows only a very low activity for native collagen, as does bromelain for elastin, but both enzymes will effect considerable structural damage against the two main connective tissue proteins given time and the right conditions. Papain is the most commonly employed protease in this process.

A concentration of about 1 mg per kg live weight is injected, although amounts vary depending on the enzyme used. There is some evidence to show that the penetration of the enzyme solution is greater in muscles with high connective tissue contents as these are generally more active and, therefore, have better vascularisation. This would seem to be a fortuitous advantage in that muscles with high collagen contents tend to be more tough and, therefore, would require greater enzyme tenderisation. However, the evidence for a beneficial effect of these enzymes in degrading connective tissue and thereby increasing tenderness is negligible. There are also associated problems in that other tissues with lower structural integrity and strength, such as liver and tongue, undergo extensive degradation of the fine connective tissue fibres and virtually disintegrate when cooked. Similarly, muscles of different textural quality undergo different degrees of proteolytic degradation and, therefore, the effects of the treatment are extremely variable and unpredictable. Some meats can become over-proteolysed and finish with a 'mushy', unpleasant paté-like character.

The other major problem is that the popular plant enzymes used in the pre-slaughter injection method are only active against denatured collagen in the early stages of cooking. At the low storage temperatures employed during conditioning papain shows little or no activity against collagen and elastin. It is only when the meat is heated that the enzymes become active. This means that the method of cooking has a major effect on the efficacy of these enzymes. If the collagen is generally well denatured but the temperature of the meat is not high enough to heat-denature the enzyme (papain and ficin are particularly heat-resistant and show very high activities at 50–60°C), then degradation will be extensive and the texture of the final product may be unacceptably 'mushy'. For example, a long slow cook at low temperatures (60–65°C) would lead to extensive proteolysis. However, a rapid high temperature cook, like, for example, frying or grilling, can lead to heat denaturation of the enzymes in the surface of the meat, some activity in the lower layers and no activity at the centre resulting in either an untenderised product or an uneven distribution of textural properties.

EFFECTS OF HANDLING, SLAUGHTER AND ONSET OF RIGOR ON MUSCLE CONNECTIVE TISSUE

There has been a considerable amount of research on the behaviour of animals during pre-slaughter handling and slaughter and the effects of such factors on carcasses during the early post-slaughter period. Much is now known about the stress factors which affect the quality of meat in terms of ultimate pH (pH_u), water-holding capacity, appearance and texture. There has also been a considerable amount of attention given to the effects of artificial treatments like the use of pre-slaughter protease injection techniques for tenderisation of meat, early post-slaughter electrical stimulation methods and specialised butchery techniques such as hot de-boning. However, very little if any research has been dedicated to the effects of such treatments on meat connective tissue. It is unlikely, considering the nature of connective tissue, in particular its stability and slow turnover, that collagen, elastin, the proteoglycans and mucopolysaccharides are at all affected by the various treatments animals may undergo in the brief time immediately pre-slaughter and during the short time before slaughter and the process of slaughter itself. The only areas in which we have any definite knowledge of the behaviour or influence of connective tissues in respect of the changes in animals and carcasses during pre-slaughter handling, slaughter and the early post-slaughter period is in terms of the effects of cold-shortening, electrical carcass stimulation and pre-slaughter protease injection (see above).

Effects of post-mortem changes on the connective tissue of meat

The quality of the connective tissue may be modified by the effects of pre-rigor mortis cold-shortening and by the 'conditioning' enzymes post-rigor.

Effects of cold-shortening on connective tissue

If muscle is too rapidly chilled after the animal is slaughtered a condition known as cold-shortening can result. This is manifested as a massive contraction of the actomyosin complex and a resultant shortening of the muscle bulk (Fig. 9.4). Cold-shortened meat has been shown to be extremely tough and so a well regulated cooling process is recommended for carcasses to avoid this undesirable effect. In the past, the toughness of cold-shortened meat was associated entirely with the contracted state of the actomyosin complex but there is some mechanical, textural and microscopical evidence that intramuscular connective tissue, and in particular collagen, may also be

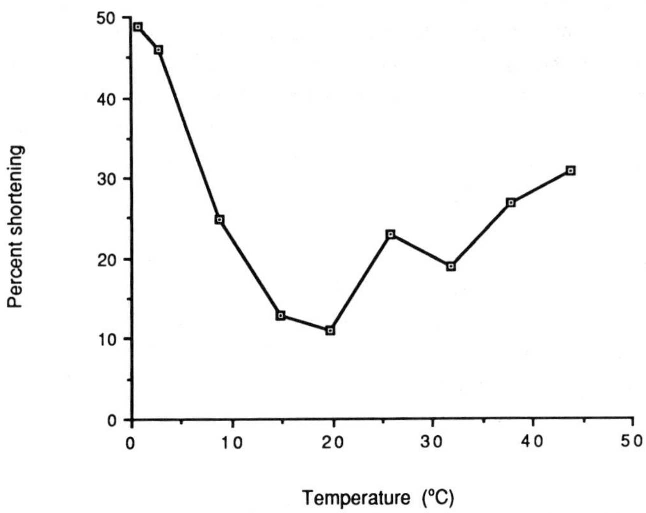

Fig. 9.4 Graph showing the relationship between the temperature used to chill pre-rigor meat and the extent of induced cold-shortening.

affected by the process and therefore contribute to the toughness of cold-shortened meat.

Comparison of perimysial collagen from normal and cold-shortened meat shows a striking difference in morphology. In the normal samples the collagen fibres are arranged in the usual sheet-like array with very little space between fibres. The collagen fibres show the usual crimped appearance (see Chapter 6) with a crimp length of about 19–20 μm. The overlayed sheets display a wide angle in the orientation of their fibres (>90°). In the cold-shortened samples there is no evidence of the crimp structure and the sheets of collagen fibres have been 'pulled' or distorted by the highly contracted muscle fibres so the angle between fibres in adjacent sheets is reduced to less than 90° (Fig. 9.5).

Cold-shortened muscle shows a gradual increase in toughness up to

(a)

Fig. 9.5 Light micrographs showing the change in perimysial fibre orientation on cold-shortening in bovine semitendinosus muscle. (a) Unstressed normal, post-rigor perimysial fibres.

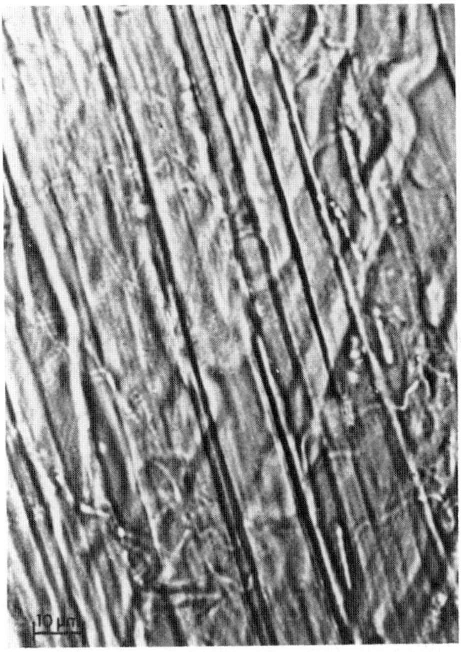

(b)

Fig. 9.5—*contd* (b) Perimysial fibres in cold-shortened meat.

about 40% contraction and a decline in toughness above 40% contraction. The changes observed in collagen in the shortened muscle may play a part in the increased toughness, although the major effect has been well established as an excess contraction of the myofibrils. It is suggested that the tensioned, more densely packed array of collagen fibres presents a different barrier to shear than the looser, untensioned arrays seen in normal meat. Such a change may well be expected to contribute to the changes in toughness of the raw meat as assessed by mechanical means (see Chapter 7) since the observations are made on the raw cold-shortened meat. However, it is possible that the density of the perimysial sheets has a direct bearing on the final textural properties of even the cooked meat as the disposition of the denatured insoluble collagen after cooking may rely considerably on its macromolecular conformation before heating. It may affect the tension generated during shrinkage when cooked, but is unlikely to affect the residual strength of the denatured perimysium, which is a primary determinant of texture.

Effect of conditioning on muscle connective tissue collagen

After the onset of rigor mortis intracellular enzymes begin the process of degradation. Several types of enzyme have been implicated in this process to date, including acid and neutral proteases and various polysaccharidases. There are two major groups of proteases, however, that have been the subject of the most intensive research. These are the cytoplasmic calcium-activated neutral proteinases and the proteolytic enzymes of the lysosome, mainly the acidic cathepsins. The first has an optimum pH of near neutrality and the latter group have optimal pHs nearer 4·0–5·0. The activity of proteolytic enzymes during the period of controlled chilled storage of the meat known as conditioning is what leads to the so-called 'resolution of rigor', i.e. the softening of the tissue so that it becomes edible, and an extension of this process is used to tenderise meat.

The activity of these proteases has been speculated to include (a) the 'clipping' of the rigor actomyosin complex to allow relaxation of the contractile apparatus, and (b) damage to the connective tissues, in particular collagen fibres, leading to a fall in the inherent strength of the collagen matrix. Both of these processes then are thought to result in a 'softening' of the respective tissues, although the softening of the actomysin complex is clearly paramount. Evidence for the former process in which intracellular proteins undergo limited proteolysis leading to the resolution of rigor have been well documented elsewhere and will not be discussed here. At this stage it is important to consider the possible effects of conditioning on collagen itself, as this would be expected to have a direct bearing on the strength of the extracellular matrix and, therefore, its role in determining texture (Chapter 8).

Several studies have attempted to determine whether or not the intramuscular connective tissue is attacked by proteases during conditioning. The difficulty in this work has been that only very limited proteolytic damage in collagen matrices can lead to large changes in the strength, rigidity and the insolubility of the connective tissue and, therefore, the amounts of solubilised collagen which may result from such limited cleavage would be very low. It has proved extremely difficult to detect these small quantities of free collagen or collagen fragments in such a way that such small variations may be distinguished from mere experimental differences.

Early work showed that, although no detectable rise in soluble collagen could be attributed to the process of conditioning, the collagen of aged meat seemed more prone to gelatinisation than that seen in fresh, unconditioned meat. Later, it was shown that some free α-chain material seemed to be associated with the conditioning process and it was suggested that excess lactic acid produced during rigor mortis might play a role in solubilising collagen cross-linked by the acid-labile aldimine bond (see Chapter 4).

In a recent study, however, the epimysium, perimysium and endomysium were separated from conditioned and unconditioned muscles and were extracted with denaturing solvents. It was shown that higher yields of soluble material could be obtained from conditioned perimysium than from the unconditioned connective tissue (Table 9.1). On average, about 1·6 times more material was released in the soluble fraction from conditioned perimysia. Hydroxyproline analysis (Chapter 12) showed that the collagen content of the soluble material obtained from conditioned perimysium was twice that of the unconditioned material.

Another interesting difference was that some muscles showed more solubilisation than others. For example, the gastrocnemius muscle only showed a 23% higher yield of soluble material when conditioned (compared to unconditioned gastrocnemius), whereas the psoas major showed a 200% increase in solubilised material. Significantly, the collagen content of these solubilised materials was higher in the psoas major than in the gastrocnemius. It appears, therefore, that different muscles show different extents of proteolytic damage to intramuscular collagen during conditioning. This may reflect differences in overall proteolytic activity within different muscles, different final pHs, different susceptibilities of collagen in different muscles or other differences. These results provide direct proof for the release of

Table 9.1
Solubility of perimysium extracted from unconditioned and conditioned bovine muscles

	Solubilised collagen (%)	
	Unconditioned	*Conditioned*
Gastrocnemius	0·8	1·7
Psoas major	0·8	10·5
Pectoralis profundus	0·4	1·3
Gluteus medius	0·6	2·3
Supraspinatus	0·4	4·6

intracellular proteases into the extracellular milieu during conditioning. It seems likely that the enzymes involved are the cathepsins of the muscle fibre lysosomes but there is no direct evidence to support this assumption. The average final (ultimate) pH (pH_u) of meat is 5·3 (Chapter 1), a level of acidity which might be expected to inhibit fully the activity of the neutral proteinases of the muscle cells and other cells such as fibroblasts. The optimum pH of the cathepsins lies in the range 2·5–4·5 but these enzymes have been shown to be active at pHs as high as 5·5.

Several cathepsins have now been isolated and characterised from a variety of tissues including muscle, although the latter is rarely used as a primary source due to the difficulty of extracting these enzymes from that tissue. Most work has been carried out on spleen cathepsins. There are at least four lysosomal cathepsins which have been denoted as cathepsins B, D, L and N. Each enzyme is genetically distinct from the rest and they have different characteristics, e.g. as stated above their optimum pHs range from 2·5–4·5. It has been suggested that the more acidic cathepsins may act in microenvironments, i.e. at a limited distance from the cells which release them in which the tonicity and pH might be controlled by secretions from that cell. It has been speculated that the pH in these localised extracellular environments may be as low as 2·5 whilst the average pH of the surrounding tissue may be as high as 7·0. Evidence for this type of mechanism comes from the direct measurement of pH in extracellular microenvironments using specially designed microprobes. Furthermore, cells have been identified and isolated which have the capacity to spread over surfaces of connective tissue structures and produce microenvironments of high acidity to promote proteolytic breakdown. The activity of cathepsins in meat connective tissue, therefore, may be a concerted attack by the various types making use of microenvironments and the general low pH of the post-rigor tissue.

Studies have shown that muscles with an unusually high pH_u (e.g. dark cutting (DFD) meat) show less proteolytic damage than those with a pH_u of 5·3–5·5. This is further good evidence for the role of cathepsins in the process. Also, it is thought that residual blood in meat may provide some further proteolytic activity as the cathepsins of leukocytes are more active than those of muscle.

Cathepsins have a powerful collagenolytic activity which is comparable to that of pepsin (see Chapter 12). The main site of attack by these enzymes on the collagen molecule is in the non-helical terminal regions. It is thought that the enzymes may act in a concerted manner to 'clip' terminal regions of collagen molecules at the surface of fibres which may or may not result in the release of some free α-chains into solution. The residual damaged fibres may then be able to swell more readily in the acidic environment of the post-rigor meat, thus allowing further penetration of the globular cathepsin molecules and more extensive damage to the fibre. There is good evidence for this mechanism in model systems using isolated enzymes and purified insoluble collagen; recently direct evidence from studies of muscle tissue have shown *in vivo* damage to perimysial collagen in conditioned meat (see below). Only one cathepsin (cathepsin L) has yet been positively identified in muscle cells by biochemical means, and then only in low quantities. The levels found are in the order chicken, pig, sheep and beef, the latter having barely detectable levels, and it should be noted that this order also correlates with the increasing conditioning times required from chicken to beef. Certainly, the use of anti-enzyme antibodies and the immunofluorescence technique (Chapter 13) should reveal more detail about these enzymes in meat in the future.

The hypothesis that only small amounts of damage are incurred by insoluble collagen during the conditioning process is borne out by the recent studies mentioned above. An investigation of the integrity of collagen at the end of conditioning by the method of one-dimensional polyacrylamide gel peptide mapping failed to reveal any damage, i.e. new peptides (Chapter 12). This technique is relatively sensitive and minor damage to regions of the collagen molecules might have been expected to show up as unusual peptides in the maps. This result seems to argue that little or no residual damage is incurred by insoluble collagen fibres in conditioned meat. However, development of a more sensitive technique has recently provided evidence of peptide cleavage. In this study the insoluble perimysial collagen from conditioned meat was compared with non-conditioned meat by two-dimensional peptide mapping. It was evident from examination of a range of muscles that the insoluble collagen from conditioned meat showed subtle proteolytic damage that was not evident from the one-dimensional mapping techniques (Fig. 9.6). It should be remembered, as fully discussed in Chapter 8, that only a limited number of 'clips' along the polymeric fibre can produce extensive changes in the physical properties following heating. Such changes are probably more readily detectable by changes in specific physical properties. Changes

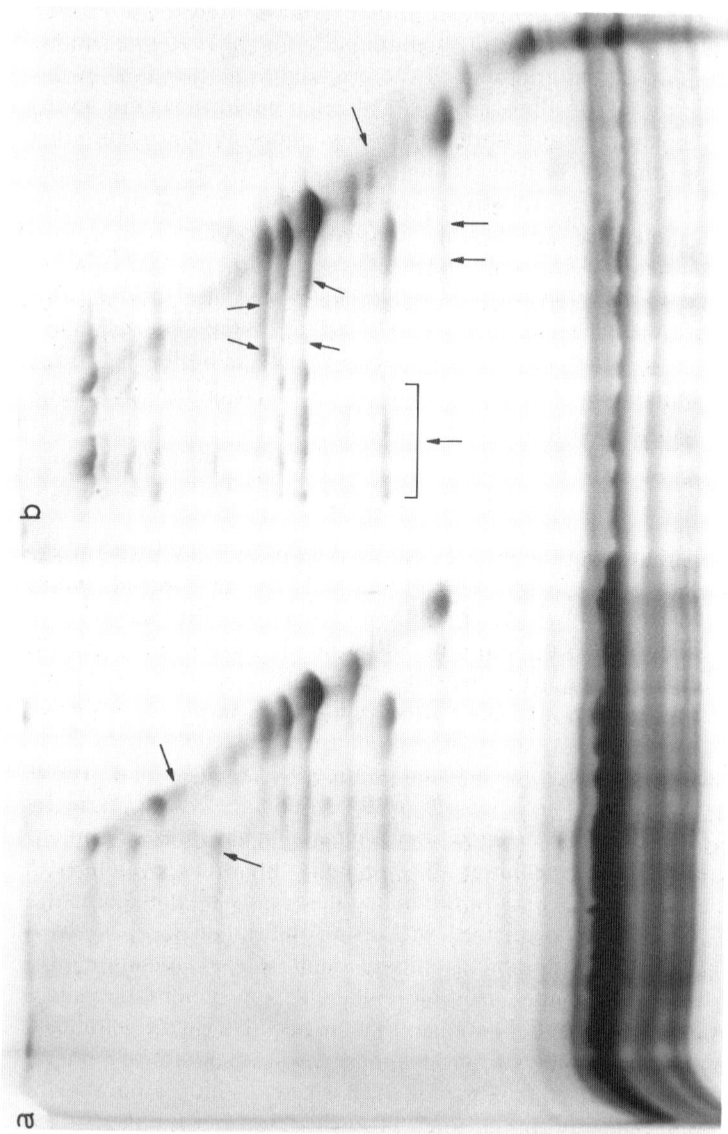

Fig. 9.6 Two-dimensional gel electrophoresis peptide mapping of insoluble perimysial collagen after CNBr and chymotrypsin digestion from (a) unconditioned and (b) conditioned muscle. The arrows highlight peptide differences between the two samples indicating the existence of subtle proteolytic damage to residual insoluble collagen after conditioning.

in the tension generated during heating and the residual strength of the denatured fibres are known to affect radically the ultimate texture of the meat. Indeed, a decrease in isometric tension has been readily observed following mild enzyme treatment of collagen fibres. A similar demonstration with the collagen fibre *in situ* in the conditioned and heated meat tends to be non-reproducible, but there is a clear trend toward a decreased tension and hence decreased shrinkage of the meat during cooking.

The increasing use of electrical stimulation to prevent cold-shortening could have an additional benefit in terms of accelerated conditioning. The rapid drop in pH resulting from electrical stimulation means that the catheptic enzymes would be acting for a short period at near body temperature rather than at chill room temperature. The activity of these enzymes against collagen will be increased at these elevated temperatures and could therefore significantly reduce the normal ten-day conditioning period.

SUMMARY

The factors influencing the amount and quality of the connective tissue in meat include age, sex, breed, type of muscle and growth rate, whilst post-mortem factors include conditioning time and the use of proteolytic enzymes. The amount of collagen in muscle is genetically predetermined, being particularly low in double-muscled animals, but the quality of the collagen depends on its turnover and growth rate of the animal.

Growth characteristics can be manipulated by nutrition and growth promoters. Rapid growth before slaughter has been shown to increase the proportion of newly synthesised collagen, and consequently the tenderness of the meat. Growth promoters are highly variable in their affect on the tenderness of meat, some increasing tenderness whilst others decrease tenderness. Details of the mode of action on the balance between synthesis and degradation of the collagen could provide a clue to the effect of these agents.

Physiological age of the animal is the major determinant of meat texture, primarily due to the maturation of the collagen fibres. Other things being equal the older the fibres the tougher the meat. At the present time there is no means of preventing maturation of the collagen fibres.

Conditioning of meat is a vital part of the process to obtain tenderness and reduce variability in quality. The major enzymes acting at pH 5·5 are believed to be the cathepsins. These enzymes act by depolymerising the collagen aggregates in the fibre, thereby decreasing the tension generated on shrinkage and the residual strength of the denatured fibres. Methods of accelerated conditioning, for example by electrical stimulation, and activation of the proteolytic enzymes, are needed by the industry to counter the cost of conditioning time.

FURTHER READING

1. Lawrie, R. A., *Meat Science,* 4th edn. Pergamon Press, Oxford & New York, 1985.
2. Bailey, A. J., Enser, M. B., Dransfield, E., Restall, D. J. & Avery, N. C., In *Muscle Hypertrophy of Genetic Origin and its Use to Improve Beef Production,* ed. J. W. King & F. Menissier. Martinus Nijhoff, The Hague, 1982, p. 178.
3. Hanrahan, J. P., *β-Agonists and their Effects on Animal Growth and Carcass Quality*. Elsevier Applied Science, London & New York, 1987.
4. Etherington, D. J., Collagen and meat quality: effects of conditioning and growth rate. In *Advances in Meat Research,* Vol. 4, ed. A. M. Pearson, T. R. Dutson & A. J. Bailey. Van Nostrand Reinhold Co., New York, 1987, 351–360.
5. Patterson, R. L. S. & Salter, I. J., Anabolic agents and meat quality: a review. *Meat Sci.,* **14** (1985) 191–220.
6. Stanton, C. & Light, N. D., The effects of conditioning on meat collagen: Part 2—direct biochemical evidence for proteolytic damage in insoluble perimysial collagen after conditioning. *Meat Sci.,* **23** (1988) 179–199.

Chapter 10

Connective Tissue in Meat Products

INTRODUCTION

Meat is eaten either as 'fresh' meat or in the form of meat products. In the former case the meat is eaten in the form of cooked conditioned muscle prepared either as a combination of muscles (joint or cut) on or off the bone, or as special cuts such as chops, chunks or mince. Meat products provide the second major class of meats consumed by man and these cover a wide range of items including simple processed meats such as cured and/or smoked products (ham, bacon) and the more complex manufactured products such as pies, sausages, burgers, re-formed meat products, patés, pastes and soups. Clearly, meat can be presented on the plate of mankind in a wide variety of forms.

All meats which are treated in some way which involves the addition of other chemicals or products or the combination or grinding of the meat and reformation of a new product may be regarded as meat products or processed meats. The processing of meat often involves many steps and relies on well-controlled manufacturing procedures. The properties of the individual proteins and macromolecules and their interactions with each other are important in the production of a meat product as they play a vital role in the emulsification, plasticity, coagulation, swelling and gelation of the final product. All these properties have an effect on the cooking characteristics and the final organoleptic attributes of the meat product.

The functional properties of a protein in a processed meat depend on its amino acid composition, molecular weight, solubility, thermal properties and the relationship of these properties to pH, temperature and salt concentration. In meat systems the pre-processing condition of the muscle such as the extent of rigor development, storage conditions and time affects the properties of the meat protein. These can be further affected by sodium chloride, polyphosphates and other chemical additives. More importantly, denaturation during cooking can fundamentally change the functional properties of a protein including its interaction with macromolecules like carbohydrates, lipids and other proteins as well as with small molecules like salts.

In earlier chapters we highlighted the difficulty of working in research with meat connective tissue components, largely caused by their intractability and low concentration in meat. Even so, 'fresh' meat has a defined structure and function and thus data derived from the study of its connective tissue can be interpreted in the light of these known parameters. However, the complex set of variables and conditions involved in the production of meat products make it even more difficult to predict the role of connective tissue components in processed meats. Although we know almost nothing of the role of proteoglycans and minor glycoproteins in the structure and function of processed meats, a limited amount of work has been done on collagen in meat products.

Another major class of meat products, the sausages, rely for their overall shape on the casings in which they are packed. These casings can be made from collagenous extracts of various types of connective tissue. The accumulated knowledge about the way collagen acts in the formation and function of sausage casings is important in terms of predicting the behaviour and reproducibility of behaviour of such products, especially during cooking. In addition, the connective tissue product known as gelatin is widely used as a food additive because of its excellent gelling characteristics. This, too, is a major product of collagen which is used, not only in meat products, but in a wide range of other foods.

In this chapter, therefore, we will consider the structure and form of comminuted meat and meat products, the role of collagen in such products and the influence of salt and other chemicals used in meat products on collagen and the structure and uses of gelatin.

MEAT EMULSIONS AND MEAT PRODUCTS

Comminuted meats, to which salts are frequently added, have traditionally been regarded as emulsions (Fig. 10.1) in which proteins

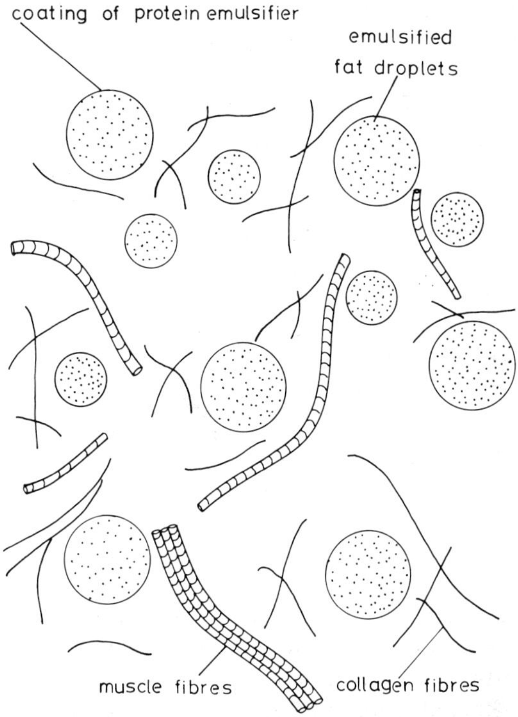

Fig. 10.1 Diagrammatic representation of a 'meat emulsion'.

encapsulate fat in a micellar structure, although there is another view that they should be considered as meat batters in which myosin gels entrap fat and other components. In this review we will consider that comminuted meats are emulsions, i.e. colloids in which aqueous sols (or colloidal dispersions) of proteins (hydrophilic phase) are mixed with fats and oils (hydrophobic phase) and stabilised in a micellar form by emulsifiers (surfactants).

Proteins are good natural emulsifiers due to their mixed hydrophobic/hydrophilic nature. As they are amphoteric, of mixed size and variable charge/hydrophobicity and are affected in terms of solubility by pH and salt concentration, their effectiveness in acting as emulsifiers in meat emulsions is variable and controllable (Fig. 10.2). Several theoretical models for the fat-binding efficiency and water-binding efficiency of proteins have been postulated and can be used to predict accurately sausage or meat product texture and consistency.

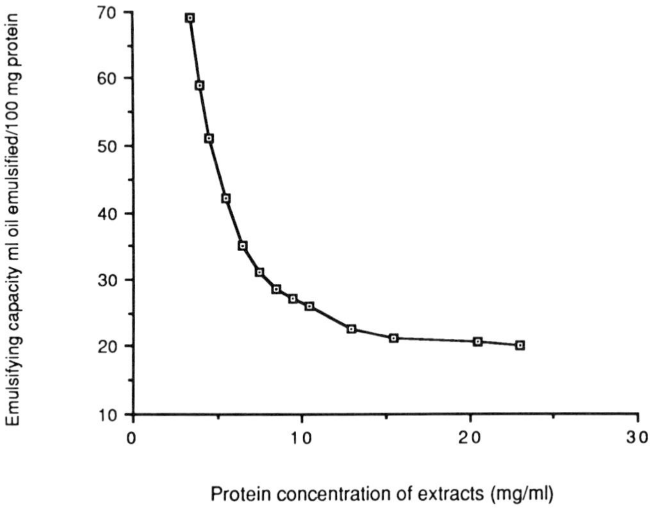

Fig. 10.2 Graph showing the correlation of emulsified oil volume and emulsifying capacity of soluble protein.

An important characteristic of meat products is their ability to bind the various components together to provide a cohesive product. The mechanisms of the binding of meat pieces are complex but are known to be affected by the extent of rigor, mechanical pre-treatment, presence of salts, temperature and time. These factors basically determine the nature of the protein extracted into the fluids in the comminuted meat which then act to bind the meat pieces together. The optimum temperature for extraction is about 7°C.

Analysis of the extracted proteins reveals myosin as the predominant component and the amount of this protein in the extracted phase increases significantly on the addition of polyphosphate to the comminute. When the binding strength is defined as the force per unit area required to pull apart bound pieces of meat in a meat product, then binding strength can be shown to be related linearly to the amount of extracted myosin. During mechanical treatment (grinding/comminution) the muscle bundles and myofibrils begin to separate; this may be an important aspect of binding since the open structure would allow ingress of the solubilised protein. The binding strength of the myosin extract increases up to a protein content of about 12%, after which the binding strength remains constant. This may be due to further increases in extraction resulting in extensive fibre disruption. Certainly, this disruption can readily be seen in the electron microscope after extensive tumbling. It would appear that a point is reached where disruption is greater than the binding capacity of the extracted protein. An important aspect of this work that should be borne in mind is that the ability to bind to the meat and the ability to form a strong gel are two different properties of proteins, and often a compromise has to be made—a rigid gel is no good if it does not bind to the meat.

More detailed analyses of the ability of the protein exudate to form a gel and bind meat have been carried out. Mixtures of actin and myosin produce stronger gels when heated than myosin alone since a three-dimensional network of fibres forms. It should be remembered that the binding is a heat-initiated reaction, no binding occurring in the raw state. Maximum binding appears to occur at about 80°C, depending on the ionic strength and pH of the mixture. Clearly the emphasis of these meat-binding studies has concentrated on myosin as this is the major component of the exudate. The sarcoplasmic proteins do not appear to have a significant influence on the binding properties.

As noted earlier, little work has been carried out on the influence of collagen on the binding, structure and properties of meat emulsions, but it is known that the texture of meat products is affected by the inclusion of connective tissue and adipose tissue. We will review the present literature below.

Collagen in comminuted meats

Collagen is present in comminuted meats and meat products either as a natural component of the connective tissue of the meat used in the process or as an additive. In the latter case the collagen may be added for specific binding/structural reasons, as in the addition of graded collagen as a binding agent or gelatin 'jelly' to pies (see below), or as a 'filler'. Filler collagen exhibits inferior properties of binding compared with higher quality collagen additives and naturally occurring meat collagen. Filler collagens are added largely to reduce the cost of a meat product by adding low cost bulk. We will discuss the role of binding (emulsifying) collagen additives in the next section.

The amount of meat collagen extracted under the conditions of pH and temperature usually involved in meat comminution is certainly very small and, therefore, collagen in protein exudates would be expected to make little contribution to the binding strength of the

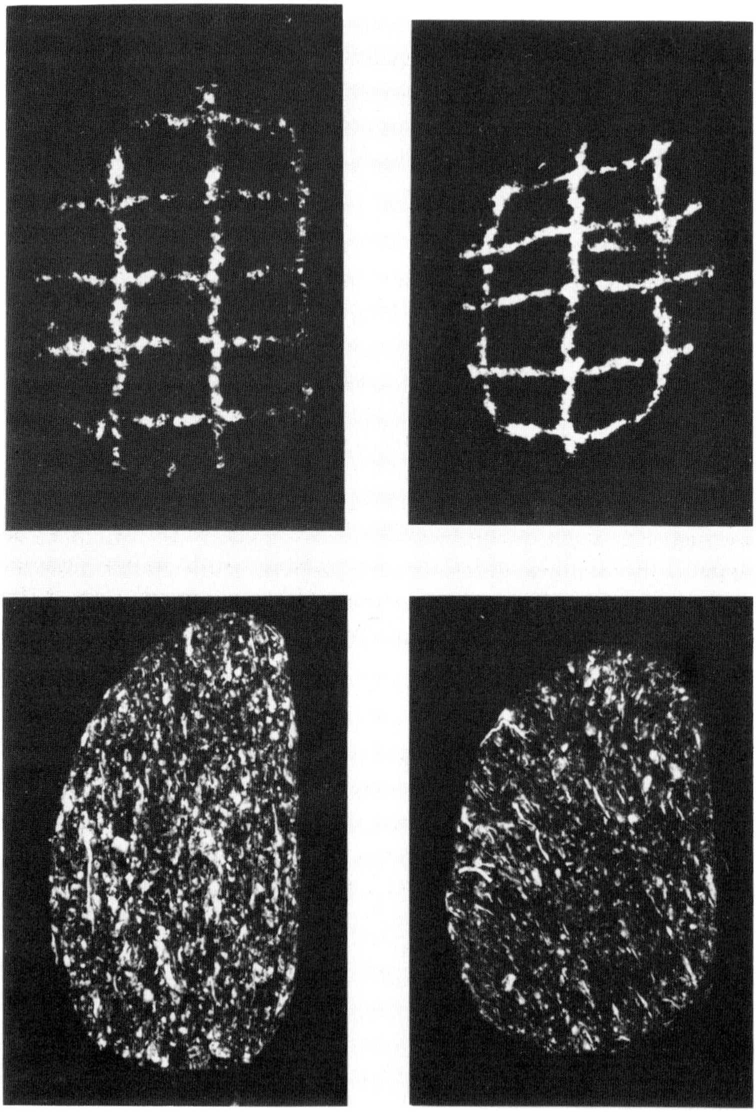

Fig. 10.3 Typical distortion of restructured steaks due to poor alignment of collagen fibres.

extracted protein gel when cooked. However, the collagen would be gelatinised (denatured to form gelatin) by a heating process over 65°C and it has been suggested that it may form pockets of fat and gelatin in products such as sausages and pies. In these cases, it is thought that the effect of collagen may be deleterious to the product quality.

The small amounts of collagen extracted in the comminution process may be important in two ways. Firstly, in affecting the rate of extraction of the myosin, and secondly, in the strength of the cooked meat pieces. Meat products made without collagen are soft, jelly-like and lack cohesion. It seems likely, therefore, that collagen plays a role in strengthening myosin gels and emulsions. This is not surprising considering the known properties of collagen and gelatin in gelation and matrix formation. As seen in Chapter 7, native insoluble collagen, when heated, undergoes denaturation but, because the chains of each triple helix are held in register by the heat-stable cross-links, the molecules quickly re-nature on cooling. This means that cooked meat products might be expected to contain a mixture of true gelatin (i.e. soluble collagen polypeptide chains) which will gel on cooling, and insoluble collagen which will melt and then re-form a coherent matrix on cooling. In both cases, smaller molecules, e.g. fat, will be entrapped. Also, intermolecular interactions with other macromolecules will be encouraged and enhanced by the opening up of the collagen triple helices and the uncovering of reactive sites such as lysine and acidic side chains.

The quality of grilling steaks re-formed from comminuted (generally flaked) meat can be badly affected by the orientation of the connective tissue in the product. Poor orientation results in excessive shrinkage producing distortion, known as 'doming', of the product and increased fluid loss (Fig. 10.3). To ensure even shrinkage the orientation of the connective tissue must be generally in the same direction. An acceptable product can be achieved by using a patty former that ensures the meat mix flowing through the machine results in an even distribution of the connective tissue.

Collagen in meat emulsions

As mentioned above, products made with comminuted meat are often referred to either as emulsions or batters. Neither is a strictly accurate definition, but however one defines 'meat emulsions', we are concerned with the functional role of collagen in the properties of these products. It is generally agreed that collagen is important in the processing of meat emulsions, in stabilising the emulsion and in providing textural properties to such products as burgers, sausages and frankfurters. It has also been suggested that collagen in meat emulsions can also affect yield, colour and even shelf-life and flavour.

To enable water and fat to mix and form stable emulsions, surfactants or emulsifiers are necessary (Fig. 10.4). These emulsifiers

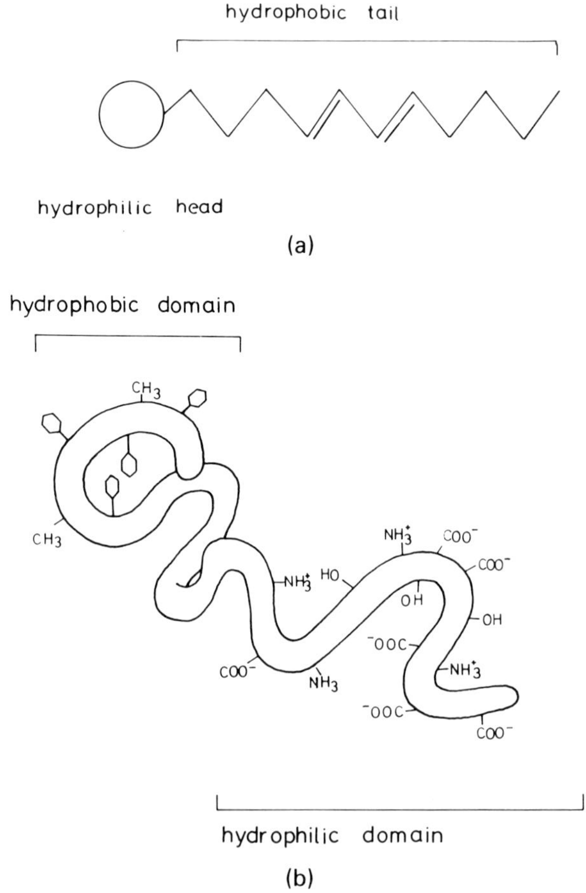

Fig. 10.4 Diagram showing the generalised structure of surfactants. (a) Small surfactant molecule, e.g. a diglyceride, (b) protein emulsifier with hydrophobic and hydrophilic domains.

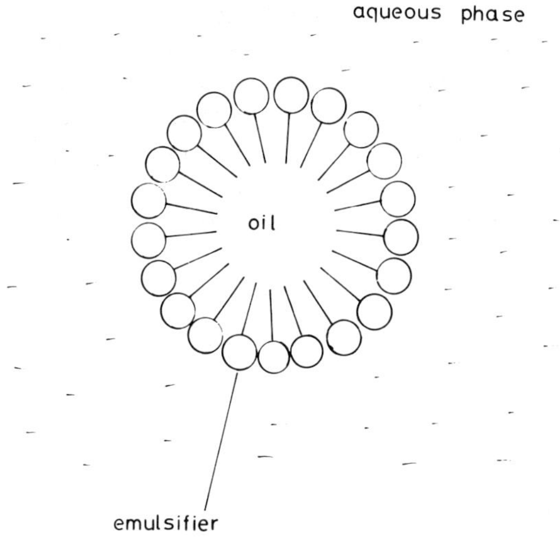

Fig. 10.4—*contd.* (c) Emulsifier surrounding fat or oil droplet in aqueous phase.

are generally partly lipophilic (hydrophobic) in order to interact with the fat, and partly hydrophilic to interact with the water phase. By interacting in this way the emulsifier concentrates at the fat–water interface, reduces the interfacial tensions between fat and water and thereby stabilises the mixture. Myosin is regarded as the major emulsifying protein in meat emulsions and if its extraction from meat pieces of the comminute is inhibited then the resulting products show the effect known as 'fat-out' or 'grease-out', i.e. a destabilisation of the emulsion resulting in the formation of fat pockets and the appearance of fat (or grease) on the surface of the product. However, collagen may also have an important role as a subsidiary emulsifier.

Many tissues and organs high in collagen content are included in products relying on a meat emulsion, as we will see below. The collagen can provide a different role depending on the temperature of processing. Collagens denature and shrink at around 65°C. As seen above, heat-denatured collagen can encapsulate fat and, on cooling, the collagen will shrink, allowing the fat droplets to coalesce. The denatured collagen (gelatin) will similarly set in separate pockets on cooling. However, it should be remembered that most of the collagen used in meat products is unlikely to solubilise as it contains high proportions of matrix cross-linked by heat-stable oxo-imine and mature cross-links (see Chapter 4); thus, it will tend to remain as gelatinous pieces.

It is unlikely that collagen becomes part of the emulsion system until heating begins. Collagen is 60% hydrophobic and hence when heat-denatured the soluble gelatin will be preferentially soluble in the fat. This high content of hydrophobic amino acids in collagen makes it likely to have unusual emulsifying abilities. In this respect, it mixes readily with oil and forms emulsions in which oil is the continuous phase. Such an effect may be expected at temperatures above 65°C in meat products. This can be a problem in products with high contents of collagen such as sausages, particularly those containing fine ground meat materials such as frankfurters. In this case it has been shown that high collagen levels (i.e. 23–33% of total protein as collagen) in frankfurters causes an imbalance of the hydrophilic/lypophilic balance (HLB) leading to 'grease-out'. These sausages, after cooking, had gelatin on their outer surfaces, a grainy inner texture and showed evidence of 'grease-out'. The appearance of the sausages was broken and curved with uneven surfaces and splitting.

The ability of gelatin to form water–oil emulsions which gel on cooling is thought to be important in some products. It is suggested that the collagen can form an ordered continuous matrix around the water and through the oil. Gelation has been defined as an aggregation during denaturation with the formation of continuous structures.

The hydrophobic nature of collagen is considered by some to be important in allowing the molecule to bind fat. If this is correct then it may provide a role for collagen in meat products, i.e. that of binding fat and leading to overall adhesion of fat to lean particles/pieces. This may be an important role in that, as we have seen above, myosin and actomyosin can act to bind protein and to some extent fat but in high fat products it may be the collagen which acts as an emulsifying intermediate to absorb excess fat. It has also been suggested that collagen (gelatin) may be able to gel in the presence of high fat concentrations, thus more effectively entrapping the fat.

Various authors have suggested maximum quantities of added collagen for meat products in terms of the effect on structure and acceptability of such products. It has been suggested that no more than 25% collagen should be present in meat blocks but experiments with beef tripe added to bologna have suggested that levels of over 15% may be unacceptable.

The fact that collagen acts as an emulsifier is now well accepted and data have been obtained showing a good correlation between meat emulsion stability and collagen content. Correlation coefficients obtained in such experiments indicate the function of collagen as an effective emulsifier. However, there is also good evidence to suggest that high levels of collagen have a detrimental effect on meat emulsions. As mentioned above, bologna sausages with 10% added beef tripe had very similar properties to control sausages, but samples with 20–40% added tripe showed increasing losses of fluid, gel-water and fat ('fat-out'), although the mechanism of such processes is not at all well understood. It is likely that the shrinkage of the collagen during cooking permits the release of the fat. Microscopic studies of these samples revealed that collagen was homogeneously distributed throughout the mix and it was postulated that the change in properties was due to a protein dilution effect whereby the effective concentration of myosin and, therefore, emulsifying/binding capacity, was reduced by increasing collagen levels.

Finally, increasing the amount of collagen in meat products is also known to affect the textural properties of the product by making it more brittle. This effect can be useful in some instances but, by and large, is regarded as being detrimental. As can be seen, much of our knowledge about the role of collagen in meat emulsions and the way it affects the quality of meat products is speculative.

SOURCES OF COLLAGEN FOR MEAT EMULSIONS

Several sources of collagen are used in the manufacture of sausages and other meat products. These include hide collagen, bone collagen, offal collagen and skeletal muscle collagen.

The corium layer of hide provides good quality collagen which can be used as an additive to coarse meat sausages up to levels of 30% by weight without detrimental effect. Isolated bone collagen has excellent properties of water and fat absorption as it has a spongy structure and is almost sterile. It is a good additive for burgers and has been shown to reduce fat exudation. It improves the textural properties of mechanically recovered meat by increasing its stiffness and can be added to almost any comminuted meat.

Offal collagen is less commonly used although body organs are a good source of connective tissue (see Chapter 1) and they represent over one-eighth of total lean meat protein in the average carcass. There have been some attempts to form texturised offal proteins by spinning techniques similar to those used for soya and other non-meat proteins, but these have not been taken up commercially and the role of collagen in such systems is not known.

The connective tissue of skeletal muscle used for comminution can be removed by pre-treatment with mechanical de-sinewing devices. This method can yield tissue low in collagen which then proves better for the production of bolognas, salamis and frankfurters. Such products show low 'fat-out' and negligible distortion on cooking. In these products therefore collagen can prove a problem and its removal prior to product formation can much improve the final product.

It remains to be seen, however, exactly how added collagen plays a role in meat products. Such knowledge would permit the formulation of a rationale for the source and form of the collagen put into the product. Much more research is needed before we understand the contribution of this and other connective tissue molecules to the structure of meat products.

EFFECTS OF VARIOUS PROCESSING METHODS ON COLLAGEN

Various processes in the manufacture of meat products require the addition of chemicals such as acids, common salt (NaCl) and polyphosphates as well as the controlled use of heat (thermal processing). Examples are the production of salted beef (corned beef in the U.S.A.), hams and bacon. Although important in the overall changes brought about in meat proteins (especially myosin) and other macromolecules, the effects of such treatments on collagen and other connective tissue components in meat products are poorly understood.

Effect of salts

The addition of salt to meat and meat comminutes enhances the solubilisation of myosin and other cytoplasmic proteins, giving rise to a sticky exudate which acts to bind and glaze chopped chunks and whole meats such as hams. Overall, the presence of salt greatly enhances the water-holding capacity of meat. The water-holding capacity of collagen, however, is not greatly affected by NaCl. Below pH 5 or above pH 8 NaCl decreases the water content of collagen, whilst between

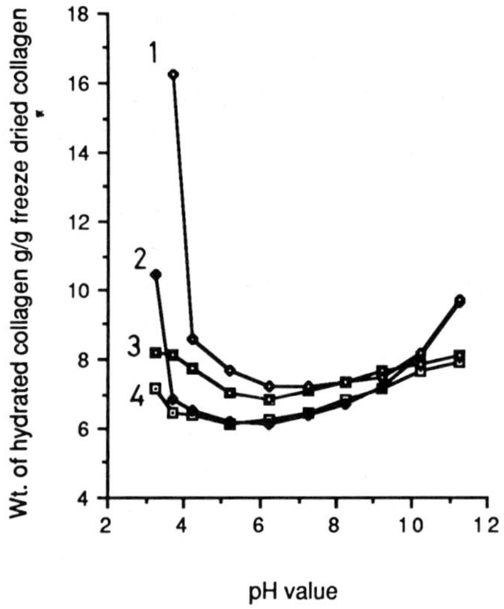

Fig. 10.5 Graph showing the effects of anions on the pH-dependent dehydration of collagen. (1) Control, (2) 0·1M NaCl, (3) 0·1M $Na_4P_2O_7$, (4) 0·1M Na_2SO_4.

pH 6 and 8 there is a small increase. High concentrations of NaCl decrease the hydration of collagen and the normal swelling of collagen in acids can be inhibited by the addition of salt.

Other salts have important effects on collagen and its water-binding properties. The effects of mono-, di- and polyvalent anions on the pH-dependent hydration of collagen is shown in Fig. 10.5. As can be seen, the water-holding capacity of collagen is minimal in the pH range 5–8 and all salts, including polyphosphates, tend to decrease its water-holding capacity. However, polyphosphates have less effect than mono- and divalent anions. In contrast, the water-holding capacity of collagen increases two-fold following thermal denaturation at 70°C. The addition of collagen to meat products which require thermal processing therefore offers functional advantage in terms of increasing the water-binding capacity of the cooked product.

Effect of pH

The hydration of collagen is at a minimum in the pH range 6–8 and only increases significantly at extreme values of the pH scale. The property of collagen matrices in taking up water at low pH (e.g. below pH 5) and to some degree at higher pHs (i.e. above pH 8) is well known. Indeed, as mentioned in earlier chapters, the decrease in the water uptake (swelling) of insoluble collagen from older animals has been attributed to the presence of increased mature cross-linking in the matrix and can be used as a measure of this parameter.

It has been shown that most of the protein components of meat swell significantly in acid during processes such as marination. In this case, meat was shown to take up to 100% its own weight during marination in acetic acid. This was associated with a concomitant decrease (up to 5–15 times) in mechanical strength as measured by shear value. It may be that in products which exhibit a relatively high acidity (e.g. salami-type fermented products where lactic acid production is promoted by the fermentation process itself), collagen swells in the same way as other proteins and may exhibit increased solubility and thus emulsifying and stabilising attributes as described above. However, little or no research has yet been carried out on this topic.

Effect of heat

The effect of heat on meat connective tissue collagen was fully discussed in Chapters 8 and 9 and on pure collagen in Chapter 7. The amount of collagen solubilised in animal tissues during cooking is related not only to the temperature of cooking, but also directly to age of the animal at slaughter and the tissue of source. Bone collagen is predominantly cross-linked by the heat-stable oxo-imine cross-link, and so even preparations from very young animals added to meat products would show little or no solubilisation on heating. Skin collagens, however, are cross-linked almost exclusively by aldimines in young animals and so collagen preparations from such sources, when added to meat products, would show high levels of solubilisation. In meat connective tissues the amount of collagen solubilised by heating is usually quite small, i.e. 10–15%, although even this small amount could have an effect on the properties of the product.

Denatured collagen is found in all cooked meat products. As discussed above, it plays an important role as an emulsifier and stabiliser, and perhaps, a binding agent; it also plays a part as a gelling and thickening agent. It is as gelatin that collagen is able to form such food gels.

GELATIN

To the biochemist gelatin is simply the heat-denatured form of any collagen. However, to the food chemist or meat scientist and in the broader industrial sense gelatin is a much more complex product. In this latter case it is the denatured and partly degraded product of animal collagen, normally derived from skin or bone. It should be recognised that gelatin is regarded as a product only of collagen, but gelatine (a word more commonly used in the industry) refers to the often coloured product obtained from animal connective tissues which contains some inorganic material and a proportion of macromolecular impurity. Though polysaccharide gelling agents may be seen by the food industry as generally more useful than gelatin (e.g. weight for weight they are more effective thickeners than gelatin), the collagen product forms good cold-setting gels under many different conditions with varying co-solutes and over a wide range of pH. The gels are generally clear and strong enough for use in foods and have the additional benefit that they melt in the mouth. Such gels form naturally to some degree in cooked meats and meat products from their natural collagen content, but gelatin as a food additive is important and will therefore be considered here.

As noted earlier in this book, the collagen found in tissues of mature animals is largely insoluble, even at high temperatures, so the extraction of this material from sources such as hide and bone requires additional chemical processing. Consequently mild techniques, resulting in little damage to the collagen α-chains, can only be used to extract gelatin from the hides of animals slaughtered at a relatively young age, e.g. pigs. Relatively mild acid treatments result in the breaking of the predominant aldimine intermolecular cross-links in the case of skin collagen and provide a high rate of extraction of good quality heat-denaturable collagen. The extracted, heat-denatured collagen is also partially purified by the removal of non-collagenous components by filtration. Such products are called Class A gelatins, have isoelectric points in the pH range 6–9, show little or no deamidation of glutamyl and aspartyl side chains and carry a net positive charge in all foods to which they are added. Naturally occurring gelatin in meat products, resulting from heating and partial extraction by comminution and other such treatments, is equivalent in properties to Class A gelatins.

Class B gelatins are produced from much more intractable tissues such as the hides of older animals or ossein (de-fatted, demineralised bone) by far more severe treatments than those used to produce Class A gelatin. In these cases, gelatin is extracted by pre-treating hides or ossein with alkali-lime for several weeks or with NaOH for several days. The swollen collagen is then washed and extracted by a series of heat treatments but up to one-quarter of the total protein contaminants are thought to be removed largely during the washing stage but the product is further clarified by filtration. Class B gelatin is almost fully deamidated and has an isoelectric point near pH 5. These gelatins are positively charged in acid foods but are negatively charged in neutral products such as meat pies.

Some work has been done on the enzymic extraction of gelatins using pronase or pepsin and such methods would seem to offer some considerable advantages, especially in terms of the effectiveness of extraction from older hides and ossein, but these techniques have not received much attention commercially and their development has not been forthcoming.

A major difference between Class A and Class B gelatins is in their molecular structure. In the former, the major molecular species are the α, β and γ components, whereas the latter is composed of a multimodal distribution of molecular weight species (as demonstrated by gel filtration chromatography). In other words, the hydrolytic nature of the extraction system is such as to lead to significant damage to the solubilised α-chains giving rise to a spectrum of lower molecular weight α-chain derivatives and higher molecular weight species. The latter may be formed by non-specific aggregation or through the extraction of multimolecular fragments from highly cross-linked collagen fibres. The nature of this mixture of components obviously varies considerably depending on the method and length of extraction.

Peptide bond cleavage during extraction appears to be somewhat limited and not entirely random. Most of the solubilisation process occurs due to breakage of intermolecular cross-links but further peptide bond cleavage is induced by heating. Thus, it seems that the majority of fragmentations are brought about during the heating processes subsequent to the limiting process.

On cooling heated gelatin solutions the re-formation of helices from separate α-chains is slow and favoured in α1-chains as opposed to α2-chains. Only aggregates of α-chains containing cross-links readily re-nature as these have in-built registration points (i.e. the cross-links themselves). The extent of helix re-formation is relatively low and dependent on concentration and temperature. The dependence of the process on solvent conditions is not great but the presence of –Gly–Pro–Pro– rich sequences in the structure of the gelatin is important as these act as nucleation points for the re-naturation. The interpenetration of domains of re-formed or partially re-formed helices leads to the formation of the gel structure and a concomitant steep rise in viscosity. This is accompanied by a high degree of organisation as triple helical zones self-assemble into a fibrillar network which can be readily observed by freeze-etching and negative staining electron microscope methods. The intermolecular bonding involved in gel formation is entirely non-covalent co-operative bonding and so is fully thermoreversible, thus explaining the ease with which gelatin gels melt on heating (for example in the mouth). The fully formed gelatin gel is classed as a partially crystalline glassy polymer and the fringed micelle model has been used to describe its structure (Fig. 10.6).

Gel strength is measured in commercial preparations by a Bloom strength, which is defined as the force in grams needed to give a stated, large distortion by a well-defined plunger to the surface of a gel of fixed dimensions. The gel to be tested (at 10°C) must contain 6·67%

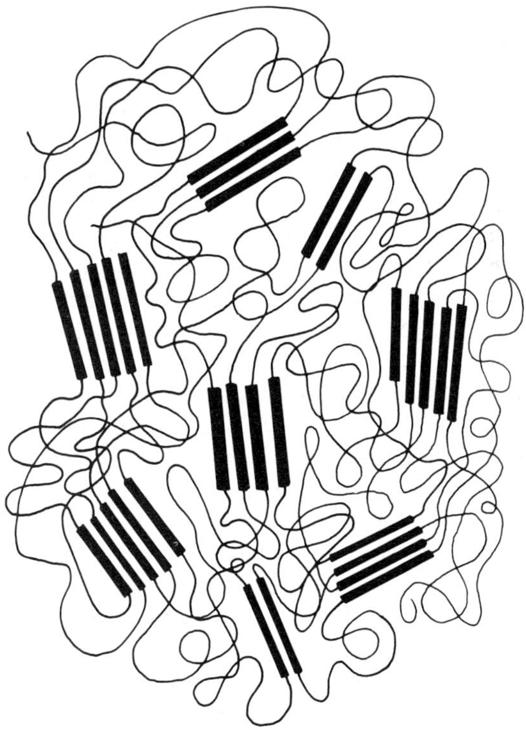

Fig. 10.6 Diagrammatic representation of the fringed micelle model of a gelatin gel.

(w/v) air-dried protein and should have been aged for 17 h at 10°C before testing. This method involves complex mechanics and is not ideal in research terms, but Bloom strength has been shown to be linearly related to the shear rigidity modulus of the gels and therefore valuable commercially.

Gelatin in food products

Gelatin is added to a wide range of foods as well as forming the major constituent of confectionery jellies and aspics. Its major use still remains in the production of gel desserts because of its melt-in-the-mouth properties, but it is also added to a range of meat products, in particular to meat pies as a jelly component.

Gelatin has also been used extensively as a stabiliser for frozen dairy products and other frozen desserts. High Bloom gelatin is added as a protective colloid to ice creams, yoghurts and cream pies. The gelatin is thought to inhibit ice crystal growth and lactose recrystallisation during storage, thus preventing undesirable graininess or 'sandiness'. The postulated mechanisms for such stabilising effects include enhancement of viscosity at low temperatures leading to decreased mobility of molecules within the freeze-concentrated matrix and a concomitant decrease in diffusion-related processes, and through its behaviour as a polymeric solute.

COLLAGEN IN SAUSAGE CASINGS

Sausage casings are made from a range of materials including natural gut, such synthetics as regenerated cellulose and thermoplastics. Collagen casings represent a small but important group. Extracted corium collagen is used to make sausage casings.

Historically, as the manufacture of sausage-type products became more automated, a more uniform casing than natural intestinal tract was clearly required. The use of natural casings (edible gut parts) from cattle, sheep and pigs has been extensively reviewed and will not be discussed here. Apart from a generally poor consumer image, they lack the uniformity in thickness and diameter necessary in automated sausage production. Regenerated collagen casings evolved using techniques similar to those previously developed for cellulose extrusion. Collagen casing manufacturing processes do not convert collagen to a solubilised product as does the gelatin extraction process; rather, they result in a much more fibrillar product which retains a relatively high degree of the native collagen fibre. The extracted collagen product is suspended in aqueous solvent and converted to an acid-swollen gel or dough which is then extruded as a tube.

The first casings of fibrillar collagen were developed in Germany in the 1920s, but only really gained popularity in the U.S.A. in the 1960s. Regenerated casings are now indispensable to modern sausage manufacture in providing product uniformity. Most casing collagen is made from bovine hides as these are easily obtainable and are relatively inexpensive. Hides, preferably from two-year-old animals, are mechanically split and the corium, which contains 85% of its dry weight as collagen, is used. Dehairing is carried out under controlled alkaline conditions normally through the use of sodium sulphide. Following this, the corium is further processed, after defatting and mechanical removal of residual adhering flesh, in lime. As in the production of gelatin, this treatment affects the properties of the collagen, especially through the deamination of the glutamyl and asparaginyl residues, cleavage of the cross-links, and dissociation of non-collagenous proteins, all of which open up the structure prior to acid swelling. The changes in the corium brought about by the sulphide-liming process also include the solubilisation and degradation of non-collagenous macromolecular components, saponification of fats, altering of the swelling properties of collagen fibres, denaturation of collagen and reduction of its isoelectric point from pH 7·8 to 4–5, the reduction of collagen T_s and the partial destruction of arginine, serine, threonine and hydroxyproline. The edible grade treated corium is then ground to a small particle size before being swollen with acid to allow the production of an extrudable 'dough'.

The so-called collagenous 'dough' produced by the alkaline extraction process is then extruded either by the wet or the dry process (Fig. 10.7). During extrusion the collagen fibres tend to align parallel to the axis of the tube emerging from the extruder. In the wet process the extruder contains an internal disc which forces the flowing gel against the sides of the extruder casing. In the dry process the extruder possesses twin counter-rotating serrated surfaces. The combination of fibre length and extruder design produces a weave-angle of cross-hatched collagen fibres. The tube of extruded collagen is then passed through a concentrated salt solution and a chamber of ammonia to reprecipitate the collagen. The swollen gel contracts to produce a film of reasonable strength which can be improved by the addition of plasticisers such as glycerine. The tube is then dried to a 10–15% water content and pleated so that a 15 m length of casing is contracted to 180–200 mm. The thickness and strength of the casing can be controlled by the amount of plasticiser and cross-linking agent added. These different properties can be used in the standard English sausage or the stronger skins required for Polish and German types of cooked meats.

A typical compositional analysis of sausage casings shows a collagen content of 50–65%, a moisture content of 13–19% and a fat content of 2·7–6·0%. The residual content is usually made up of additives such as glycerine, sorbitol, cellulose and carboxymethyl cellulose already

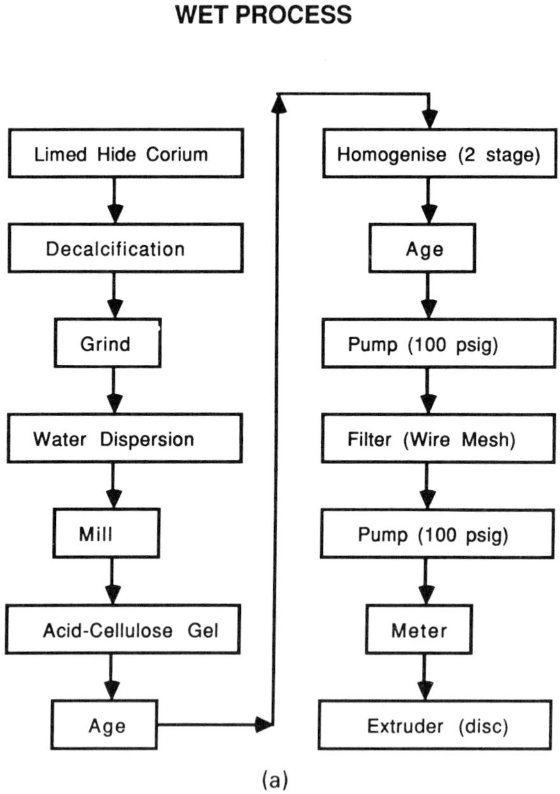

Fig. 10.7 Flow diagrams showing the (a) wet and (b) dry methods for fibrillar collagen extraction from hide for the manufacture of sausage casings.

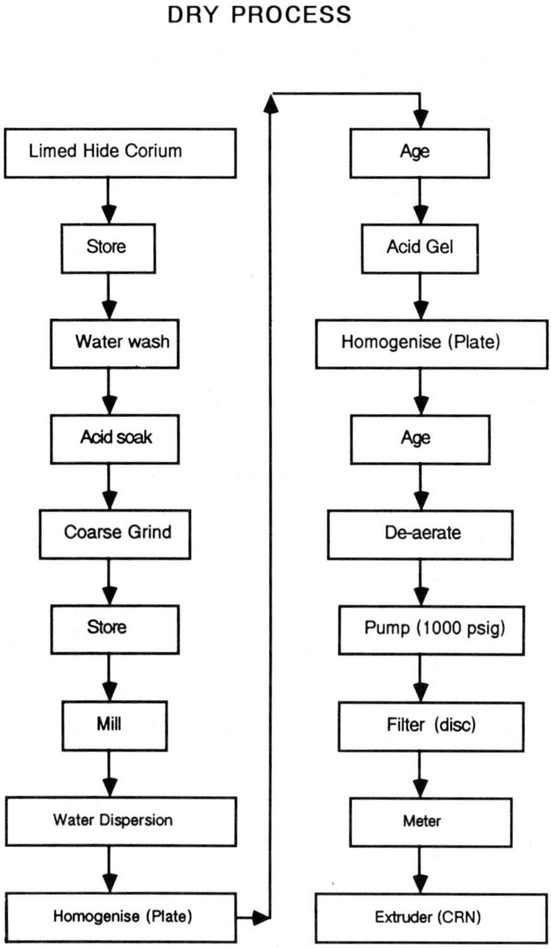

(b)

Fig. 10.7—*contd.*

mentioned. Tube film thicknesses vary considerably depending on the use required but for standard sausages thickness is usually 1·1–1·4 mm. Break tensile strength for collagen casing films varies from 600–914 g longitudinally and 275–1164 g transversely, with dry extrusion films generally exhibiting slightly higher strengths. The elasticity of such films has been shown to fall in the range 25–57%. The appearance of the films in the electron microscope (TEM) is characteristic. Films made from the dry extrusion process tend to contain much more of the large fibre bundle material whilst those made from the wet process are seen to contain better dispersed and much more fibrillar material (i.e. smaller units). Both types of film exhibit obvious cross-weave type structures of the collagen fibre/fibril network.

SUMMARY

Meat emulsions and batters are used to make many meat products including sausages, pies and burgers. Myosin is the main protein emulsifier in these emulsions but collagen also has a role as an emulsifier. Because of its hydrophobic nature it is able to bind high quantities of fat. This can be an advantage and it has been suggested that collagen may act as an adhesive between fatty and lean particles and may absorb excess fat. However, it can also cause problems. Excess collagen added to meat products such as sausages can cause the formation of pockets of gelatin and fat on cooking, 'grease-out', splitting in sausage casings, a brittle filling and gelatin on the surface of products. Collagen is a useful additive for meat products, however, and is used as a filler. It can be purified from hide and bones for such purposes.

Salts, pH and heat all affect the water-holding capacity (WHC) of collagen. Sodium chloride decreases its WHC in the pH ranges below 5 and above 8 but polyphosphates decrease its WHC in the range 5–8. pH also alters the WHC of collagen and at low pH (below 4–5) insoluble collagen swells considerably and takes up water. Heat causes about a two-fold increase in WHC.

Gelatin is denatured collagen. It may be formed merely by heating native collagen *in situ* or it may be purified from a range of tissues, notably hide and bones, using mild acid extraction, enzyme extraction or more commonly, sulphide and alkali extraction. Two grades of gelatin may be extracted; Class A which is high grade, relatively undamaged molecular material and Class B which is extracted by harsh means and consequently has a range of molecular weights and altered properties. All gelatin is soluble and able to form gels on cooling from hot solution and thus it is an important food additive.

A dispersed but fibrous collagen product made from hides and other sources is insoluble material which is then re-aggregated and used to make sausage casings. This is an important component in the manufacture of sausages.

FURTHER READING

1. Karmas, E. (ed.), *Sausage Casing Technology*. Noyes Data Corporation, NJ, 1974.
2. Stainsby, G., Gelatin gels. In *Advances in Meat Research,* Vol. 4: Collagen as a Food, ed. A. M. Pearson, T. R. Dutson & A. J. Bailey. AVI, New York, 1987.
3. Jones, N. R., The uses of gelatin in edible products. In *Science and Technology of Gelatin,* ed. A. G. Ward & A. Courts. Academic Press, London, 1977.
4. Ledward, D. A., Gelation of gelatin. In *Functional Properties of Food Macromolecules,* ed. J. R. Mitchell & D. A. Ledward, Elsevier Applied Science, London, 1986, Ch. 4.
5. Hamm, R., Post-mortem changes in muscles affecting the quality of comminuted meat products. In *Developments in Meat Science,* Vol. 2, ed. R. A. Lawrie. Applied Science Publishers, London, 1982.
6. Hermansson, A. M., Some physicochemical aspects of the structure formation of proteins. In *Biochemical Aspects of New Protein Food,* ed. J. Alder Nissen, B. O. Eggum, L. Munck & H. S. Olsen. FEBS, Copenhagen, 1977.
7. Price, J. F. & Schweigert, B. S. (eds), *The Science of Meat and Meat Products*. Van Nostrand Reinhold, San Francisco, 1971.

Chapter 11

Collagen in the Diet and Nutritional Properties of Meat Collagen

INTRODUCTION

Apart from the natural enjoyment and satisfaction of eating we need food to obtain fuel for metabolic reactions and the essential nutrients which are components and participants in those reactions. Human nutritional requirements are not particularly complex. In brief, we need elements of our diet as an energy source—protein, carbohydrate or fat can serve this purpose and the efficiency with which these nutrients can be used for the production of energy is fat > carbohydrate > protein. We need vitamins as enzymic co-factors and metabolic components and we need mineral elements for the activation of metalloproteins, the maintenance of salt homeostasis, the activity of nerves, muscles and multiple chemoreceptors and for the construction of bones and teeth and other connective tissue elements. Finally, we need proteins as a supply of amino acids, the building blocks of all new proteins synthesised in the body to replace routine losses of tissues and organs.

Meat connective tissue contains fat in that a certain amount of lipid material is associated with the perimysium. This varies from muscle to muscle and with the overall fatness of the animal from which the meat is obtained. Our dietary need for lipid is very low (it has been estimated to be as low as 2% by weight) and there is only one known essential fatty acid (linolenic acid). Indeed, the major dietary constraints on fat levels concern the link between high lipid intake and heart disease and obesity. However, even though this is currently a major area of concern and research in nutrition, especially in respect of meat, since connective tissue is not a major source of dietary lipid we will not consider this subject in this book. Connective tissue in meat is composed, as we have seen in earlier chapters, largely of various collagens and only small amounts of elastin and proteoglycans. Therefore, the main contribution of these tissues to the human diet is through the contribution of collagen to the protein intake.

High protein foods like meat provide dietary protein as a source of indispensable essential amino acids, i.e. those amino acids which we cannot synthesise for ourselves. Dietary protein also provides an adequate level of non-specific nitrogen. These nutrients are needed for such processes as the synthesis of all bodily proteins including all the enzymes of the body, the synthesis of all the structural components of the body such as skin, bone, blood vessels, muscles and hair, the synthesis of protein hormones and (because proteins form the component enzymes of all the synthetic processes of the body) the synthesis of all macromolecules including carbohydrates and fats.

Because any protein can provide non-specific nitrogen, the nutritional quality of a protein in the diet is judged by its amino acid content and, specifically, its content of the required essential amino acids. However, the way a protein is digested by the body and the amino acids assimilated and utilised have a profound influence on the nutritional availability of that protein. For this reason, all proteins can be assayed for nutritional value in a variety of ways, each of which reflects a different aspect of that protein's nutritional function. These parameters are known as biological value, digestibility, protein efficiency ratio and net protein utilisation. Before discussing the contribution of meat connective tissue components to human nutrition it would be useful to briefly review the meaning and use of these various parameters of nutritional value.

This chapter, then, will assess the nutritional value and status of collagen and to some extent, that of other meat connective tissue glycoproteins and proteoglycans, in terms of their contribution to our diet and their biological usefulness.

Biological value of proteins

The biological value (BV) of a protein is primarily dependent on the level of the essential amino acids in it as well as on their bioavailability. Bioavailability indicates the ease with which usable amino acids can be obtained from a dietary protein. Thus, the ease with which a protein is digested to its component amino acids determines the bioavailability of those amino acids. Similarly, any amino acids in a protein which undergoes non-reversible reactions during preparation and cooking prior to consumption will be biologically unavailable.

The essential amino acids are leucine (Leu), isoleucine (Ile), lysine (Lys), methionine (Met), phenylalanine (Phe), threonine (Thr), tryptophan (Trp), valine (Val) and histidine (His) (only needed in young humans). If any one of these essential amino acids is not present in adequate quantity in a protein it limits the utilisation of other amino acids in that protein so that they will have been wasted as new protein is not synthesised. A protein that almost entirely lacks any one essential amino acid is biologically useless. Waste amino acids are not easily stored by the body and so such a protein is ultimately excreted as waste nitrogen. The measurement of excreted waste nitrogen in experimental animals (normally rats) provides the basis of determining the biological value of proteins. The value of a protein is normally measured against a theoretical value of 100%. Egg protein has the highest biological value known, i.e. 94%. Meat protein has a measured biological value varying from as low as 62% (veal) to as high as 85–90%. Fish fillet has a BV of 75–90% depending on the source of the sample. The BV of rice is about 86%, whilst that of corn is only about 40%. It is generally agreed that a BV over 70% is required for a protein to be nutritionally useful to humans.

Biological value used to be defined by comparing the composition of a protein to a high quality reference protein such as egg protein. The current preferred approach, however, is to compare the essential

Table 11.1
Reference amino acid patterns based on nutritional requirements for various age groups

Amino acid	*Infant*	*2–5 years*	*10–12 years*	*Adult*
Histidine	26	19	19	16
Isoleucine	46	28	28	13
Leucine	93	66	44	19
Lysine	66	58	44	16
Methionine/Cysteine	42	25	22	17
Phenylalanine/Tyrosine	72	63	22	19
Threonine	43	34	28	9
Tryptophan	17	11	9	5
Valine	55	35	25	13

Data from FAO/WHO/UNO (1985). Figures indicate mg amino acid per g protein based on protein intakes of 0·99 (10–12 years) and 0·75 (adult) g protein per kg body wt per day.

amino acid content of a protein with the known required levels of essential amino acids for humans. The FAO/WHO/UNO (1985) developed a table of levels of essential amino acids (called a pattern) based on the requirements of all age groups and given in terms of safe levels of protein intake for four different age groups, infant, 2–5 years, 10–12 years and adult (Table 11.1). For the infant the composition of mothers' milk was used to define the level of essential amino acids. Essentially they took the highest level of each specific amino acid believed to be required for each age group, tabulated the levels so obtained for all the essential amino acids, and then proposed these levels as a reference pattern for scoring protein sources.

Protein digestibility

The nutritional value of a protein must be considered in the light of the way in which that protein and its constituent amino acids are used by the body. The first step in protein utilisation by the body is the hydrolytic degradation of the protein in the digestive tract to its component amino acids. This process involves acid conditions in the stomach, the stomach acid protease pepsin and the group of proteases secreted into the small intestine from the pancreas and the cells of the duodenum, i.e. trypsin, chymotrypsin, pancreatin, elastase, etc. (see Scheme 11.1). The digestibility of a protein is the degree to which a particular protein is degraded to usable amino acids in this process. Protein digestibility is determined in the rat and is defined as:

$$\text{Digestibility} = \frac{\text{N in diet} - \text{N in faeces} - \text{N in Metabolism} \times 100}{\text{Nutrient in diet}}$$

Protein efficiency ratio

Protein efficiency ratio (PER) values are used to reflect both the level of essential amino acids in a protein or proteinaceous food as well as bioavailability of those amino acids. Thus, PER takes account both of the BV of a protein and its digestibility. It also considers the effects on bioavailability of an essential amino acid of non-reversible chemical reactions of that amino acid prior to consumption of the protein. The protein efficiency ratio (PER) is, therefore, defined as the weight gain in an animal relative to protein consumed in a specific period using the same level of protein in the diet. Thus:

$$\text{PER} = \frac{\text{Weight gain by test animal}}{\text{Weight of protein consumed}}$$

	Location and comment
Mastication of meat connective tissue collagen or collagen component of meat products	Mouth. Little or no structurally induced damage to collagen matrices
Warming to body temperature 37°C	Mouth, oesophagus
Mixing with 0·001–0·01M HCl	Stomach, pH 2–3
Swelling of collagen fibres	Acid-induced swelling
Acid/heat-induced denaturation	
Degradation of native molecules at N- and C-termini releasing whole triple helices	Attack by proteolytic enzyme pepsin secreted by stomach lining
Degradation of released and denatured molecules to large fragments	
Neutralisation and increase of pH to 7·5–8·0	Passage of chyme from stomach to upper small intestine (duodenum)
Action of enzymes of small intestine on residual fibrous collagen, released whole molecules and proteolytic fragments of collagen	Trypsin, chymotrypsin, elastase, and pancreatin secreted from pancreas
Adsorption of collagen amino acids into blood	Through lining of small intestine

Scheme 11.1 Flow diagram showing the proposed route and mechanism of digestion of collagen in the human alimentary canal.

Low PER values are due to low levels of essential amino acid, poor distribution, low protein digestibility or loss of essential amino acids through reaction prior to consumption.

Net protein utilisation

The net protein utilisation (NPU) and net protein ratio (NPR) values of the diet are also utilised as measures of nutritional status. This is because the PER has serious limitations as a protein quality indicator as it does not consider the maintenance requirement of the test animals used in the assay. This principal drawback can be removed by including in the measurements a second group of experimental animals on a protein-free diet. The weight loss in this group is equivalent to the contribution of protein to body maintenance over the experimental period and can be thus added to the weight gain of the test group. The NPU is identical to NPR except that the calculation is based on body nitrogen rather than weight. NPR can, therefore, be defined as:

$$\text{NPR} = \frac{\text{Weight gain} + \text{weight loss by protein free groups}}{\text{Protein consumed}}$$

NUTRITIONAL PROPERTIES OF COLLAGEN

Having defined the major terms used in the determination of the nutritional quality of a protein it is now possible to consider collagen in the light of our current knowledge of its amino acid composition and other parameters.

Biological value of collagen

Judged by the standards outlined above collagen has a very poor nutritional value, i.e. because it lacks tryptophan its BV is effectively 0. Because of this it is generally believed that collagen is of negligible nutritional significance. However, a closer consideration of the facts leads us to the conclusion that this statement is not strictly true. Although not a viable positive contributor to the nutritional value of dietary protein, collagen could be seen as a negative contributor. In other words, because collagen has a unique amino acid composition in that it contains no tryptophan and low levels of the other essential amino acids, it may be possible that this protein could 'dilute out' the BVs of other better quality proteins. Hence, it is of interest to know

Table 11.2
Comparison of the FAO/WHO standard for essential amino acid requirements in standard proteins with the actual content of these amino acids in collagen and muscle

Amino acid	*Protein*	*Collagen*	*Muscle*	*Collagen/muscle* 30:70
	(*mg of amino acid per g*)			
Histidine	19	7	34	26
Isoleucine	28	16	48	38
Leucine	66	34	81	66
Lysine	58	30	89	72
Methionine/Cysteine	25	8	40	31
Phenylalanine/Tyrosine	63	28	80	64
Threonine	34	19	46	38
Tryptophan	11	—	11	7
Valine	35	25	50	41

whether the inclusion of large amounts of collagen to meat products (see Chapter 10) could significantly decrease the nutritional value of their proteins.

The essential amino acid content of collagen represents 6–22% of its total amino acids compared to the reference pattern which represents 36% in the 'ideal' protein. There is clearly a large reduction in the levels of almost all of these components (Table 11.2). In contrast, muscle tissue provides an excess of all the essential amino acids. Mixtures of collagen with muscle can, therefore, to some extent, balance out the deficiency in collagen. The addition of up to 30% collagen only caused a deficiency in tryptophan down to 7 mg per g protein compared to the values of 11 mg per g protein as the standard. Overall there is a reduction in the levels of total essential amino acids of about 15–20%. These compositional analyses clearly suggest that, contrary to the general belief, significant amounts of collagen can be added to meat products without substantially affecting the nutritional value of the product in terms of total essential amino acids.

Digestibility of collagen

The figures for the biological value and nutritional significance of collagen are theoretical in that they are obtained following complete hydrolysis of the protein. However, *in vivo* the bioavailability of the amino acids may not be 100% since this depends on the digestibility of the protein as discussed above.

Because of the intractable nature of insoluble collagen it has been assumed in the past that this material will have a relatively low digestibility in humans. Indeed, it has been suggested that the insolubility of collagen and hence its probable non-digestibility could lead to it acting as dietary fibre similar to plant wall components. This suggestion was based on the possible substitution of collagen for the lack of plant fibre in the diet of healthy eskimos. Collagen could, therefore, play an important dietary role unrelated to its biological value as a protein. However, recent work on the digestibility of collagen has shown it to be high and indeed to be equivalent to other meat proteins and thus it seems unlikely that it would have a role as dietary fibre.

Rat model *in vivo* studies of meat protein show a digestibility of 96% whilst the digestibility of gelatin is 89%. Both figures are close to that found for casein. Trials with fibrous collagen which had been fully heat denatured before being fed to rats showed digestibility equivalent to that of gelatin, i.e. 90%. It seems that digestion is essentially complete irrespective of the age of the animal and source of the collagen. However, these figures make two important assumptions—the equivalence of human and rat digestive systems and that all dietary collagen will be fully denatured by heat before being consumed. Indeed, values obtained for protein quality from rat assays do not provide consistent relationships where parallel or similar studies have been carried out with humans. Certainly, collagen is digested by man as shown by the increased excretion of hydroxyproline following the feeding of gelatin and high collagen-based meat products. However, it is impossible to estimate from such data the total extent of breakdown of the protein in the human gut.

As discussed in other chapters, collagen swells in acid conditions and under such circumstances is highly digestible by pepsin (Chapter 12). The primary site of activity of this stomach enzyme is at the N- and C-terminal regions of the molecule to release whole triple helices from matrix collagen. The conditions in the human gut, i.e. 0·001–0·01 M HCl (pH 2–3) and 37°C, would encourage denaturation for the released triple helices which would then act as good substrates for further pepsin attack (Scheme 11.1). One might predict therefore, that the bulk of collagen passing out of the stomach in the chyme would be in the form of triple helices, damaged helices and fragments of variable size (amino acids to large polypeptides). Once partially degraded or even just denatured to gelatin, collagen is readily digested by most proteinases. Thus, in the small intestine the action of trypsin, chymotrypsin, pancreatin and elastase should further degrade the majority of collagen to amino acids or very small peptides (Scheme 11.1).

Limited studies on the action of protease mixtures and sequential protease attack on highly cross-linked insoluble collagen have shown that this substrate is readily digested to amino acids at 37°C, although there is some evidence of a very small amount (probably less than 5%) of resistant material that remains after digestion with a molecular weight in excess of 2500 (approximately 25 amino acids). It is supposed that this material is of a highly hydrophobic nature and may derive from the mature cross-link region of the molecule. On the theoretical basis outlined above, therefore, it would be expected that collagen would show a high digestibility under the conditions in the human stomach.

Recent studies using casein as a reference have demonstrated up to 95% digestibility of insoluble collagen (bovine achilles tendon collagen) in rats. However, the digestibility of the insoluble collagen could be reduced to 71% by the administration of omeprazole, an inhibitor of gastric secretion. This result suggests that even a limited acid environment in the stomach at 37°C will lead to sufficient swelling and denaturation of insoluble collagen to allow intestinal enzymes such as trypsin, chymotrypsin, pancreatase and elastase to digest the protein. The researchers proposed that the limited acid environment would lead also to the breakage of aldimine cross-links which would enhance the digestion process by allowing more swelling and denaturation.

In the same study it was shown that the *in vitro* solubilisation of collagen by pepsin in HCl was pH dependent. Below pH 2·5 even highly insoluble bovine achilles tendon collagen was 90% solubilised by pepsin. Above pH 2·5 the solubility of the achilles tendon collagen fell rapidly so that at pH 3·0 it was less than 50% soluble. Gelatin was nearly 100% soluble at all pHs up to 5·0. This data indicates that the solubility of intractable matrix collagen is highly pH dependent but, taken with the rat data, suggests that in most cases this material will be at least 90–95% solubilised by gut pepsin. Thus, although human gut pH is variable depending on the individual examined and his or her dietary consumption, it seems fair to conclude that insoluble collagen will be effectively digested by most people. Indeed, limited studies on humans fed a high collagen diet showed that collagen was as well digested as muscle protein—about 80%.

The protein efficiency ratio (PER) of collagen

The PER values of meat products with collagen content up to 40% fall between 2·4 and 3·2. None fall below the limit of 2·4, i.e. 80% of the PER for casein, which is deemed to be the level for a healthy diet. Clearly high collagen content does not significantly affect the nutritional value of the product.

In the U.S.A. a minimum PER of 2·5 and a minimum essential amino acid content of 32% have been specified for most fabricated foods. Increasing the collagen content of meat products from 10% to 30% reduced the PER value of those products from 3·0 to 2·5. This led to the conclusion that the addition of up to 30% collagen would not have a significant effect on the protein nutritional status of such products. Similarly, using varying amounts of collagen in meat bioassays it has been possible to calculate the following linear regression equation for the influence of collagen on meat product PER:

$$\text{PER} = -0{\cdot}0220 \times \%\ \text{collagen in product} - 3{\cdot}1528$$

From this equation it is possible to calculate a maximum allowable collagen content in meat products of 28·5% given a PER of 2·5. In other words, there is no danger of seriously affecting the nutritional quality of a meat product by increasing the collagen content up to 28·5%.

Net protein utilisation (NPU) of collagen

Increases in collagen in meat or meat products also decrease the value of NPU (NPR). The recommended level, for weaning foods, is an NPU of 65. Early work, which has since been corroborated, showed that even very poor quality muscle samples with very high collagen content had acceptable NPUs, e.g. shin with 23·6% collagen had an NPU of 69. Top quality meat such as beef fillet steak (2·5% collagen) has been shown to have an NPU of 82. The NPU of meat has been shown to vary from 62–78 with an average of 74.

The levels of collagen in sausages and similar meat products in the U.S.A. and Europe varies from 15–35%; thus, as far as NPU (with an apparent tolerance up to 36% collagen) is concerned there should be no problem from such products. In some countries, e.g. Russia, nutritionists recommend the inclusion of collagen to improve protein digestion and it is considered to have nutritional and medical benefits.

The addition of collagen to formulated meats to produce added value products by restructuring clearly lowers the nutritional value of the product. However, up to 30% collagen can be added to meat products before the minimum PER of 2·5 is approached. Collagen, if used within limits in a balanced protein diet, should not decrease the nutritive value of that diet.

Fortification of high collagen content meat products

The nutritional value of any diet can be improved by the direct addition of the deficient essential amino acids (protein fortification). Although this approach seems simple and potentially effective there are two main disadvantages of this method. There is good evidence that, compared to dietary protein amino acids, there are differences in bioavailability of added free amino acids due to the variation seen in the uptake of these across the intenstinal wall compared with amino acids from dietary protein. In addition, the bioavailability of the added free amino acids may be radically affected by undergoing Maillard-type reactions during processing and more especially cooking. A typical example is lysine which is known to take part in Maillard reactions with reducing sugars to form Schiff's base compounds which

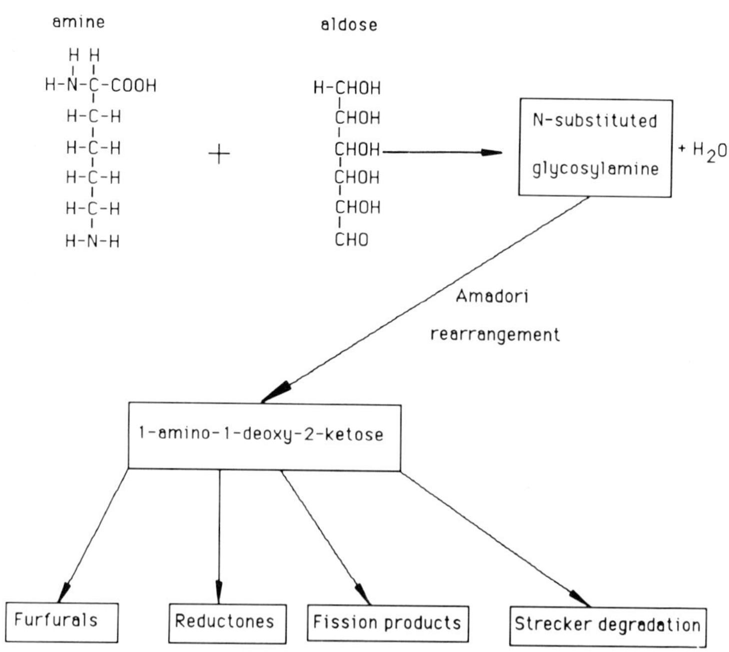

Fig. 11.1 The Maillard reaction and Strecker degradation.

subsequently undergo Strecker degradation to non-reversible products, thus rendering the lysine unavailable (Fig. 11.1).

Many free amino acids have their own taste and thus may affect the flavour of the product to which they are added. For example, methional and dimethyl disulphide, which are formed from free methionine in processed foods, have undesirable flavours. Finally, many free amino acids are readily soluble and could leach out of the product during processing.

These problems can be overcome to some extent by the addition of amino acids coupled to carriers (a method which has been proved in some plant proteins) or by the addition of other proteins rich in the deficient limiting amino acids. However, no data are available for the covalent coupling of essential amino acids to collagen. Further work is required in this area if collagen is to be made a nutritionally respectable protein.

NUTRITIONAL PROPERTIES OF OTHER MEAT CONNECTIVE TISSUE COMPONENTS

Collagen is the major component of meat connective tissues and, as such, makes the main contribution of those tissues to the overall nutritional quality of meat and meat products. However, it is worth remembering that these tissues also contain variable amounts of other components such as elastin, glycoproteins and carbohydrate-containing proteoglycans. As elastin has such an unusual amino acid composition, as discussed in Chapter 5, with a low content of several essential amino acids (Table 11.2) it would appear, like collagen, to have a low nutritional value. On the other hand the glycoprotein and proteoglycans would be expected to have high nutritional values. However, the individual contribution to nutrition of these components has not been much researched, especially in connection with meat connective tissues, as they represent such a small proportion of the total tissue weight and are not added in any quantity, as is collagen, to meat products or other foods.

LEGISLATION

Connective tissue is present, to widely varying extents, in the raw material of the industry from flesh, which includes muscle and adipose tissue, skin and sinew, and offal, such as liver, kidney and heart, and are considered as contributing to the 'meat content'. The collagen content of these tissues ranges from 90% in tendon to about 2% in muscle and less than 1% in liver.

It is clear from this variability that the collagen content of meat requires definition. Indeed, all meat products are subject to legal requirements, codes of practice and in the U.K. are defined in The Meat Products and Spreadable Fish Products Regulations 1984 as: 'the flesh, including fat, and the skin, rind, gristle and sinew in amounts naturally associated with the flesh used of any animal or bird which is normally used for human consumption'.

Clearly connective tissue is regarded as an ingredient of meat and meat products, although the definition of how much is associated with the flesh is debatable. Although there is only about 2% associated with most lean meat (Table 11.3), the meat product regulations allow the addition of (e.g. pork rind) up to about 10%, which is assumed to be that naturally associated with the flesh used. In this case the 10% is taken as the connective tissue associated with untrimmed meat cuts which include the tendon and thick epimysial tracts (Table 11.4 gives the average collagen content of various bovine cuts). The addition of cooked rind has the added advantage that during cooking it takes up 50% more water than before it is added to the product. Similarly,

Table 11.3
Collagen content of trimmed muscles

Muscle	*Bulls*	*Cows*	*Steers*
M. obliquus externus abdominis	1·13	1·13	1·03
M. obliquus internus abdominis	0·60	0·61	0·58
M. rectus abdominis	0·84	0·70	0·59
M. transversus abdominis	1·13	0·90	0·86
M. pectoralis profundus	0·92	0·98	0·89
M. pectoralis superficialis	1·76	1·63	1·46
M. serratus ventralis	0·94	0·84	0·73
M. brachiocephalicus	1·86	1·50	1·58
M. biceps brachii	1·30	1·03	0·99
M. longissimus capitis et atlantis	0·90	0·97	1·00
M. splenius	0·86	0·78	0·77
M. sternocephalicus	1·45	1·70	1·52
M. complexus	0·82	0·86	0·87
M. longissimus thoracis	0·50	0·52	0·51

Table 11.4
Average collagen content of bovine cuts

Cut	*Average collagen content* (%)	*Range* (% *wet fat-free*)[a]
Masseter	20	10–30
Shin	12	6–16
Neck	12	7–20
Clod	10	4–16
Brisket	10	4–20
Flank	9	3–20
Chuck and blade	9	5–15
Diaphragm	8	4–18

[a] Wet fat-free connective tissue is defined as hydroxyproline content × 37.

Average for most beef cuts is 10% although some, e.g. masseter, are high at 20%. The recommendation of 10–20% for connective tissue in beefburgers, for example, is seen to be reasonable from the figures given in the table.

because of the higher nitrogen value of collagen (18·7%) than muscle (17%), the addition of collagen tends to increase the apparent lean meat content of meat products.

The level of connective tissue in meat products is highly variable and can range from 10–35%. The Food Standards Committee report on meat products recommended that new regulations on the composition of meat products should limit the connective tissue as that associated with the lean meat content, to 10%. The enforcement of regulations defining the level of connective tissue will depend on the determination of collagen, the major component, by hydroxyproline determinations (Chapter 12). The proportion of hydroxyproline in collagen varies with the collagen type but since the major component is Type I collagen with only about 20% Type III, it is reasonable to take a value close to that for Type I (Table 11.5). The value generally used is 14·0.

Within the European Community new meat product regulations are being adopted to provide more informative labelling. Instead of requirements for a statutory minimum they will be based more on ensuring that the labelling includes a declaration of meat content and maximum added water and connective tissue content.

Table 11.5
Hydroxyproline content of different collagen types

Collagen type	*% Hydroxyproline*	*Tissue*
I	13·1	Skin, tendon, bone
II	15·0	Cartilage
III	17·4	Vascular system
IV	16·6	Basement membrane

SUMMARY

Meat is an important source of protein in the diet and the major meat proteins provide a good balance of the essential amino acids needed by man. The overall nutritive value of meat protein is high when compared to either reference proteins (i.e. international standards set for essential amino acid content) or good quality protein like egg protein. The connective tissue components of meat, however, are unusual in that the two major proteins, collagen and elastin, have uncharacteristic amino acid compositions which lack or are low in several of the essential amino acids. This makes them nutritionally poor and could influence the overall quality of meat if they were present in high amounts. However, these components are not present in fresh meat at a high enough level to affect the overall quality of the tissue. Meat products can contain a substantial amount of added connective tissue which is normally high in collagen. This may be expected to lower the nutritional standard of such products and this is indeed observed. However, up to 30% collagen can be added to meat products without seriously affecting their nutritional quality, i.e. dropping their PER or NPU below unacceptable levels. For this reason legislation exists to limit the amount of collagen and connective tissue that can be added to such products in order to ensure their nutritive value.

FURTHER READING

1. Ashgar, A. & Hendrickson, R. L., Characteristics of collagen in food systems. In *Advances in Food Research,* Vol. 28. Academic Press, New York, 1982, pp. 231–372.

2. Lee, Y., Elliot, J. G., Rickansrud, D. A. & Hagberg, E. C., Predicting protein efficiency ratio by the chemical determination of connective tissue in meat. *J. Food Sci.*, **43,** 1978, 1359.
3. Mauron, J., The analyses of food proteins, amino acid composition and nutritive value. In *Proteins in Human Nutrition,* ed. J. W. G. Porter & B. A. Rolls. Academic Press, New York, 1973.
4. Reutersward, A. L., Asp, N.-G. & Bjorak, J., Protein digestibility of pig skin and bovine tendon in rats. *J. Prod. Technol.*, **20,** 1985, 745.
5. Whitmore, R., Booth, A., Naghski, J. & Swift, C., Digestibility and safety of limed hide collagen in rat feeding experiments. *J. Food Sci.*, **40,** 1975, 101.
6. FAO/WHO/UNO, *Energy and Protein Requirements,* WHO Technical Report Series 724. WHO, Geneva, 1985.

Chapter 12

Preparation and Analysis of Meat Connective Tissues and Collagen Types

INTRODUCTION

The last 20 years have witnessed considerable advances in the techniques used to purify and analyse connective tissue, particularly collagen. As has been stated earlier, the major collagens of meat are Types I, III, IV and V and so this chapter will be restricted to the proven methods for the investigation of these types of collagen as well as gross analysis of whole connective tissue. First, we will describe the methods available for isolation and purification of meat connective tissues. Secondly, methods for the purification of the procollagens of Types I, III, IV and V, neutral salt-soluble and acid-soluble collagen, pepsin-solubilised collagen, and insoluble collagen will be discussed. The routine analysis of pure collagens by techniques such as amino acid analysis and peptide mapping will be dealt with in the third part of this chapter. This will be followed by a description of chemical methods for analysing and quantitating collagen and elastin cross-links and quantitating and assessing total collagen and elastin content in meat samples as well as estimation of the different types of collagen present.

SEPARATION OF EPIMYSIUM, PERIMYSIUM AND ENDOMYSIUM

As only a small fraction of the whole muscle is represented by the connective tissue it has proved difficult to separate each of the hierarchical domains. Earlier methods utilised milled, deep frozen muscle to filter out gross connective tissue; this can be useful for total connective tissue work. Another method used controlled homogenisation of the tissue followed by graded sieving. This technique has now been improved and modified to allow good separation of the connective tissue domains.

Following careful dissection of the muscle from the carcass, the *epimysium* may be dissected away from the surface and stored for later use at −20°C. Samples of the muscle (usually 50–100 g wet weight) are then homogenised in a Waring-type blender at full speed for 10 s in 100–200 ml of ice-cold 50mM $CaCl_2$. The sample is then filtered through a graded 1 mm sieve. The collected gross connective tissue is termed the crude *perimysium* and the material passing through the filter is the crude *endomysial* preparation. Examination of the latter sample under the light microscope shows that it consists of some fibrous insoluble material but is mainly made up of short tubes of endomysium full of contractile protein. This material may then be further washed in hypotonic buffers as described in Scheme 12.1 until all the actomyosin is removed and the undamaged endomysial tubes may be used for further work.

Further purification of the collagenous components of epimysium, perimysium and endomysium is possible by taking the separated fractions and washing in chaotropic reagents, acid solutions or detergents. Perhaps the most successful of these is the use of 1% (w/v) sodium dodecyl sulphate (SDS) in low ionic strength solution at pH 7·4. In this case the tissues are washed by continuous stirring at 15–20°C for 30–60 min time periods in the solvent with collection of the residual solids by centrifugation after each wash until a stable collagen content is reached. The method generally yields clean, white tissue masses with collagen contents of 60–95% by weight.

EXTRACTION, SEPARATION AND PURIFICATION OF COLLAGENS

Procollagens

Varying amounts of the different procollagens may be extracted from any collagenous tissue dependent on the total amount of collagen in that tissue and the rate of cleavage of the extension peptides at the N- and C-terminal ends of the procollagen molecules.

Type I procollagen is processed to collagen molecules very rapidly

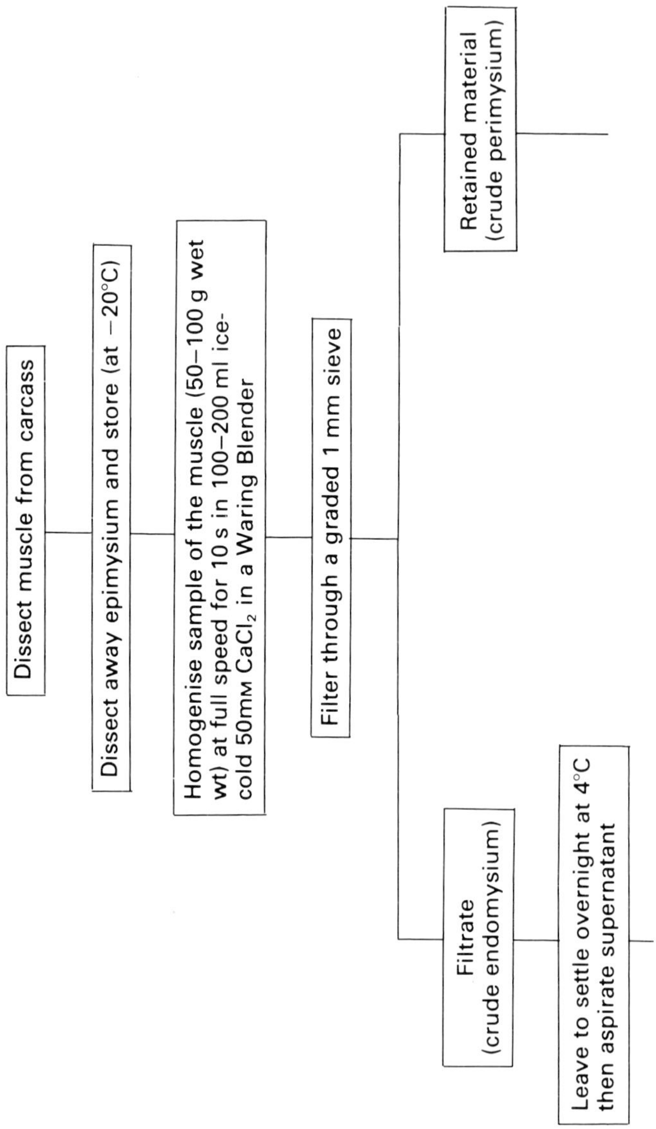
Dissect muscle from carcass
Dissect away epimysium and store (at −20°C)
Homogenise sample of the muscle (50–100 g wet wt) at full speed for 10 s in 100–200 ml ice-cold 50mM $CaCl_2$ in a Waring Blender
Filter through a graded 1 mm sieve
Retained material (crude perimysium)
Filtrate (crude endomysium)
Leave to settle overnight at 4°C then aspirate supernatant

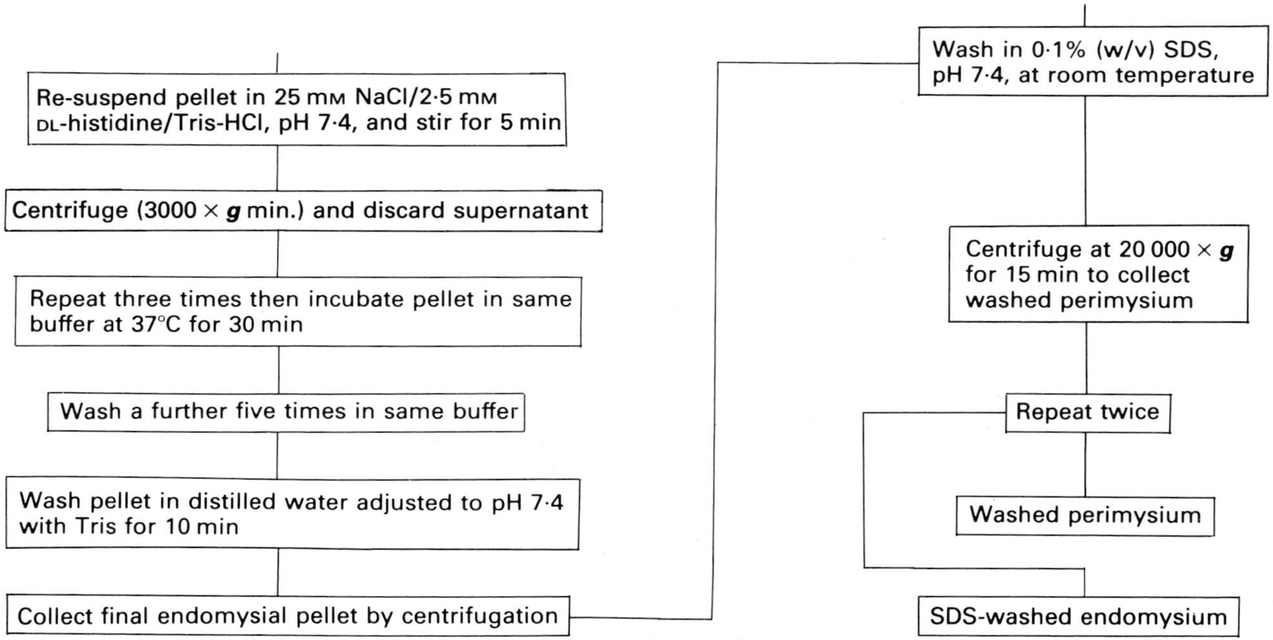

Scheme 12.1 Flow diagram showing the preparative procedures for the purification of epimysial, perimysial and endomysial fractions from meat.

Homogenise or mince connective tissue at 4°C in 10–20 vol. 0·15M NaCl, 0·05M Tris-HCl, pH 7·4, with protease inhibitors

↓

Filter through muslin then through Celite

↓

Add ammonium sulphate to 30% saturation

↓

Collect precipitate by centrifuging at 15 000 × ***g*** for 20 min

↓

Wash pellet twice with buffer containing NaCl added to 20% (w/v) and discard soluble material

↓

Extract pellet twice with 330 ml of buffer containing 1M NaCl and pool supernatants (70 000 × ***g*** for 15 min)

↓

Add NaCl to final concentration of 20% (w/v) and collect precipitate by centrifuging at 70 000 × ***g*** for 15 min

↓

Suspend pellet in water and dialyse exhaustively vs water

↓

Adjust to 0·2M NaCl, 0·05M Tris-HCl, pH 7·6, and pass through DEAE-cellulose (4 × 2·5 cm) equilibrated in same buffer

↓

Collect unbound material, dialyse and adjust to 0·02M NaCl, 0·03M Tris-HCl, pH 7·5, 2M urea and pass through DEAE-cellulose in same buffer at 8°C with a gradient of 0·02–0·2M NaCl over 600 ml

↓

Pool and dialyse pro-collagen Types I and III and store in 0·05M acetic acid frozen in liquid nitrogen

Scheme 12.2 Flow diagram showing the preparative procedures for the purification of procollagens Types I and III.

and is consequently present in most tissues in only very low concentrations. Type III procollagen has a slower processing rate and a higher proportion is present in tissue and is therefore slightly easier to prepare. However, neither variety is present to any great extent in mature or old muscle and by far the best source of these precursors is foetal skin or placenta. The procedure for their preparation described below is illustrated in Scheme 12.2. The extracting buffer (Tris-buffered saline) is supplemented with protease inhibitors to prevent non-specific cleavage of procollagen propeptides during preparation. The procedure consists of salt fractionation steps followed by a diethylaminoethyl (DEAE)-cellulose ion exchange chromatography step to remove acid protein contaminants. Finally DEAE–cellulose chromatography in 2M urea at low ionic strength is used to separate the procollagens (I and III) from the collagens (Types I and III) (Fig. 12.1).

Types I and III collagen are not bound to the DEAE-cellulose and elute in the void volume. Type I procollagen is then eluted prior to Type III procollagen by means of a linear salt gradient from 20–200mM NaCl over a total volume of 600 ml. Column effluent is monitored at 230 nm and fractions are pooled, dialysed against water, lyophilised and stored at 4°C.

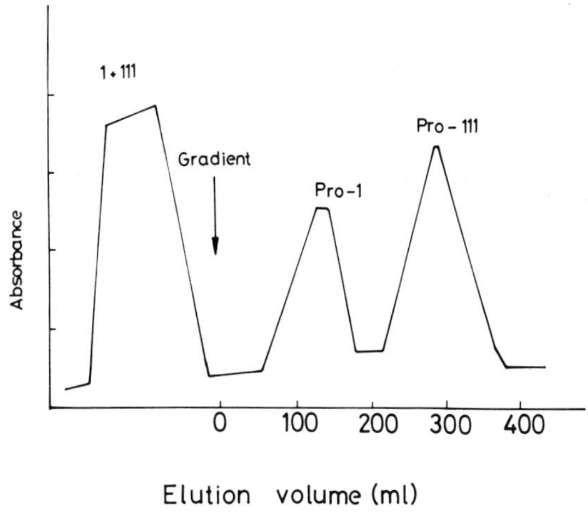

Fig. 12.1 Typical elution profile from DEAE-cellulose chromatography of Types I and III collagens and their procollagens.

Neutral salt-soluble collagen

Very little or no collagen in mature muscle is soluble in neutral salt solution (1M NaCl, 0·05M Tris-HCl pH 7·4). However, small quantities of this non-cross-linked collagen can be extracted from homogenised foetal or new-born muscle samples or specific tissues, e.g. guinea pig skin. The tissue (briefly pre-washed with water) should be well homogenised in the neutral salt solution at 4°C (maximum 50 g wet weight of tissue to 100 ml extractant) using a pre-chilled Waring-type blender. Usually, one or two 10-second treatments at full power are sufficient to disintegrate the tissue fully. It is preferable to apply several short treatments in the homogeniser, inspecting the sample at intervals, rather than one long treatment in order to avoid clumping of the collagen. A single filtration step (as described above) after homogenisation and before neutral salt extraction is advised in order to remove the bulk of cellular protein. The homogenised sample should be stirred continuously at 4–15°C overnight before collecting the supernatant by centrifugation. The extraction should be repeated under similar conditions. The collagen extracted consists almost entirely of Type I with some Type III and can be readily further purified by salt fractionation (see below).

Acid-soluble collagen

Collagen cross-linked only by the acid-labile aldimine can be solubilised by extraction with 0·5M acetic acid. The best sources of such material are placenta or foetal skin but small quantities can be extracted from the tendons and intramuscular connective tissue of foetuses or very young animals. The reason that only small amounts of collagen can be extracted from tendon and muscle is the prominence of the acid stable oxo-imine cross-link in these tissues. The major product from these sources is Type I collagen.

Simple homogenising in 0·5M acetic acid at 4°C with repeated extractions at this temperature can yield large amounts of Type I collagen; Types III, IV and V are largely unaffected and remain insoluble. Pure Type I collagen can be obtained by salt fractionation as described below. To avoid digestion of collagens by endogenous proteases it is advisable to add inhibitors to the acetic acid, although the presence of acid-labile peptide bonds in the C-terminal non-helical region of the $\alpha 1(\mathrm{I})$-chain means that the product of acid extraction is always reduced in size by the loss of between 10 and 15 residues from the C-terminal end.

Pepsin-soluble collagen

Collagen can be extracted from most foetal, pre-mature and mature connective tissue by solubilising the source tissue, after preliminary washing with salt, acid, chaotropic or detergent solutions, with the enzyme pepsin. The action of this protease in releasing whole, soluble collagen molecules from the insoluble matrix has been discussed earlier and is illustrated diagrammatically in Fig. 12.2. Virtually all protocols for this method use a substrate-to-enzyme ratio of 10:1 at temperatures of 4–15°C for times of 4–72 h. In general, a 24 h treatment at 10°C is applicable to muscle connective tissue specimens although placenta, a much better source, needs longer incubation at the same temperature, i.e. up to 72 h. The treatment is carried out in 0·5M acetic acid (pH 2·5) and insoluble material is collected and stored or discarded after centrifugation (15 000 × *g* for 20 min). The various collagen types can then be precipitated from the clear supernatant as described below.

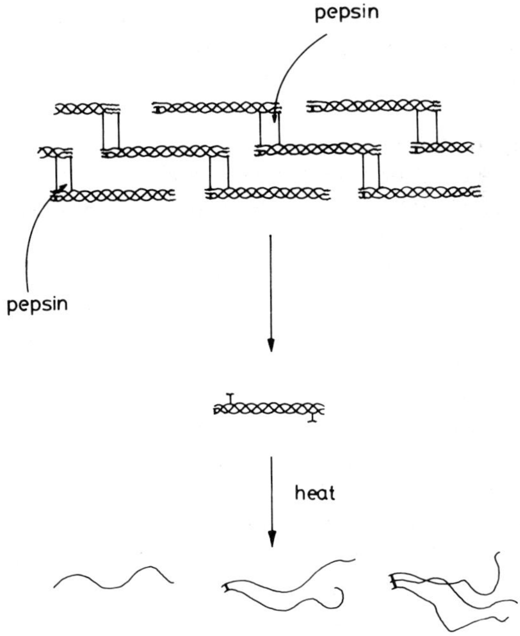

Fig. 12.2 Diagrammatic representation of the action of the proteolytic enzyme pepsin on matrix collagen to yield a soluble molecular fraction and subsequent heat denaturation.

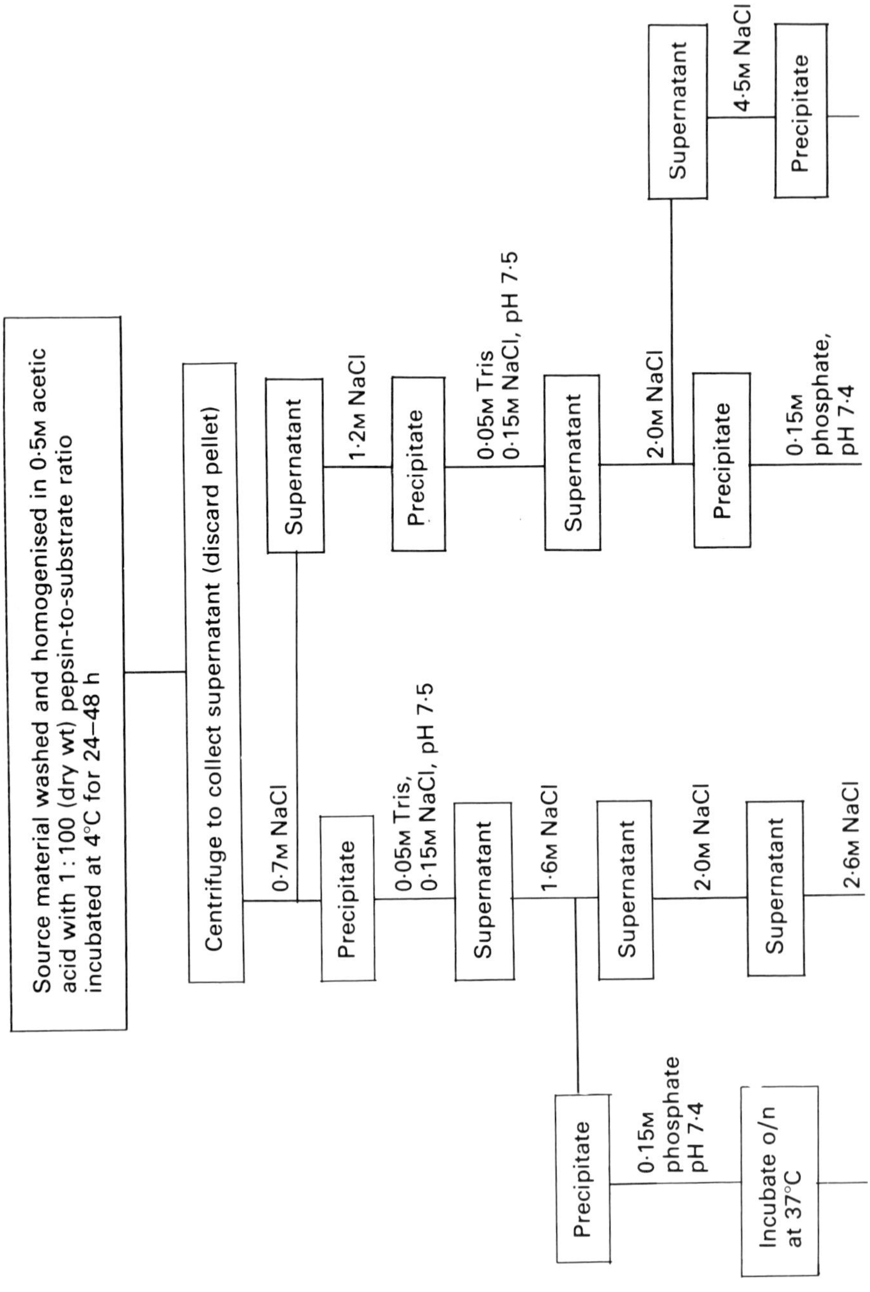
Source material washed and homogenised in 0·5м acetic acid with 1:100 (dry wt) pepsin-to-substrate ratio incubated at 4°C for 24–48 h
Centrifuge to collect supernatant (discard pellet)
0·7м NaCl
Supernatant
1·2м NaCl
Precipitate
0·05м Tris 0·15м NaCl, pH 7·5
Supernatant
2·0м NaCl
Supernatant
4·5м NaCl
Precipitate
Precipitate
0·15м phosphate, pH 7·4
Precipitate
0·05м Tris, 0·15м NaCl, pH 7·5
Supernatant
1·6м NaCl
Supernatant
2·0м NaCl
Supernatant
2·6м NaCl
Precipitate
0·15м phosphate pH 7·4
Incubate o/n at 37°C

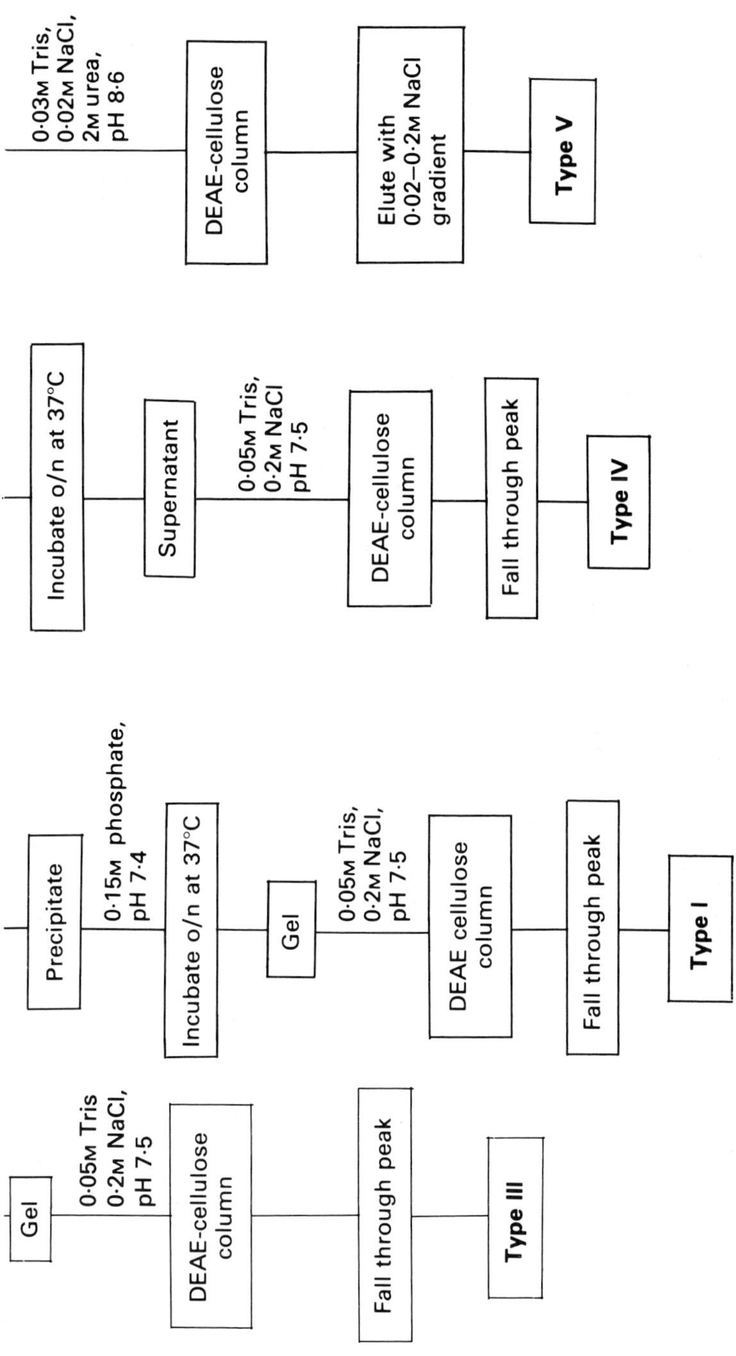

Scheme 12.3 Flow diagram showing the preparative procedures for the purification of Types I, III, IV and V collagens from meat connective tissue or any alternative connective tissue source.

Residual insoluble collagen

Acid and neutral salt-insoluble collagen in both foetal and adult muscle connective tissue can be readily prepared by homogenising the sample at 4°C (in a blender or Polytron homogeniser) in 0·5M acetic acid containing 8M urea and repeatedly extracting with the same buffer until no further protein is extracted. The resulting pellet should be thoroughly washed with water and contains predominantly Type I and Type III collagen and small amounts (5-10%) of Types IV and V collagen (plus some elastin from certain muscles). Preparations of insoluble collagen are generally only made for studies of cross-linking or to act as substrates for specific proteolytic enzymes.

Separation of native collagen types

Differential salt precipitation of collagen types

The cold (4°C) supernatant after pepsin digestion or 0·5M acetic acid extraction is made 0·7M in NaCl, left at 4°C for 18–24 h and the precipitated collagen is collected by centrifugation. The precipitate is dissolved in 0·5M NaCl, 0·05M Tris-HCl, pH 7·5, at 4°C and dialysed against the same buffer. By increasing the salt concentration the various types of collagen can now be salted out of solution and collected as precipitates. Intractable precipitates can be redissolved in 0·5M acetic acid and then dialysed against neutral salt solution if

Table 12.1

Summary of salt concentrations required to precipitate different collagen types and behaviour on heat gelation and DEAE-cellulose chromatography

Collagen type	*Molarity of NaCl*	*Effect of heat*	*Behaviour on DEAE-cellulose*
I	2–2·6	Forms gel[a]	Not bound[b]
III	1·6	Forms gel	Not bound[b]
IV	2·0	Soluble	Bound[c] eluted with gradient (0·02–0·2M NaCl)
V	4·5	Soluble	Bound[c] eluted with gradient (0·02–0·2M NaCl)

[a] From neutral salt solution (0·15M NaCl; 0·05M Tris-Cl, pH 7·4).
[b] Equilibrated in 0·05M Tris-HCl, 0·2M NaCl, pH 7.4.
[c] Equilibrated in 0·03M Tris-HCl, 0·02M NaCl, 2·0M urea, pH 8·6.

further fractionation is necessary. Scheme 12.3 summarizes the procedures adopted to purify Types I, III, IV and V by salt fractionation from a pepsin-solubilized supernatant of muscle connective tissue (or any other collagenous source). The technique of heat gelation can be used to separate partially purified Type IV or V collagen from contaminating Types I and III. A final DEAE-cellulose chromatography step in 0·2M NaCl as described in Scheme 12.3 assures the purity of the final preparation. Table 12.1 summarises the concentration of salt necessary at neutral pH to precipitate the different collagen types and their behaviour on heat gelation and DEAE-cellulose chromatography. The purity of all collagen preparations should be checked by sodium dodecyl sulphate (SDS)-polyacrylamide gel electrophoresis (see below).

Ion-exchange chromatography

The separation of specific collagen types on an ion-exchange column under non-denaturing conditions has proved possible in a limited number of cases. CM-cellulose chromatography at acid pH has been used to separate Type IV collagen from Type VII and Type V. Separation is generally achieved by elution with 0·4M sodium acetate at pH 4·8 at a column temperature of 4 to 8°C.

CM-Cellulose chromatography at alkaline pH has also been used to separate Type I and Type V. The column is eluted with Tris buffer at pH 8·0.

DEAE-cellulose column chromatography has also been used to isolate collagen types normally difficult to separate, for example, Type V from Type IV. The collagens are eluted from the column using Tris buffer at pH 8·6 and a column temperature of 15°C.

HPLC

A method has also been reported for the high performance liquid chromatography (HPLC) gel filtration separation of the native molecular forms of Types I and III collagens. The method is particularly useful in removing the last trace of Type III from Type I prepared by salt precipitation. It is anticipated that these methods may be scaled up to the preparative level and that both Types IV and V will also be resolved in the near future.

Resolution of the individual α-chains of collagen types

As described above, native collagens may be prepared by salt fractionation. It may be necessary, however, to separate and investigate the component α-chains of the various collagens. The purified collagen is denatured, usually in buffered 4–6M urea at pH 4·8, by heating above 42°C for 30–60 min. The separate α-chains of the collagen obtained after denaturation are then purified chromatographically.

Ion-exchange chromatography

The α-chains of Types I, III and IV collagen molecules can be separated by carboxymethyl (CM)-cellulose chromatography in 0·05M sodium acetate buffer, pH 4·8, at 42°C with a salt gradient (0·02–0·2M NaCl)—see, for example, Types I and III in Fig. 12.3.

It should be noted here that the α-chain composition of Type IV collagen has been, until recently, controversial. It is clear now that two Type IV α-chains exist and several different protocols, variations of the standard pepsin solubilisation, have been adopted in their preparation. These include the use of reduction and alkylation, re-pepsinisation and gel filtration on agarose gels. Type IV procollagen has been purified from a mouse tumour and from cultures of amniotic fluid cells. The various degradation products purified after pepsin digestion of basement membrane material can also be produced by pepsin treatment of procollagen made in cell culture.

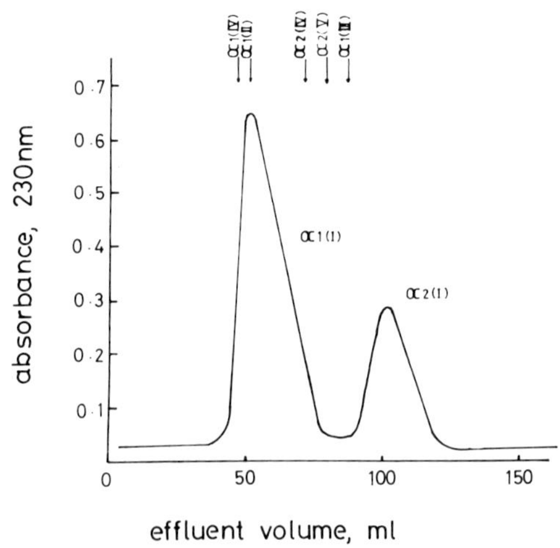

Fig. 12.3 CM-cellulose chromatography of individual α-chains from purified collagens.

Type V collagen $\alpha1(V)$ and $\alpha2(V)$ chains can be separated after denaturation, by CM-cellulose, phosphocellulose, or hydroxy-apatite chromatography. However probably the most popular method is that of DEAE-cellulose chromatography. Native Type V is denatured in 0·025M Tris–HCl, 4M urea, pH 8·6 for 20 min at 45°C and then applied to a column (2·5 × 11 cm) of DEAE-cellulose equilibrated in the same buffer at 42°C and eluted with a gradient of NaCl (0–0·1M) in a total volume of 1000 ml.

Gel filtration

The technique of gel filtration may also be used to separate denatured collagen components of different molecular weight. The method is therefore generally used to separate α, β and γ and higher components. In general the collagen is heat denatured (at 60°C for 20–30 min) in a reagent such as 1M CaCl/0·05M Tris-HCl, pH 7·4, before chromatography on an agarose gel equilibrated in the same buffer. Components may be separated in this way on agarose A1·5m or A5m, or on Sepharose 6B or 4B (Fig. 12.4).

HPLC

Recent developments in HPLC technology have offered hope for the evolution of extremely rapid, high resolution separations of different

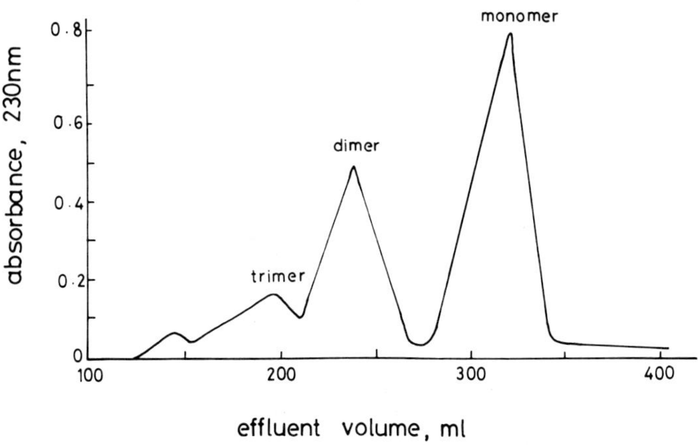

Fig. 12.4 Separation by molecular weight difference of monomer, dimer and trimeric components of collagen α-chains after denaturation by agarose gel filtration chromatography.

denatured α-chains. The main advantage offered by such methods is that of speed, as most fractional salt separations described above take days or weeks to complete.

Large pore (30 nm and 50 nm) reverse-phase HPLC supports with hydrophobic side chains (octyl, octadecyl, diphenyl and cyanopropyl) can be utilised with volatile pyridine buffers at pH 4·5 using post-column fluorescent detection to resolve the $\alpha 1$(I) and $\alpha 2$(I) chains of Type I collagen at denaturing temperatures. The method is classical reverse-phase chromatography carried out on a support which allows full permeation of the collagen chains into the stationary phase. The resolution of denatured chains of Types I and III collagen on 30 nm and 50 nm silica derivatised with octadecyl and cyanopropyl groups can also be achieved using UV-transparent solvents. The advantages of these methods are that UV detection may be used and undesirable solvents such as pyridine are avoided.

ROUTINE ANALYSIS OF PURE COLLAGENS

Collagen preparations are generally checked for purity by amino acid analysis, SDS-polyacrylamide gel electrophoresis or by CNBr or proteolytic peptide mapping.

Amino acid analysis

Purified collagens or muscle connective tissue samples are hydrolysed in sealed glass tubes in redistilled 6M HCl at a concentration of 1 mg ml^{-1} and at 110°C for 24 h. After hydrolysis the amino acid mixture is taken to dryness either passively under vacuum or by rotary evaporation at 50°C. The residue is dissolved in water and dried again before being dissolved in 0·001M HCl and analysed by routine ion exchange amino acid analysis with ninhydrin detection. The number of residues per 1000 residues in Types I, III, IV and V collagen are given in Table 2.2. Modifications of the procedure for quantitation of oxidised cysteine and methionine residues follow standard methods.

Amino acid analysis by HPLC has become more popular in recent years owing to its rapidity and convenience. The dansyl (DNS) or phenylthiohydantoin (PTH) pre-derivatisation methods are commonly used. In these cases the dry amino acid mixture obtained after hydrolysis is taken up in an appropriate buffer, the derivatising agent is added and the mixture incubated. The DNS or PTH amino acid derivatives are then separated by reverse-phase HPLC on 3 or 5 μm spherical particulate supports with high efficiencies in solvents buffered near neutrality (e.g. 0·05M sodium phosphate, pH 6·2). Typically, a complete separation of derivatised amino acids can be achieved in 25–35 min.

SDS-polyacrylamide gel electrophoresis

Samples of pure collagen are dissolved in SDS at 60°C for 30 min to a final concentration of 5 mg ml^{-1}. Most workers use a buffer containing 2% (w/v) SDS, 0·125M Tris-HCl, pH 6·8, 10% (v/v) glycerol, and 0·005% (w/v) bromophenol blue marker dye. Twenty-five to 100 μg of protein are loaded in wells in a pre-cast SDS-polyacrylamide slab gel (generally containing 5% (w/v) acrylamide and 0·167% (w/v) *NN′*-methylenebis-acrylamide). Some laboratories use the borate buffer system in casting and running these gels but the Tris-HCl/Tris-glycine system is now used more frequently. In this case the acrylamide is dissolved prior to polymerisation in 0·375M Tris-HCl, pH 8·8, containing 0·1% (w/v) SDS and is run at 100–200 V with anodic and cathodic buffers of 0·05M Tris-glycine (6 g Tris, 28·8 g glycine per litre), pH 6·8, containing 0·1% (w/v) SDS.

Protein bands are visualised by staining the gel with Coomassie Brilliant Blue R250 dissolved to 0·1% (w/v) in methanol/acetic acid/water (5:1:5) and destained at 20–45°C with several changes of 7·5% (v/v) acetic acid containing 5% (v/v) methanol. More sensitive stains (2- to 50-fold increased sensitivity) have been developed using silver for both proteins and carbohydrate as well as for specific saccharides attached to collagenous chains. A typical staining profile obtained with Coomassie Blue after electrophoresis of Types I, III, IV and V is shown in Fig. 12.5.

Types III and IV collagens both contain interchain disulphide bridges and so migrate at the top of the gel if these bonds are not chemically reduced. On addition of 2% (v/v) 2-mercaptoethanol to samples of Types III collagen prior to electrophoresis a band is seen migrating in the position of or near to the α1(I) chain of Type I collagen. This represents reduced α1(III) chains. In mixtures of Types I and III collagen this makes assessment of the amount of Type III collagen present difficult and so an interrupted reduction method is used. 2-Mercaptoethanol (5 μl of 30% (v/v) in SDS-buffer) is added to the required wells on the gel between 10 and 15 min after the start of the run. The α1(III) chains now migrate significantly more slowly than

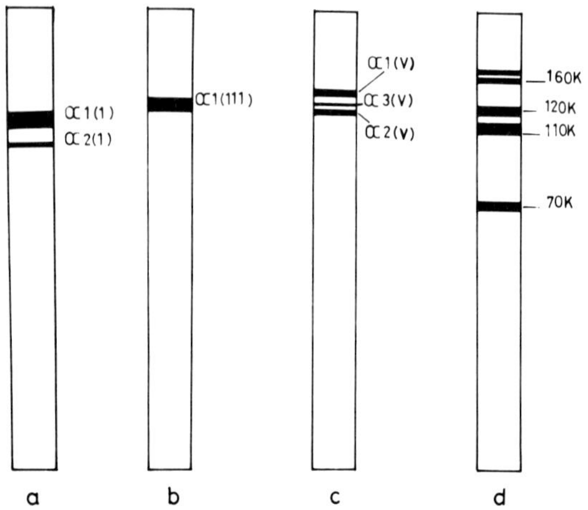

Fig. 12.5 Separation of molecular components of (a) Type I, (b) Type III, (c) Type V and (d) Type IV collagens by SDS-polyacrylamide gel electrophoresis.

the $\alpha1$(I) chains. Reduced, pepsin-solubilised Type IV collagen appears as several bands on gels; molecular weights of 160 000, 120 000, 100 000, 70 000 and 55 000 have all been reported, depending on the extent of pepsin digestion. However, careful digestion and extracts from tissue culture reveal two bands with apparent molecular weight of 120 000 ($\alpha1$) and 95 000 ($\alpha2$). Type V collagen, purified from muscle connective tissue by pepsin solubilisation, is not mercaptoethanol sensitive and the three chains migrate on gels with molecular weights slightly in excess of 100 000 (typically around 115 000).

It is clear that the major chains of all four types (I, III, IV and V) can be successfully separated by SDS-polyacrylamide gel electrophoresis and this provides a useful qualitative analytical technique. The method can be made semi-quantitative by use of a scanning spectrophotometer or densitometer to estimate the material in each stained band.

Peptide mapping techniques

The most commonly used peptide mapping method utilises CNBr and SDS-polyacrylamide gels. As mentioned in earlier chapters, CNBr cleaves proteins selectively at methionine residues; with their limited numbers of methionines, collagens give relatively simple sets of peptides which may be resolved by SDS-polyacrylamide gel electrophoresis. The sample of pure collagen or muscle connective tissue is dissolved or suspended to a final concentration of 10 mg ml^{-1} in 70% (v/v) formic acid at 25–30°C and an equal weight of solid CNBr is added. Cleavage takes place over a period of 4–6 h at this temperature. For convenience and ease of handling the toxic reagent it is possible to use a solution of CNBr in acetonitrile (2 g ml^{-1}) so that a very accurate addition of the compound can be made by volume. After the reaction is finished the solution is reduced in volume by rotary evaporation at 40°C to several millilitres and diluted to the original volume with water. This procedure is repeated three or four times to remove excess reagent and acid before the sample is lyophilised.

Freeze dried CNBr peptide mixtures are dissolved at 5 mg ml^{-1} concentration in the SDS buffer described above and are separated on 10% (w/v) acrylamide gels containing 0·34% (w/v) *NN'*-methylene-bisacrylamide and 0·1% (w/v) SDS with a stacking (loading) gel containing 3% (w/v) acrylamide and 0·1% (w/v) *NN'*-methylene-bisacrylamide in the Tris-Cl/Tris-glycine buffer system described above. Peptides are visualized by Coomassie Blue staining and can be assessed quantitatively by densitometric scanning. Typical peptide maps are shown in Fig. 12.6. It should be noted that Types I and III CNBr peptide maps contain well-characterised linear uncleaved peptides and the same is almost certainly true of Types IV and V, although the CNBr peptide components of these collagens have not yet been fully characterised. Each peptide map is unique and can be used to verify the type of collagen purified.

Increased resolution of some components in collagen CNBr peptide mixtures can be achieved using gradient gel electrophoresis. In this method exactly the same procedures are followed as those described above but the separating gel is formed as a gradient of polyacrylamide ranging from 10–30% (w/v).

Other peptide mapping techniques for collagen have been reported, particularly two-dimensional (2D) CNBr-SDS-polyacrylamide gel electrophoresis and 2D protease peptide (e.g. chymotrypsin, V8 protease) maps. Both methods rely on the ability to cleave collagens specifically and reproducibly into a limited number of fragments and then to separate those fragments using two different electrophoretic methods sequentially with the same sample. In most cases the sample is subjected to isoelectric focussing in the first dimension. This involves running the sample on a gel containing a gradient of pH formed by the

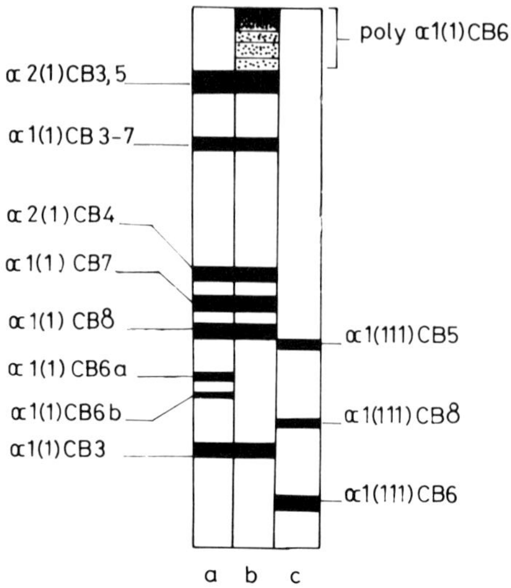

Fig. 12.6 Separation of CNBr peptides of different collagen types by SDS-polyacrylamide gel electrophoresis. (a) CNBr digest of Type I collagen from young tissue, (b) CNBr digest of Type I collagen from old/mature tissue showing polymeric peptide components (see Chapter 4), and (c) CNBr digest of type III collagen from young tissue.

inclusion of specific chemicals and normally in the presence of urea. After separation and without staining this first gel is overlaid at right angles to the first direction of electrophoresis on the second gel, normally a standard electrophoresis gel. Now, a current is applied and the second dimension separation is carried out. These procedures allow much more detailed separations of collagen fragments but involve much longer times than standard electrophoresis.

HPLC

CNBr peptide mapping by the SDS-polyacrylamide gel technique is efficient and reliable but recently it has been supplemented by a very rapid HPLC method which allows the resolution of very small amounts of peptide mixtures on reverse-phase columns. The method uses UV-transparent solvents and so accurate quantitation can be made on eluted peaks. Because the solvents used are volatile, samples can be recovered by fraction collection after HPLC resolution and dried in 2–3 h by vacuum centrifugation.

HPLC methods are also used effectively to separate and identify fragments of collagen derived from proteolytic cleavage by enzymes such as trypsin, chymotrypsin and thermolysin. These methods have been successfully applied in the sequencing of collagens but can equally be used for the analysis of collagens in various situations.

CM-cellulose separation of CNBr peptides

This method has been used extensively as a preparative technique in the production of collagen CNBr peptides. It consists of loading the CNBr peptide mixture in 0·02M citrate buffer, pH 3·6, containing 0·02M NaCl and 4M urea on a short (10 × 2·5 cm) CM-cellulose column equilibrated in the same buffer at 42°C. Peptides are eluted with an NaCl gradient (0·02–0·17M) over 1000 ml. The technique has been used analytically but has been overtaken by the more rapid and reliable gel electrophoresis method. However, it is still the primary method for bulk purification of collagen CNBr peptides.

Carbohydrate analysis

By far the most common form of analysis of carbohydrates required by collagen chemists is the observation of hydroxylysine glycosides released during alkaline hydrolysis. The protein is suspended or dissolved at 1 mg ml^{-1} in 0·2M NaOH and is hydrolysed in a sealed polypropylene vial at 110°C for 48 h. The cooled hydrolysate is neutralised with 6M HCl and desalted on Sephadex G-10 or Biogel P2 equilibrated in water. All material eluting before the salt peak is collected, rotary evaporated to dryness and taken up in 0·001M HCl prior to standard amino acid analysis. The migration position of both

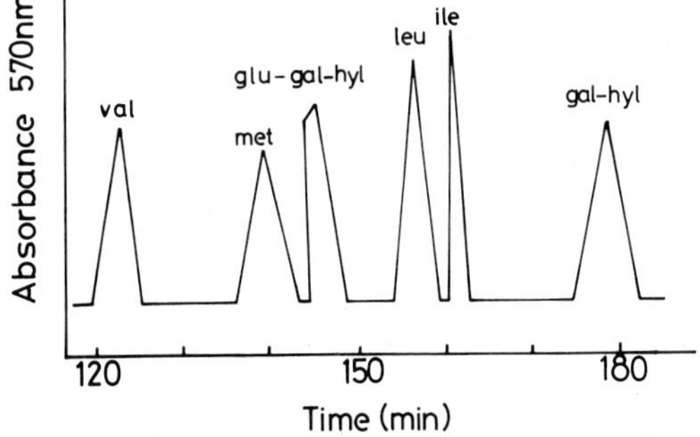

Fig. 12.7 Ion exchange amino acid analysis of hydroxylysine glycosides glucosylgalactosyl-hydroxylysine (Glu–Gal–Hyl) and galactosyl-hydroxylysine (Gal–Hyl) on the amino acid analyser in citrate buffers with ninhydrin detection.

glycosides is shown in Fig. 12.7. Detailed carbohydrate analysis of collagens and procollagens is best carried out by classical gas–liquid chromatographic methods which are well established and will not be described here. Using these techniques the structure and sequence of the polysaccharide moieties such as the mannose derivatives attached to the non-helical propeptide regions can be determined. Simpler and less expensive but only qualitative methods include paper or thin layer chromatography and electrophoresis.

Estimation of amounts of Types I, III, IV and V collagen

At present, estimation of Types I and III collagen in tissues can be made reasonably accurately by using the CNBr/SDS-polyacrylamide gel technique. Such quantitations have been of considerable value in allowing an accurate biochemical assessment of the proportion of these collagens in various meat samples before and after treatments such as cooking. Purified muscle connective tissues are well suited to the application of this method as they generally contain 90% or more of their total collagen complement as Types I and III and often show greater than 95% solubility in CNBr or following pepsin digestion. Obviously, the effectiveness and accuracy of any method of this kind depends on the total solubilisation of the tissue in question, and in most cases CNBr is the most effective reagent.

Using the pepsin extraction method native collagen molecules are released from the matrix and can be separated by SDS-polyacrylamide gel electrophoresis as described above. Using the interrupted reduction method leads to unequivocal identification of the Type III chain and its subsequent quantitation relative to Type I.

Using the CNBr digestion technique the sample of muscle connective tissue is finely chopped with a scalpel and suspended either in 70% (v/v) formic acid or 0·5M acetic and is homogenised to give a fine suspension. CNBr (an equal weight) or pepsin (1:10 enzyme to substrate) is added to the formic acid suspension or acetic acid suspension respectively and incubation carried out as described in previous sections. After digestion and gel electrophoresis the peptide bands are visualised with Coomassie Brilliant Blue. Gels can be either dried under vacuum on a porous polyethylene drying slab and scanned by reflectance densitometry or scanned directly (in a specially adapted cuvette or as slabs on a flat bed scanner) by transmission densitometry. The higher molecular weight peptide bands for Types I and III are well separated; α1(I) CB8 and α1(III)CB8 (or α1((III)CB5) are generally chosen for quantitation. None of these peptides is involved in cross-linking and so almost 100% of each can be expected to migrate in the correct position despite the age and extent of cross-linking of the sample. As Type I collagen contains only two $\alpha 1$ chains (i.e. two $\alpha 1$(I)CB8 per molecule) as compared to Type III with three chains (i.e. three $\alpha 1$(III)CB8 per molecule), the quantitation should be corrected by multiplying the arbitrary figures obtained for Type I by a factor of 1·5.

It should be noted that this method reveals only the ratio of Type I to Type III, but can be adapted for an approximate calculation of the absolute amount of each type. If enough sample is available a hydroxyproline assay can establish the total amount of collagen per unit weight of the muscle connective tissue and so, using the calculated Type I-to-Type III ratio and knowing the amount of sample loaded on each gel track, absolute amounts of both types can be estimated.

This method is useful and readily available to most workers as it requires little development and the apparatus is relatively cheap. By using internal standard CNBr peptide mixtures on each gel and by careful handling it is possible to obtain results to an accuracy of better than 10% providing the Type III-to-Type I collagen ratio exceeds 0·1. However, it is not possible to quantitate Types IV and V by this method due to the very low proportion of these collagens present in the muscle connective tissue and the current lack of knowledge of their peptide patterns. To overcome this problem various laboratories are currently developing immunoassays to CNBr peptides of Types I, II, III, IV and V collagens. The method involves purifying a specific CNBr peptide from each collagen type, raising poly- or monoclonal antibodies, purifying the antibody preparation and using it in inhibition assays for collagen quantitation.

QUANTIFICATION OF COLLAGEN TYPES BY IMMUNOASSAY

Each collagen type is a distinct gene product (as explained in Chapter 2). Specific antibodies to collagen types in the species of interest are raised in donor animals as follows. Collagen samples, purified and checked both quantitatively and qualitatively by the methods cited above are dissolved in a neutral isotonic buffer such as phosphate-buffered saline (PBS: e.g. 0·15M NaCl, 0·02M sodium phosphate buffer, pH 7·4) by stirring at 4°C for several days at a concentration of 1–2 mg ml^{-1}. After dissolution the solutions are centrifuged to remove any undissolved material and the resulting solutions are combined (normally at a ratio of 1:1) with Freund's Complete Adjuvant for the first injection into donor animals and with Freund's Incomplete Adjuvant for subsequent injections. Antibodies to the injected collagen type are produced in a primary immune response by the donor and subsequently at higher levels in the secondary response (Scheme 12.4).

The researcher can harvest antibodies raised in this way from laboratory animals such as rabbits, or larger donors such as goats and sheep, by simply humanely bleeding the donor and collecting the immune serum from the blood by centrifuging out the cellular fraction. The immune serum contains many other proteins apart from the range of active anti-collagen antibodies (A_c) but, nevertheless, the serum can be shown to contain the active A_c fraction by routine immunological tests such as enzyme-linked immunosorbent assay (ELISA) or radioimmunoassay (RIA).

Once harvested, the serum can be treated to remove the unwanted protein fractions and to purify the specific A_c fraction. This is usually achieved by a method which selects out the highly specific antibodies (A_s) which react only with the specific, genetically distinct collagen type which has been used as the antigen to raise the antibody fraction. This method is known as immuno-absorption. In this technique each of the genetic types of collagen must be purified and then linked to an insoluble chromatography material such as agarose beads. This controlled insolubilisation or immobilisation of the collagen types can be done relatively easily by first activating the agarose to make it reactive with amine groups. Once the collagens are immobilised to agarose beads these materials can be packed into columns and equilibrated in physiological buffers (Scheme 12.5).

Although there are several methods of choice for using these immuno-absorbent columns the immune serum is generally cycled, at low temperature (4–10°C), around columns containing immobilised forms of all the collagen types except the one of interest (i.e. the one used to immunise the animal in the first instance). In this way, any antibodies in the A_c fraction of the serum which cross-react with the other collagen types will be absorbed onto the columns. Once this step has been carried out the immune serum is cycled through a column containing an immobilised form of only the collagen of interest (i.e. the original antigen). Now the A_s fraction will be absorbed to the column. After sufficient cycling to allow full absorption, the column is washed free of serum proteins with clean physiological buffer. The A_s

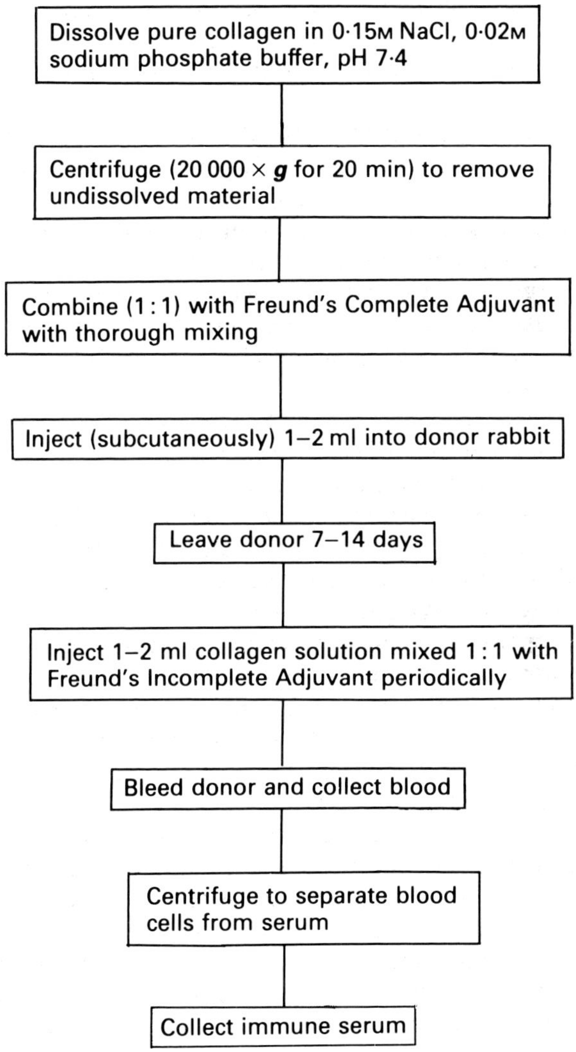

Scheme 12.4 Flow diagram showing the procedures for raising antibodies to purified collagens in donor animals (e.g. rabbits).

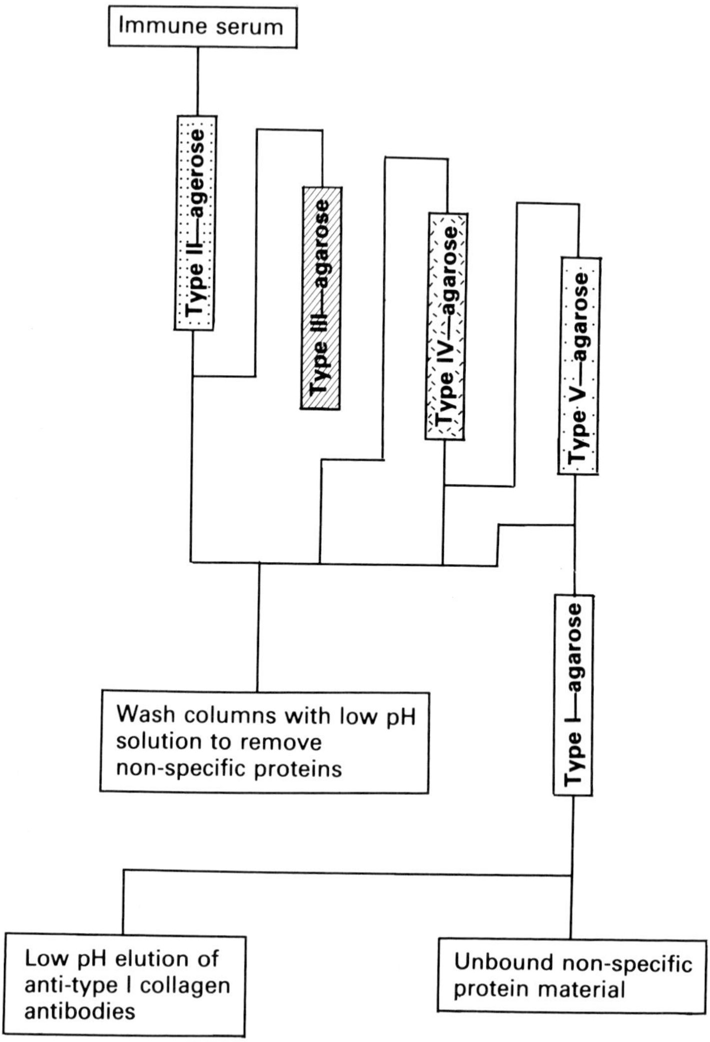

Scheme 12.5 Flow diagram showing the preparative procedures for the purification of anti-collagen antibodies using adsorption chromatography on agarose-collagen affinity columns.

fraction is then eluted from the column with a short, low pH wash and can then be stored at physiological pH at 4°C in the presence of a bacteriostat such as sodium azide for months (Scheme 12.5).

The A_s fraction which contains antibodies which will only react with the collagen type desired can now be used in assays for the collagen type of interest. Although RIA can be used, the method of preference, because it does not use radiolabels, is that of ELISA. In this method, a known quantity of the collagen of interest is coated on a plastic plate at a known concentration. The A_s fraction is applied to the plate and will bind to the collagen substrate. A second antibody is now applied which recognises the first A_s fraction (e.g. anti-rabbit IgG raised in goats if the A_s fraction is raised in rabbits). This second antibody is pre-labelled with an enzyme such as horseradish peroxidase which uses a substrate that is converted to a coloured product. After washing the plate the substrate for the enzyme is applied and the colour intensity produced by the reaction is measured as an estimate of the extent of enzyme, and therefore second and first antibody, present. By inclusion of serial dilutions of the unknown sample containing the collagen type of interest at the second step (with the A_s fraction) the inhibition of the binding of A_s to the collagen substrate on the plate can be measured and, therefore, the amount of unknown in the sample can be calculated.

Clearly, this method relies on the unknown collagen being soluble and not cross-reacting with the antibodies in A_s. This means that either pepsin-soluble fractions from meat connective tissue must be used or, if the sample is intractable to pepsin digestion, CNBr fractions. In the latter case the antigen will be denatured and be only a fragment of the original collagen molecule. In this case the A_s fraction must recognise the denatured site and therefore must be specifically raised against CNBr fractions or be a monoclonal antibody with a high degree of sequence specificity.

Preparation of monoclonal antibodies

The difficulties encountered using polyclonal antibodies, such as cross-reactivity, limited quantities available and the unavoidable variation between batches, can be overcome to a certain extent by using monoclonal antibodies.

The diversity of antibodies produced in animals is due to each lymphocyte producing a different antibody. Cloning of individual lymphocytes would produce a single antibody, but this has not proved possible. However, the production of a single antibody has been produced indirectly by fusing lymphocytes with myeloma cells. The resulting hybridoma cells can be propagated indefinitely *in vitro,* thereby producing a continuous supply of highly specific monoclonal antibodies.

The myeloma cell lines most readily available are of mouse origin, hence to produce hybridomas mice must be immunised with the type of collagen to which an antibody is required, generally with several injections of native collagen at two-week intervals. The collagen is initially given subcutaneously in complete Freunds adjuvant and subsequently injected with incomplete adjuvant or without adjuvant and administered intraperitoneally. About 200 μg of collagen are used per injection. The mice are bled after the last injection and the serum tested for antibody activity. The lymphocytes are then removed from the spleen of the mouse by agitation of pieces of spleen or by flushing out the cells.

The myeloma cells are then fused with the mouse lymphocytes obtained from the immunised animal, using polyethylene glycol to promote the fusion. The hybridomas can then be selected by plating the entire cell suspension into a large number of culture wells containing HAT medium. Unfused myeloma cells cannot survive in HAT medium, but the hybridoma cells survive and grow and can therefore be easily selected out from the culture wells. The hybridomas synthesising anti-collagen antibodies are identified by an ELISA assay. This narrows down the number of culture wells of interest and these are cloned by limiting dilution. The productive clones are grown and the antibody harvested from the spent culture medium in which the hybridomas are growing, generally by precipitation with ammonium sulphate.

A higher yield of antibodies can be obtained by injecting the hybridoma cells into the peritoneal cavity of a host mouse and then harvesting the ascites fluid.

QUANTIFICATION OF TOTAL MEAT COLLAGEN

The measurement of total collagen content in meat, as in most other tissues, has been carried out for some years in most laboratories by means of the method which uses the oxidative reaction of free hydroxyproline with a chromophore. Hydroxyproline is found in only a very few proteins in very small quantities (i.e. less than 1% in elastin, Clq and acetyl choline esterase) but is present in collagen at levels in excess of 13% by weight. This makes hydroxyproline an ideal marker for collagen quantitation.

The hydroxyproline content of the collagens varies from 13% by weight in Type I to 17·4% in Type III (Table 11.5). However, as Type I collagen tends to predominate in meat connective tissue, a figure of 14% is taken for the calculation of collagen content from hydroxyproline. The method requires total hydrolysis of all the connective tissue to liberate free hydroxyproline prior to quantitation. Once in the free imino acid form there are three methods for the quantitation of hydroxyproline.

Colorimetric method

The colorimetric technique is based on reaction of oxidised hydroxyproline (i.e. as a pyrrole) with *p*-dimethylaminobenzaldehyde (PAB—Ehrlich's reagent). In this method the acid hydrolysate of meat connective tissue or whole meat (see earlier) is oxidised with chloramine-T. There are many variations available on the technique which allow the quantitation of hydroxyproline in samples of various kinds, but the major method is based on that of Stegemann as modified by Woessner (Scheme 12.6). In this method, a 2 ml sample of the acid hydrolysate diluted with water to contain 1–5 μg of hydroxyproline is added to 1 ml of 0·05M chloramine-T solution (made by mixing 20 ml of a 35% solution of chloramine-T in water with 30 ml ethylene glycol monomethyl ether and 50 ml of a buffer containing 50 g citric acid monohydrate, 12 ml glacial acetic acid, 120 g sodium acetate trihydrate and 34 g sodium hydroxide per litre of water and adjusted to pH 6·0). Oxidation proceeds at room temperature for 20 min at which time excess chloramine-T is removed by addition of 1 ml of 3·15M perchloric acid with shaking. After 5 min but not more than 10 min 1 ml of PAB solution (20% w/v in ethylene glycol monomethyl ether) is added with shaking and the mixture is incubated for 20 min at 60°C. After cooling under tap water the absorbance of the mixture is read at 560 nm and compared with a curve constructed from hydroxyproline standards ranging from 0–5 $\mu g\,ml^{-1}$.

The method has been automated and a reliable assay procedure developed with a throughput of many samples per hour. There are some drawbacks with the technique, however, which should be

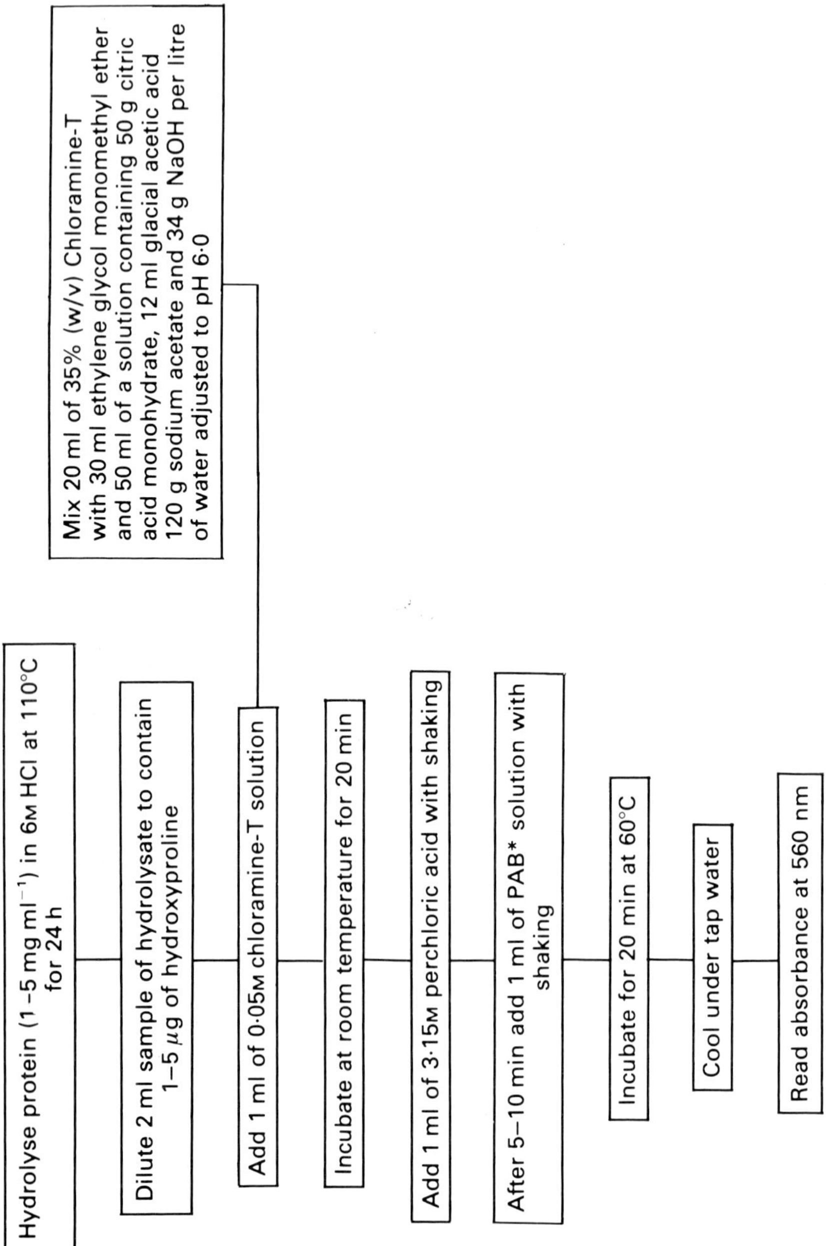

* 20% (w/v) *p*-aminobenzoic acid in ethylene glycol monomethyl ether.

Scheme 12.6 Flow diagram showing the procedure for hydroxyproline estimation using the classical colorimetric method.

mentioned here. Some tissues may contain significant amounts of elastin, e.g. ligaments, arterial walls and skin. In the case of muscle the M. semitendinosus has been shown to contain up to 40% of the connective tissue as elastin. As elastin itself contains 1% hydroxyproline this may lead to discrepancies in the quantitation of collagen from hydroxyproline data. In the case of tissues suspected of having a high elastin content, therefore, the total elastin should also be calculated from the procedures described below and used to correct the initial collagen figure.

In addition, in meat where the collagen levels are low (see Chapter 8) the colour yield from the hydroxyproline assay can be significantly increased by non-specific chromophores which are present in the meat. This can lead to some inaccuracies in the method which need to be accounted for. As stated above, several modifications of the technique exist but the British Standards Institute method minimises this interference and is the preferred method for routine analysis of meat samples.

Ion exchange method

Traditional ion exchange separation on columns of sulphonated epoxy resins in bead form is perhaps the best known of these methods. The acid hydrolysate is dried *in vacuo* and taken up in 0·001M HCl. A sample is applied to a short column (12–20 × 1 cm) of analyser resin equilibrated in 0·2M sodium citrate buffer, pH 2·9, at 60°C. The sample should contain enough hydroxyproline to give a 50% deflection on the recording device (normally in the range 200–500 nmol). The column is pre-calibrated with a standard containing a known quantity of hydroxyproline. The sample hydroxyproline is eluted with the sodium citrate buffer in about 15–20 min. Hydroxyproline is detected as its reaction product either with ninhydrin or with a post-column fluorescence technique using fluorescamine or *o*-phthalaldehyde. The assay can be modified and speeded up by use of different buffers and reducing column length and size, but generally only two to three samples can be handled per hour. The method, however, is extremely accurate and reproducible.

HPLC method

An alternative method of hydroxyproline determination relies on amino acid analysis after pre-column derivatisation with a chromophore or fluorophore such as dansyl chloride and subsequent separation by reverse phase high performance liquid chromatography (RPHPLC). Hydroxyproline estimation in meat samples using this method (Scheme 12.7) involves the following steps. Acid hydrolysates are rotary evaporated to dryness at 55°C three times from HPLC grade water. Samples are then dissolved in 1 ml of the RPHPLC Buffer B

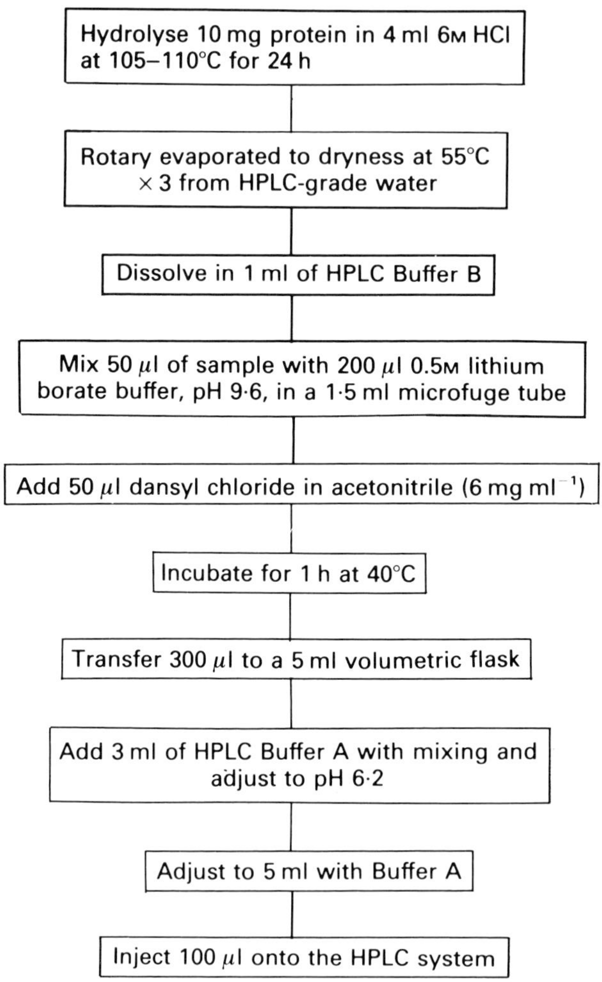

Scheme 12.7 Flow diagram showing the procedure for hydroxyproline estimation using the pre-column derivatisation/high performance liquid chromatography method.

(see below) and transferred to microfuge tubes. Samples can be kept frozen at −20°C if necessary.

Fifty microlitres of each sample is mixed with 200 μl 0·5M lithium borate buffer, pH 9·6, and 50 μl dansyl chloride (DNS-Cl) in CH_3CN (6 mg ml^{-1}) in a 1·5 ml microfuge tube wrapped in foil. The mixture is incubated for 1 h at 40°C. The dansylated material (300 μl) is then transferred to a 5 ml volumetric flask, mixed with 3 ml of RPHPLC Buffer A (see below) and adjusted to pH 6·2 by careful addition of 0·1M HCl. The final volume is adjusted to 5 ml with Buffer A. One hundred microlitres of each sample is injected into the HPLC system.

HPLC analysis is carried out on a 15 × 0·46 cm column of Rainin 5 μ C_{18} reverse phase material (or equivalent high grade HPLC material) at 20°C and a flow rate of 1·5 ml min^{-1}. Solvent A should contain 0·02M sodium phosphate buffer, pH 6·2, with 5% (v/v) CH_3CN and Solvent B should contain 95% CH_3CN with 5% of the phosphate buffer. A gradient is used to elute the DNS-amino acids as follows: 15% Solvent B for 5 min, 15–40% Solvent B for 8 min, 40–50% Solvent B for 7 min, 50–70% Solvent B for 5 min, 70–100% Solvent B for 5 min, 100–100% Solvent B for 2 min, 100–15% Solvent B for 5 min. Detection is by optical density at 254 nm and fluorescence detection with an excitation filter of 368 nm and emission filter of 500 nm.

The method is extremely reproducible and sensitive (pmol amounts of hydroxyproline can be detected and less than 1 nmol easily quantified) and can be made very rapid by the reduction in length of the gradient cited above. Typical elution times for DNS-hydroxyproline are 7–10 min with a column regeneration time of 8–10 min. Thus, the analysis of between three and four samples per hour is possible and with an autosampling device attached to the HPLC analyser up to 96 samples a day can be accurately processed. The method can also be used with whole meat samples without undue interference of non-collagen derived components.

QUANTIFICATION OF TOTAL MEAT ELASTIN

The quantification of elastin in meat is carried out by the destructive hydrolysis of all other insoluble materials except elastin. After this process the elastin, which remains insoluble during hydrolysis, can be weighed as a freeze-dried insoluble residue and its original concentration in the meat back-calculated. Two methods are available to carry out this determination—hydrolysis in 0·1M NaOH or by cyanogen bromide digestion. The former technique relies on the chemical resistance of elastin to alkaline hydrolysis and the latter to the fact that elastin is one of the few insoluble proteins which contain no methionine (the amino acid attacked by cyanogen bromide) and, therefore, will not be solubilised by that reagent.

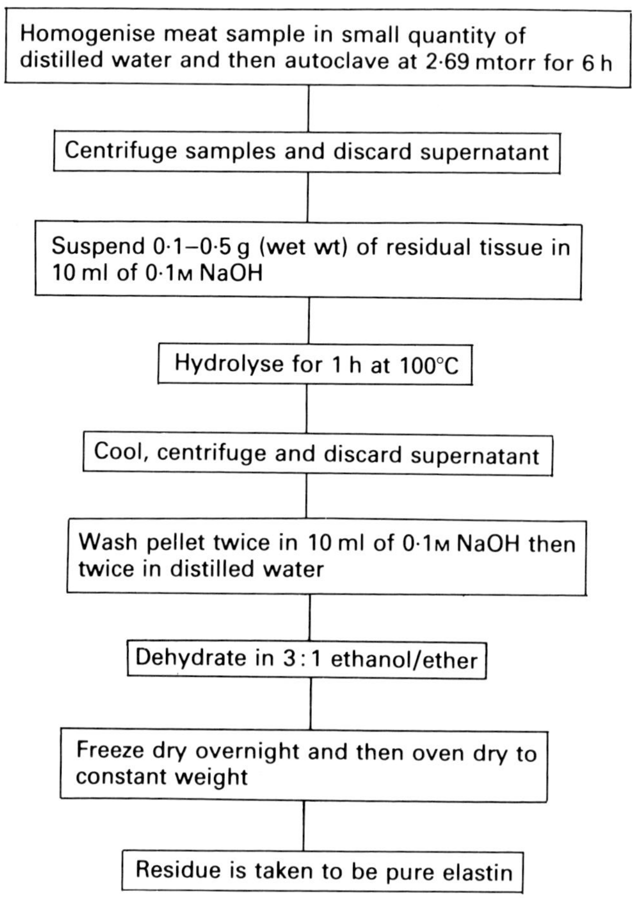

Scheme 12.8 Flow diagram showing the procedure for elastin estimation using the alkaline hydrolysis method.

In the alkaline hydrolysis method (Scheme 12.8) the meat samples are homogenised in a minimal quantity of distilled water and then autoclaved at 2·69 mtorr for 6 h. The samples are centrifuged and the supernatants of soluble proteins and gelatinised collagen are discarded. Then 0·1–0·5 g (wet weight) of the residual tissue is suspended in 10 ml of 0·1M NaOH and hydrolysis is carried out for 1 h at 100°C. Samples are cooled and the residues are separated from supernatants (which are discarded) by centrifugation. The pellet is washed twice in 10 ml of 0·1M NaOH and twice in distilled water before being dehydrated in 3:1 ethanol/ether and transferred to a weighing bottle. The samples are freeze dried overnight and then oven dried to constant weight. The residue weight is taken to be the weight of elastin in the original sample.

The method has some variations to cut down the manipulation and time factors but is found to be generally very reliable. Amino acid analysis of the acid-hydrolysed elastin residue gives a good indication of the effectiveness of the method as elastin has such a unique amino acid profile (see Chapter 5). However, the method is lengthy and somewhat laborious. The cyanogen bromide method provides a quicker and simpler alternative (Scheme 12.9). The samples are homogenised in a volume of ice-cold 50mM $CaCl_2$. Insoluble tissue is collected by brief centrifugation and the supernatant, containing a good deal of the soluble fraction of meat, is discarded. The pellet is washed once more with two volumes of distilled water prior to suspension in 70% (v/v) formic acid and the addition of CNBr as described above. Digestion with CNBr is normally carried out at 20°C for 4–8 h. However, older tissues, i.e. meat samples from mature animals, might prove to be less easily digested and so a temperature of 40°C is recommended in these cases. The digestion should always be done in a fume cupboard due to the volatile and poisonous nature of CNBr.

After digestion the sample is diluted ten-fold with distilled water and rotary evaporated to the original volume to remove excess CNBr. Insoluble material is collected by brief centrifugation and the pellet, after a wash with water, is freeze and oven dried as described for the NaOH method above. Once again, the residue weight is taken to be the weight of elastin. The effectiveness of the method can be easily checked by acid hydrolysis of the residue and amino acid analysis to ensure the purity of the elastin pellet.

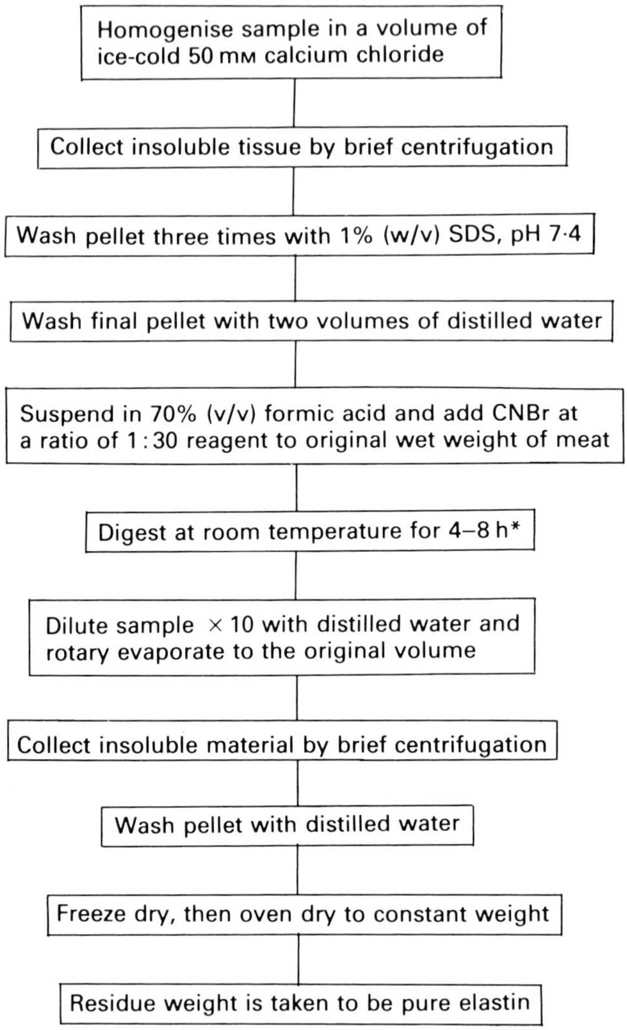

* It may be necessary with intractable tissues to carry out digestion at elevated temperature (40°C)

Scheme 12.9 Flow diagram showing the procedure for elastin estimation using the CNBr digestion method.

ANALYSIS OF REDUCIBLE COLLAGEN CROSS-LINKS

In the past, the mode of investigation of the reducible cross-links has been through the use of tritiated borohydride to reduce cross-linked compounds, thereby incorporating a radioactive label into them and making it possible to identify them. It should be remembered that cross-links exist as the unreduced form in collagen, although we shall confine ourselves to a discussion of their artificially reduced products. The method of borohydride reduction of isolated chains, soluble collagen, reprecipitated fibres, or suspensions of tissue follow a similar general procedure (Scheme 12.10).

The sample is dissolved or suspended in 0·9% (w/v) NaCl to a final concentration of 2–20 mg ml^{-1}. The solution is usually buffered at neutral pH with 0·02M phosphate although this leads to the base-catalysed production of an artifact, as we shall see later. With meat samples, where collagen concentration is very low, it is advisable to pre-wash the connective tissue after homogenisation as described above. The use of sodium dodecyl sulphate (SDS) for this pre-wash is to be recommended provided excess detergent is removed from the connective tissue pellet before the sample is chemically reduced.

Tritiated potassium borohydride, diluted with the non-radioactive reagent to give a final specific radioactivity of approximately 10 mCi $mmol^{-1}$, is added directly to the sample with stirring to give a substrate-to-reagent ratio of 30:1 (w/w) based on collagen dry weight. The reaction is allowed to proceed for 1 h at room temperature, after which time it is terminated by the addition of acetic acid to pH 3·0. Exhaustive dialysis against water or centrifugation and washing with excess water removes unreacted borohydride and reaction by-products and the sample may then be freeze dried or used immediately for analysis.

Reduced components are analysed after hydrolysis of the sample. The hydrolysis may be carried out under acid (redistilled 6M HCl) or alkaline (2M NaOH in polypropylene bottles) conditions at 110°C for 24 or 48 h respectively. The hydrolysate is dried *in vacuo* and, after desalting by ion exchange chromatography if necessary, may be analysed using a variety of methods. During preparative procedures the hydrolysed sample should be subjected to a preliminary chromatographic step. For example, when up to 10 g of starting material is to be used the hydrolysate should be passed through a column (90 × 5 cm) of Sephadex G-10 or BioGel P-2 equilibrated in 0·1M acetic acid.

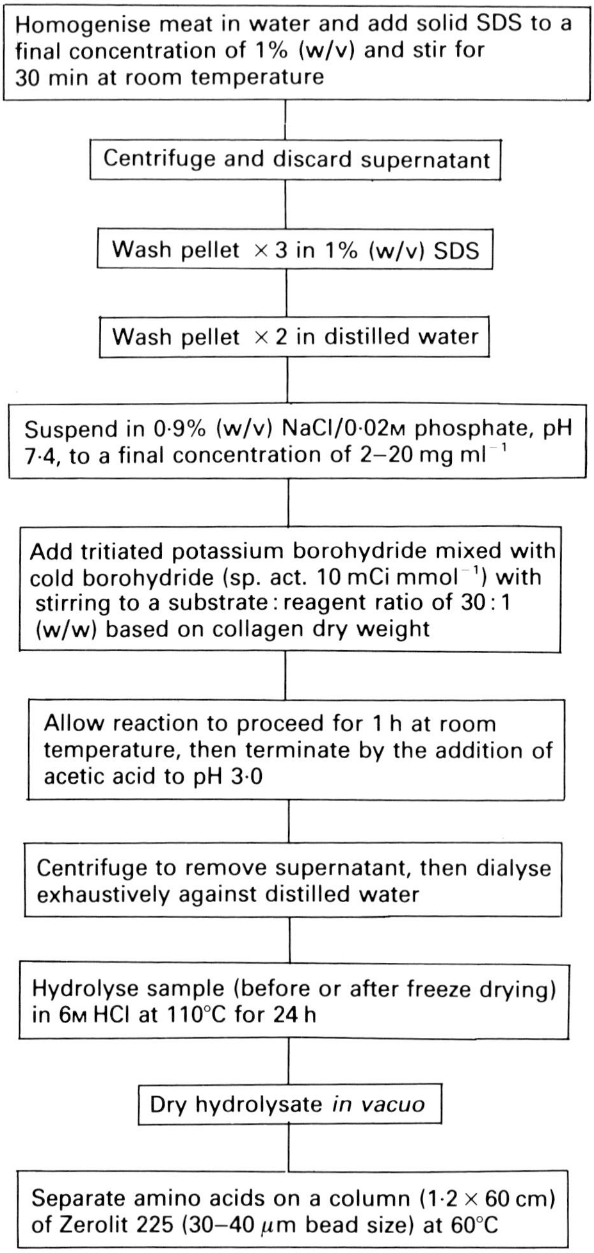

Scheme 12.10 Flow diagram showing the procedure for production of radioactively labelled reducible collagen cross-links using the classical ion exchange chromatography method.

The tritiated reduced cross-linked compounds elute before the bulk of the other amino acids owing to their larger size. If more than 10 g of the material is to be hydrolysed a displacement chromatography method may be used (up to 500 g). However, this procedure is lengthy and complex, and for 50–100 g of material several preliminary gel filtration steps might be regarded as easier and less time-consuming.

All the reduced hydrolysis products may be separated and identified using a single-column ion exchange system. The method employs a modified Technicon autoanalyser with a column (1·2 × 60 cm) of Zerolit 225 (30–40 μm bead size) maintained at 60°C by a water jacket. The resin is washed with 2M HCl and converted to the pyridinium form by stirring in 2M pyridine for 24 h prior to equilibration in 0·1M pyridine-formic acid buffer, pH 2·9. The column is eluted with a gradient formed by running 1M pyridine-formic acid buffer, pH 5·0, into a mixing vessel containing 350 ml of the starting buffer (Fig. 12.8). The flow rate is 60 ml h^{-1} and fractions are collected every 5 min. At the end of a run the column is regenerated by elution with 2M pyridine and then starting buffer.

If ninhydrin detection of eluted amino acids is required a stream-splitting device may be used at the base of the column to divert one-fifth of the eluate to the detection system. The pyridine should be routinely redistilled from ninhydrin (2 g $litre^{-1}$) to avoid interference in the detection process from non-specific amine contaminants. The expense of a continuous automated ninhydrin system may be obviated by locating ninhydrin-positive peaks as follows. Aliquots from fractions are spotted on to Whatman No. 1 paper, and peaks are located by spraying with ninhydrin (0·25% (w/v) in acetone). Initial standardisation of a column must be made, however, by analysis of fractions by classical chromatographic methods and the use of pure preparations of individual amino acids.

Fractions are analysed for radioactivity by counting a sample (0·2 ml) in 3 ml of Insta-Gel or an equivalent scintillation fluid. On an analytical scale this method is found to be adequate to separate all the major components found in reduced collagen, although the peak obtained should be run through a second system for verification of identity. A typical chromatograph is shown in Fig. 12.9. On a larger scale, components isolated from the Zerolit 225 column may be further purified by elution in 0·35M sodium citrate buffer, pH 5·25, from a 60 cm modified amino acid analyser column maintained at 56°C. Isolated components are then desalted by adsorption on a

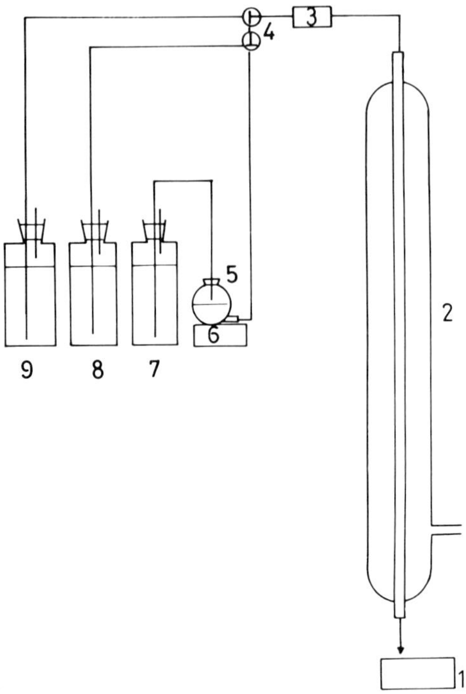

Fig. 12.8 Diagrammatic representation of the apparatus used for the ion exchange chromatographic separation of radiolabelled collagen reducible cross-links. (1) Fraction collector, (2) jacketted chromatography column at 60°C, (3) pump (4) taps, (5) mixing flask, (6) magnetic stirrer, (7) limiting buffer, (8) starting buffer, (9) 2M pyridine for washing column.

column (5 × 0·5 cm) of Amberlite GC-120 washed with water and displacement with 2M NH_3.

Minor radioactive compounds are identified as the two reduced cross-link precursors 5,6-dihydroxynorleucine and 6-hydroxy-norleucine, the acid hydrolysis by-product of the latter, 6-chloro-norleucine, the reduced intramolecular cross-link (reduced aldol), and lysinonorleucine. Early eluting material (Fraction A) was characterised and shown to contain the reduced condensation products of glucose or mannose with lysine or hydroxylysine—hexosyllysine and hexosylhydroxylysine, respectively. Fraction 11 has been named histidinohydroxymerodesmosine and fully characterised but has been shown to be an artifact of the reduction reaction. The compound is

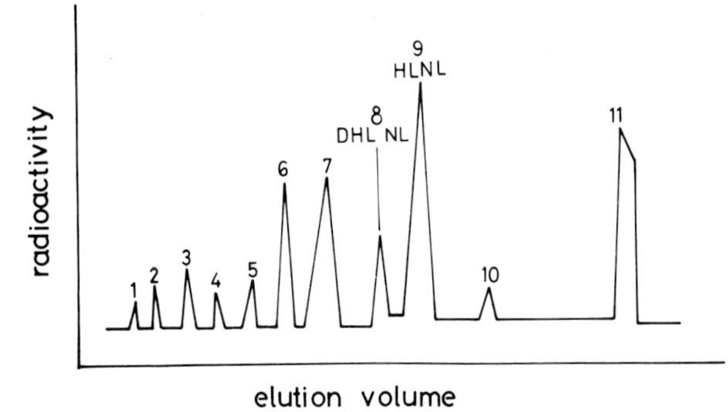

Fig. 12.9 Diagrammatic representation of a typical chromatograph obtained using ion exchange chromatography to separate and quantify radiolabelled collagen reducible cross-links. (1) 5,6-dihydroxynorleucine, (2) 6-hydroxynorleucine, (3) reduced aldol, (4) 6-chloronorleucine, (5) hexosylhydroxylysine, (6) hexosyl-lysine, (7) anhydro-derivative of hexosyl-lysine, (8) hydroxylysinohydroxynorleucine, (9) hydroxylysinonorleucine, (10) lysinonorleucine, (11) histidinohydroxymerodesmosine.

thought to be formed during the reaction with borohydride by a base-catalysed Michael addition of histidine across the double bond of an intramolecular aldol.

The two reduced cross-link compounds correspond to the material chromatographing under peaks 8 and 9. The first of these peaks contains dihydroxylysinonorleucine (8), and the second contains hydroxylysinonorleucine (9). Confirmation of the identity of these components must be made by using a different analytical system, for example ion exchange chromatography in 0·35M sodium citrate buffer at pH 5·25 on the amino acid analyser. The disaccharide derivative of dihydroxylysinonorleucine co-chromatographs with the reduced aldol on both pyridine-formate and citrate systems. These compounds may be resolved by gel filtration on BioGel P-2 (140 × 1·6 cm) in 0·1M acetic acid.

Periodate oxidation

Further confirmation of the nature of the isolated reduced cross-link compounds can be obtained by carrying out a Smith degradation with periodate. One volume of sodium periodate (5 mg ml^{-1} in 0·1M sodium citrate buffer, pH 5·5) is added to 3 volumes of the suspected reduced cross-link in the same buffer (1·5 mg ml^{-1}). After 5 min the pH is raised to 7·5 with 5M NaOH and the products are reduced by the addition of excess KBH_4. After 30 min the pH is adjusted to 3·0 with 6M HCl and the solution can then be applied to an amino acid analyser. Dihydroxylysinonorleucine yields predominantly proline and some norvaline, whereas hydroxylysinonorleucine yields lysine and proline.

The relatively low resolution of the pyridine-formate system and the presence of co-chromatographing hexosyllysine derivatives on the citrate system makes accurate one-step quantitation of cross-links in tissue hydrolysates difficult. Quantitation based on the incorporation of radioactivity cannot be used owing to the difficulty in assessing the exact extent of tritiation per cross-link in each reduced sample.

Two new methods have recently been developed for use with untritiated samples using high performance liquid chromatography (HPLC). The first method uses aminopropyl bonded silica for normal phase high performance liquid chromatography (NPHPLC). In this method, in which the hydrolysate is loaded in high organic phase concentrations and eluted with an increasing polar phase concentration, the reduced cross-links elute late from the column and can be quantitated by post-column detection with ninhydrin or a more sensitive amine reactive reagent such as *O*-phthalaldehyde or fluorescamine. In the second method, standard analysis using the dansyl-derivatives of amino acids is modified to allow separation of the DNS-derivatives of reduced cross-link components. This is the preferred technique as the cross-links can be quantitated directly from their fluorescence as they elute from the column. Both methods have the advantage that they allow the quantitation of reducible cross-links of collagen in very small samples.

For analysis by NPHPLC the sample is reduced and hydrolysed as described above and re-suspended in water and neutralised with dilute NaOH and loaded on a 25 × 0·46 cm column of aminopropyl bonded silica (5–10 μm) at a flow rate of 2 ml min^{-1} and at room temperature. Detection of amino acids is at 205 nm so all solvents must be high spectroscopy grade. Amino acids are eluted with a gradient formed from Solvent A (10mM potassium phosphate buffer, pH 4·3) and Solvent B (CH_3CN/water in a ratio of 50:7) as follows: 80% Solvent B for 15 min, 80–60% Solvent B for 40 min, 60% Solvent B for 10 min, 60–50% Solvent B for 6 min, 50–80% Solvent B for 9 min (Fig. 12.10).

Amino acids can be more sensitively detected by post-column derivatisation with *o*-phthalaldehyde or Fluoram and tritium-reduced cross-links can also be detected by scintillation counting of fractions

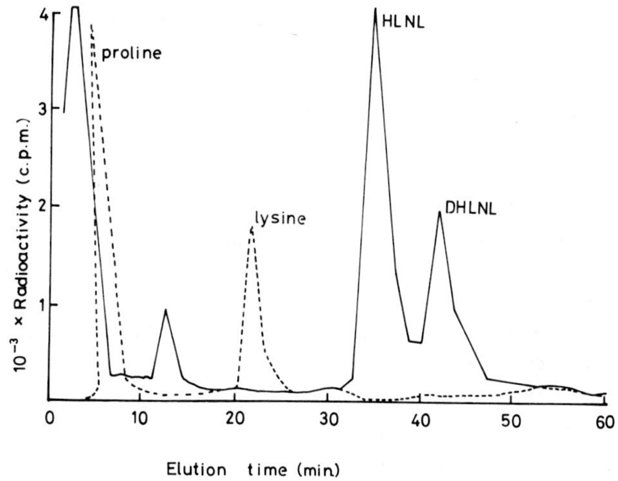

Fig. 12.10 Typical chromatograph obtained using normal phase high performance liquid chromatography to separate and quantify the reduced radiolabelled aldimine (HLNL) and oxo-imine (DHLNL) cross-links.

collected during HPLC. The method is rapid and relatively sensitive (reduced oxo-imine and aldimine compounds elute in about 55 min), but the aminopropyl columns are not stable and resolving power is lost over 20–50 runs.

For analysis by RPHPLC, the acid hydrolysed sample is labelled with dansyl chloride as follows. A sample of the hydrolysate equivalent to approximately 15 μg is dissolved in 0·5M lithium borate buffer, pH 9·5, and a solution of dansyl chloride (6 mg ml^{-1} in CH_3CN) is added so that the final molar ratio of dansyl chloride to hydrolysate is about 16:1 and the ratio of aqueous to organic solvent is 3:1. The sample is dansylated at 60°C for 40 min or at 35°C overnight in the dark. After dansylation the pH is adjusted to 4–7 with orthophosphoric acid and the volume made up to 500 μl with HPLC grade water.

The filtered or centrifuged sample is then loaded on a 25 × 0·46 cm column of high grade 5 μm C_{18} RPHPLC material equilibrated in 10% Solvent B (see below). Chromatography is carried out at 20°C and a flow rate of 1 ml min^{-1}. The elution gradient is formed with 10mM lithium phosphate buffer, pH 6·3 (Solvent A) and far-UV grade CH_3CN (Solvent B). The gradient is 10–35% Solvent B for 40 min, 35% Solvent B for 15 min, 35–80%, Solvent B for 3 min, 80% Solvent

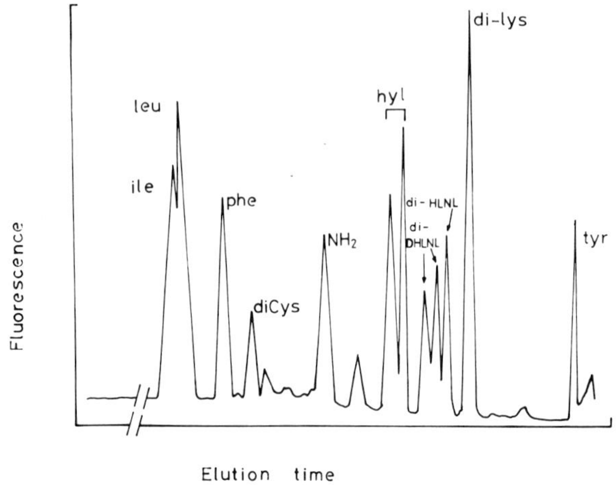

Fig. 12.11 Typical chromatograph obtained using reversed phase high performance liquid chromatography to separate and quantify the reduced aldimine (di-HLNL—di-dansyl derivative) and oxo-imine (di-DHLNL—two isomers of the di-dansylated component) cross-links.

B for 2 min, 80–10% Solvent B for 5 min. Dansylated cross-links are detected with UV at 254 nm and fluorescence analysis of the eluate as described above. A typical elution profile is shown in Fig. 12.11.

ANALYSIS OF MATURE COLLAGEN CROSS-LINKS

There are few reports in the literature of the purification and assay of putative mature cross-link compounds and only two or three compounds have been isolated from acid hydrolysates of mature insoluble collagen as discussed in Chapter 4. Of these only one, pyridinoline, can be relatively easily purified. This compound may also now be assayed using a rapid HPLC technique.

The compound has a molecular weight in excess of 400 and so can be separated relatively easily from the bulk of amino acids (molecular weight range 90–180) by adopting the large-scale gel filtration method described in the section in this chapter on isolation of reduced cross-link components. Fluorescence at 400–410 nm with excitation at 295 nm or simple UV detection at 295 nm may be used to monitor for the compound. The fluorescent peak in the molecular weight range

300–600 is pooled and may be further purified by ion exchange chromatography on a 30 × 1 cm amino acid analyser column run under standard conditions in citrate buffers (pH 2·9–4·65) with the last buffer extended for 40 min. The compound is pooled and de-salted on a short column of Sephadex G-10 in 0·1M acetic acid and may be stored dry at −20°C. Other techniques for the purification of the compound include the original method, but these can be lengthy and not always as successful as that described above. The compound isolated by either method shows a characteristic peak on the amino acid analyser and an acid–base shift in UV absorption from 295 nm in acid (0·1M HCl) to 325 nm in alkali.

A recently developed HPLC method allows the separation and assay of pyridinoline from whole connective tissues without prior purification. This method relies on the use of fluorescence detection of the compound after reverse phase HPLC. The acid hydrolysate of a collagen or connective tissue sample is dried and dissolved in a solvent containing 1% (v/v) heptafluorobutyric acid (HFBA). The sample (normally 20–50 μl) is loaded onto a 25 cm column of octadecyl derivatised (C_{18}) silica (3–5 μm) at room temperature in the same solvent and the pyridinoline and other amino acids are eluted with a gradient of 21–25% Solvent B (see below) over 20 min at a flow rate of 1 ml min^{-1} (Solvent A is 0·01M HFBA containing 5% (v/v) CH_3CN; Solvent B is CH_3CN containing 0·01M HFBA). The pyridinoline elutes as a sharp peak late in the chromatogram (about 12 min) and may be accurately quantified from its fluorescence intensity through comparison with a known standard.

An immunoassay procedure that is extremely sensitive (pmol level) has also been developed for the quantitation of pyridinoline. The method uses antibodies raised in donor animals to haptenised pyridinoline (conjugated to albumin) in an inhibition assay, as mentioned below for desmosines.

ELASTIN CROSS-LINKS—THE DESMOSINES

Desmosine and isodesmosine are the main covalent cross-links of elastin (see Chapter 5) and are present in sufficient quantities to be detected directly on amino acid analysis of acid hydrolysates. The isomers elute in the region of tyrosine. Insoluble elastin fibres also contain a proportion of dehydro-lysinonorleucine and allysine aldol, the molecular precursors of desmosine. The latter two compounds are readily assayed after routine tritiated borohydride reduction, as described earlier for collagen cross-linking compounds. All these cross-link components are large and may be concentrated in solution prior to amino acid analysis from hydrolysates by preliminary gel filtration (on a suitable support such as Sephadex G-10 or Fractogel 100) and collection of the high molecular weight components.

More recently two new methods, which are extremely accurate and relatively rapid, have emerged for the quantification of the desmosines. These are immunoassay methods and high performance liquid chromatography techniques. Enzyme-linked immunosorbent assays (ELISA) and radioimmunoassays (RIA) have both been described for desmosine. In both instances antibodies to desmosine (conjugated to albumin as a carrier) are raised in donor animals such as rabbits as described earlier in this chapter. No pre-purification of anti-desmosine antibodies is needed as, in both cases, an inhibition assay procedure is used. The assays are sensitive down to the 1–2 pmol level and are useful as multiple samples can be assayed simultaneously.

The HPLC methods evolved for desmosine assay utilise standard ion exchange HPLC or separation of pre-derivatised amino acids by reverse phase HPLC. The latter method involves the use of dansyl chloride to label the amino acids in an elastin or tissue hydrolysate prior to separation on a high performance C_{18} column ($25 \times 0{\cdot}46$ cm, 5 μm particle size). The dansylated sample (see earlier discussion) is loaded onto the column which is equilibrated at room temperature in Solvent A (25mM NaH_2PO_4 and 25mM acetic acid mixed with CH_3CN in the ratio 86:14). Chromatography is carried out at a flow rate of 1 ml min^{-1}. A gradient for the elution of dansylated amino acids is produced by the addition of Solvent B (pure CH_3CN) as follows: 0% Solvent B for 5 min, 0–8% Solvent B for 5 min, 8–8% Solvent B for 15 min, 8–45% Solvent B for 30 min, 45–58% Solvent B for 5 min, 58–0% Solvent B for 5 min, 0–0% Solvent B for 20 min. Dansylated amino acids including desmosine are detected with UV at 254 nm and fluorescence analysis of the eluate is carried out as described above.

SUMMARY

Methods are described for the separation and purification of the three main connective tissue domains of meat. Techniques for the purification of procollagens, the various main genetic collagen types, insoluble collagen and collagen α-chains are also described. Routine laboratory methods for the analysis of collagen by amino acid analysis, carbohydrate analysis and separation of whole chains by SDS-polyacrylamide gel electrophoresis are examined. The controlled degradation of collagens to their discrete CNBr fragments is discussed and the subsequent peptide mapping by SDS-polyacrylamide gel electrophoresis, HPLC or CMC-ion exchange chromatography is also described. Methods for the estimation of Types I, III, IV and V collagen are described and new immunoassay methods for the estimation of different collagens in mixtures are mentioned briefly.

The quantification of meat collagen by the classical colorimetric method is described in full, as are the slower but more reliable methods of hydroxyproline estimation by amino acid analysis and HPLC. Elastin quantification using the CNBr and NaOH methods for extracting this intractable protein are discussed. The analysis of collagen reducible cross-links by classical ion exchange methods after their reduction with radioactive borohydride are described. A new method for the faster and more reliable measurement of these bonds by reverse phase HPLC is also described. The estimation of elastin cross-links (desmosines) by ion exchange amino acid analysis and by the new methods of immunoassay are discussed, as is the application of HPLC and immunoassay techniques to the quantification of pyridinoline.

FURTHER READING

1. Bailey, A. J. & Light, N. D., Analysis of collagen cross-links. In *Methods in Enzymology,* Vol. 82A, (ed. L. W. Cunningham & D. W. Frederiksen). Academic Press, New York and London, 1980.
2. Negro, A., Garbisa, S., Gotte, L. & Spina, M., The use of reverse phase liquid chromatography and pre-column derivatization with dansyl chloride for quantitation of specific amino acids in collagen and elastin. *Anal. Biochem.,* **160,** 1987, 39–46.
3. Gunja-Smith, Z., An enzyme linked immunosorbent assay to quantitate the elastin cross-link desmosine in tissue and urine samples. *Anal. Biochem.,* **147,** 1985, 258–264.
4. Robins, S. P., A radio-immunoassay for the collagen cross-link pyridinoline. *Biochem. J.,* **207**, 1982, 617–620.

Chapter 13

Physical and Sensory Methods of Analysis

INTRODUCTION

Early work on the structure of muscle connective tissue using light microscopy was supplemented by elegant scanning and transmission electron miscoscopy. This work has been further extended by research on the fundamental structure of collagen molecules by specialised electron microscopy methods like rotary shadowing. All these studies have been carried out in parallel with work on the mechanical properties of meat which has produced such diverse approaches as classical stress–strain tests, fracture mechanics and shear tests. Mechanical tests are often carried out in concert with more subjective analyses such as sensory taste panels. Recent developments of these techniques, including the refinement of the quantitative descriptive analysis method and the emergence of new or more powerful computer-based statistical analysis packages for the treatment and interpretation of sensory panel data, has allowed a much more sophisticated use of these methods.

This chapter will discuss these physical methods available for the analysis of meat connective tissue including microscopy, mechanical testing and sensory panelling.

LIGHT MICROSCOPY

The varied and simple techniques of light microscopy have proved invaluable in learning about the gross structure of meat and meat products and in elucidating some of the changes that occur in meat during processing. There are many well described methods of sample preparation and staining and it would be impossible to deal with them all in detail here. However, the major stains used for collagen, elastin and proteoglycans will be discussed.

Classically, samples of fresh or conditioned muscle are obtained and blocks 1 cm thick are fixed for a minimum of 72 h (and sometimes as long as 1 week) in 10% (v/v) formalin buffered with 0·05M sodium phosphate buffer, pH 6·5. Alternatively, samples may be fixed in 1·15–40% (v/v) glutaraldehyde buffered at pH 7·4 with either Sorenson's phosphate or 0·1M cacodylate buffer for 3 h. After fixation samples are briefly washed in water to remove excess superficial formalin (Scheme 13.1).

Separated or microdissected connective tissues from the fixed samples can be viewed on glass slides. This technique has allowed the visualisation of perimysium and perimysial–endomyial junctions. In this case, the fixed muscle is restrained and individual muscle fibres or muscle fibre bundles are teased from the cut ends of the block. Residual endomysial tubes attached to fine sheets of perimysium can then be carefully cut from the end of the block, washed in water and dehydrated in graded alcohol before embedding in Araldite or another suitable resin.

Phase contrast microscopy has been used to study isolated collagenous meat connective tissue without further treatment of fixed and Araldite embedded samples. Thick sections are cut on a standard ultramicrotome and mounted on glass slides in a suitable mountant (e.g. Permount) prior to viewing by phase contrast microscopy.

The method of choice in several recent studies of meat connective tissue has been the use of fast (cryo-) frozen blocks of untreated material. This method avoids the use of any fixatives or waxes/resins and is thought, therefore, to maintain a high degree of structural integrity in the tissue. The method has drawbacks but it is the method of choice for immunolocalisation studies. Wax embedded, Araldite embedded or free mounted connective tissue or meat samples can be stained for various connective tissue components using a variety of stains.

Collagen

Collagen can be differentiated using the classical Van Gieson stain and the trichrome methods of Masson or Mallory. The Van Gieson stain uses acid fuchsin in aqueous solution combined with picric acid which acts

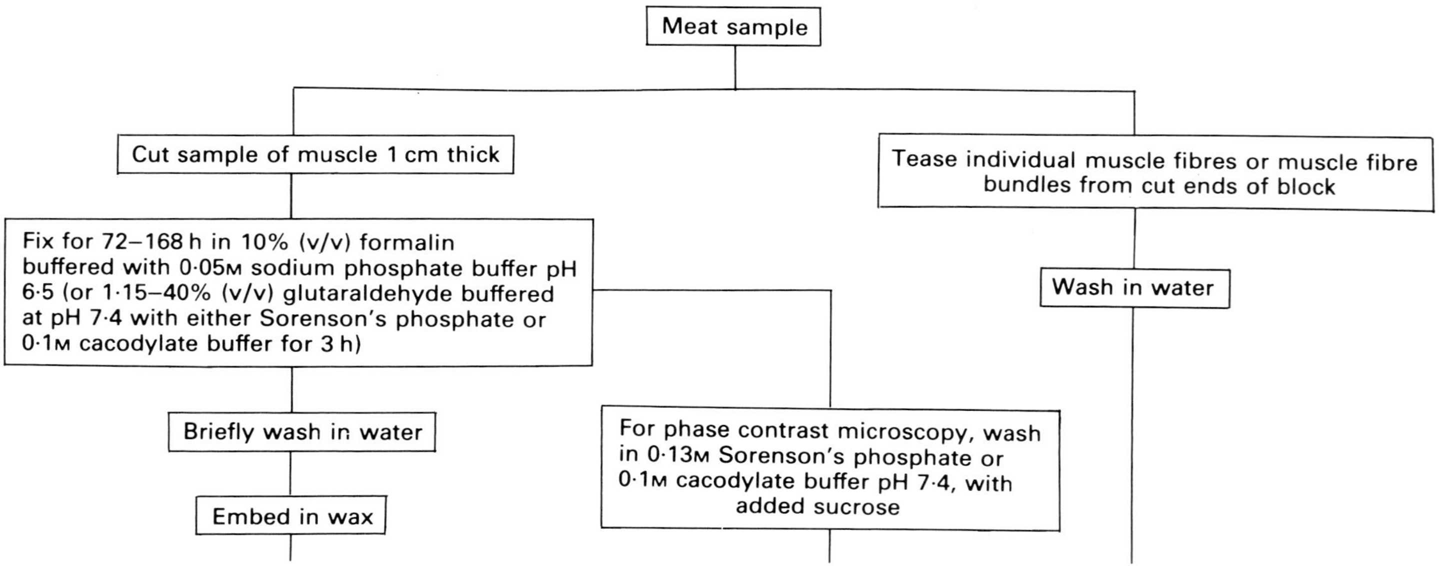
Meat sample
Cut sample of muscle 1 cm thick
Tease individual muscle fibres or muscle fibre bundles from cut ends of block
Fix for 72–168 h in 10% (v/v) formalin buffered with 0·05M sodium phosphate buffer pH 6·5 (or 1·15–40% (v/v) glutaraldehyde buffered at pH 7·4 with either Sorenson's phosphate or 0·1M cacodylate buffer for 3 h)
Wash in water
Briefly wash in water
For phase contrast microscopy, wash in 0·13M Sorenson's phosphate or 0·1M cacodylate buffer pH 7·4, with added sucrose
Embed in wax

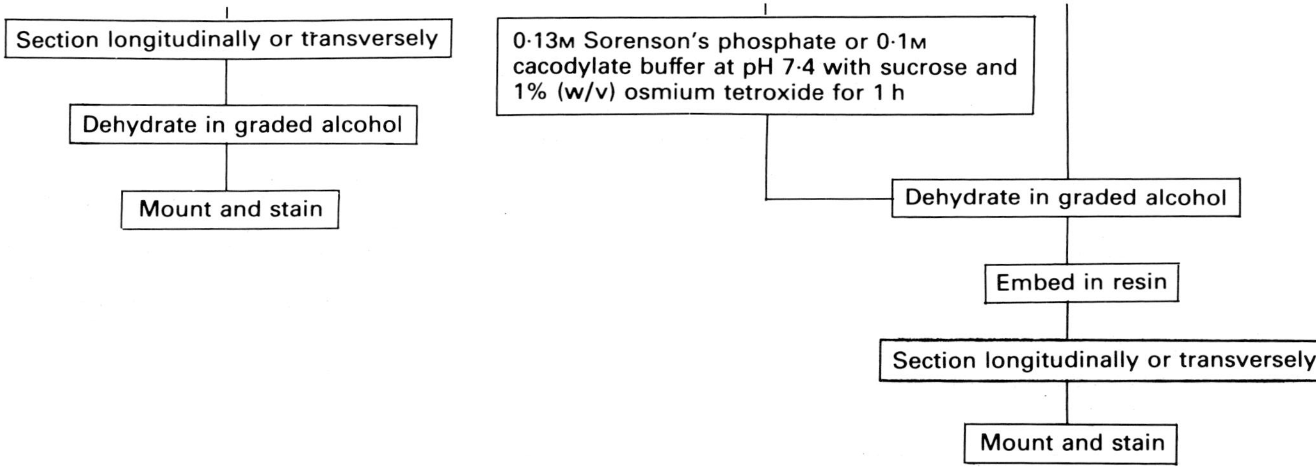

Scheme 13.1 Flow diagram showing procedures for the preparation of samples of meat or meat connective tissue for classical light microscopy.

as a cytological stain. Cell nuclei are stained with iron–haemotoxylin complex known as Weigert's Haemotoxylin. The trichrome stains use aniline blue or light green to demonstrate collagen fibres. An improved version of the Van Gieson technique for the differentiation of collagen uses Sirius Red F3BA in place of acid fuchsin and gives better staining of both coarse and fine collagen fibres. It also stains reticulin and basement membrane and has less tendency to fade than other stains. This method, which is more sensitive than traditional stains, is certainly to be recommended for higher power work on meat connective tissues. The Periodic Acid Foot method has also been used for collagen staining, although this is less commonly employed as a routine stain.

Reticulin

Several of the silver staining methods are useful for the visualisation of the so-called reticulin fibres associated with the basement membrane of muscle (as mentioned in earlier chapters, these fibres are probably Type III collagen). A good standard method is that described by Gomori which uses an ammoniacal silver stain. A cleaner presentation of the stained elements is obtained if the silver-stained preparation is toned with gold chloride.

Elastin

Weigert's Haemotoxylin and Verhoeff's Van Gieson (VVG) have been used for the demonstration of elastin. Another method, described by Margolena and Dolnick, uses orcein, aniline blue and Orange G in acid solution to differentiate between collagen and elastin fibres. The Verhoeff and Weigert methods for elastin staining readily show the presence of large elastic fibres but some difficulties have been encountered with these stains in differentiating fine, young elastic fibres. A combination of Victoria Blue 4R and ethyl violet has been shown to overcome these problems and give good definition of fine elastic fibres; this method is therefore recommended for more sensitive work.

Glycosaminoglycans

Glycosaminoglycans (GAG) can be differentiated using the periodic acid-Schiff (PAS) method or with Alcian Blue. One method combines Alcian Blue with PAS and immunoperoxidase for the simultaneous demonstration of keratan and neutral and acid mucosubstances. Another method demonstrates the distribution of sulphonated GAG

Table 13.1
Microscopical stains and their uses

Stain	*Application*	*Colour*
Van Gieson	Differentation of collagen (fibrous)	Purple
Trichrome (Masson or Mallory)	Differentiation of collagen (fibrous)	Blue/green
Sirius Red F3BA	Differentiation of collagen	Red
Periodic Acid Foot	Differentiation of collagen (fibrous)	Red
Gomori silver stain	Reticulin fibres	Black
Weigert's Haemotoxylin	Large elastin fibres	Black
Verhoeff's Van Gieson	Large elastin fibres	Black
Orcein, aniline blue, Orange G	Differentiation of collagen and elastin	—
Victoria Blue 4R/ethyl violet	Young elastin fibres	Violet
Alcian Blue	Glycosaminoglycans	Blue
Periodic Acid Schiff	Glycosaminoglycans	Pink
^{125}I cationic ferritin	Sulphonated glycosaminoglycans	—

in basement membrane associated material of the kidney using ^{125}I-labelled cationic ferritin as a probe. This enabled light and electron microscopic autoradiography to be performed. A summary of stains and their uses with connective tissue components is shown in Table 13.1.

Immunochemical staining

The use of purified antibody serum (see Chapter 12) to localise the collagen type of interest in muscle is done by using it as a stain. The muscle sample is sectioned to a thickness of 5–7 μm, after snap freezing it in liquid nitrogen at −20°C in a cryotome. This method, rather than the classical technique of tissue sectioning of wax or resin embedded samples, ensures that all the antigenic sites will remain in their native state and therefore will still be reactive with the antibody preparation. The sectioned samples are taken up onto microscope slides and a drop of the diluted A_s fraction is placed over the sample. After incubation of the A_s fraction with the section, it is washed free of unbound antibody. Antibodies which have reacted with tissue collagen are then located by using a second antibody which is itself labelled with a fluorescent or coloured compound. In this case, this second antibody is specific for the IgG from the species used to raise the A_s fraction. In other words, if the original immune serum was obtained from immunised rabbits then the second antibody must be an anti-rabbit IgG antibody labelled with a chromophore or fluorophore.

The most popular label is fluorescein, a brilliant green fluorescent substance. Fluorescein-labelled antibodies to rabbit, goat, sheep and other IgGs are now available commercially. Other labelling systems include the red fluorophore rhodamine and several enzyme-linked systems such as horseradish peroxidase which, when coupled to the second antibody, can be used *in situ* to deposit coloured reaction products at the site of antibody binding within the tissues.

The results obtained with this highly specific and powerful tool are seen in the light microscope with the use of fluorescence optics. The collagens stained with specific antibody fractions and fluorophore-labelled IgG (the sandwich technique) show up as bright or diffuse green areas, depending on their concentration, distribution and, most importantly, the avidity of the particular antibody preparation. Because each antibody preparation will have different avidities for the tissue collagens it is directed against, and because these avidities cannot be measured in the actual staining procedure, it is impossible to quantitate the collagen present by image processing methods. The method is therefore purely qualitative. In the future, the use of monoclonal antibodies for such staining methods, which are specific for only one antigenic site on any one protein, may allow such accurate quantitation, or at least semi-quantitation, in tissue sections.

ELECTRON MICROSCOPY OF MEAT CONNECTIVE TISSUE

The two main methods of electron microscopy relevant to meat tissue studies are transmission electron microscopy (TEM) and scanning electron microscopy (SEM).

Transmission electron microscopy

In general, for TEM, samples of meat can be prepared by fixation, washing and Araldite embedding as described above for light microscopy. There are several methods available for fixation and resin embedding, and any may be chosen for use with muscle tissues. High resolution microscopy usually requires better fixation than that obtained using the simple formalin methods cited above.

Common fixatives for TEM are glutaraldehyde (2% (v/v)) and Karnovsky's fluid (containing 3% (v/v) glutaraldehyde and 1·5% formaldehyde). The fixative solutions also contain sucrose or other modifiers of osmolality, e.g. 6·9% (w/v) sucrose with 2% (v/v) glutaraldehyde to give an osmolality of 480 milliosmols. Fixatives should have a pH in the range 7·2–7·5 for meat and buffers such as phosphate and cacodylate are commonly used. It is recommended that fixation should be carried out initially for a short period of time at 4°C to prevent autolytic activity in muscle samples, but then at higher temperatures (20–30°C) to accelerate the process and preserve fine structural detail which can be lost by slow fixation. Post-fixation is normally carried out in osmic acid followed by dehydration as described above.

Fixed, dehydrated and embedded meat samples are sectioned on an ultramicrotome to thicknesses of approximately 75 μm, mounted on either carbon-coated or uncoated grids, and usually stained with 2% (w/v) uranyl acetate in 50% (v/v) methanol for 15 min. Sections are then post-stained with lead citrate for 2 min prior to electron microscopy. The technique allows a detailed analysis of the axial banding along the fibre.

Negative staining

This technique is applicable to suspensions of homogenised tissue, reprecipitated fibres and segment long spacing (SLS) crystallites (side-to-side aggregates of collagen molecules induced by addition of ATP). The suspension of fibres is mixed with an equal volume of silico- or phospho-tungstic acid (2% (w/v)) at pH 8·4 and applied to electron microscope grids covered in a carbon-coated collodion film. After leaving to settle, excess liquid is removed and the grids allowed to dry. Electron microscopy of the fibres reveals the 67 nm axial periodicity whilst analysis of SLS crystallites allows accurate determination of the molecular length, and that of reconstituted CNBr peptides.

Immuno-gold labelling

As an extension of the antibody fluorescence labelling of tissue components in the light microscope heavy metals can be attached to antibodies and used for location studies in the electron microscope. Ferritin and peroxidase have been used but the current method of choice appears to be antibodies labelled with gold. The technique can be improved by enhancing the gold with silver.

Sections are incubated with a monoclonal antibody to the specific collagen of interest for 2 h at room temperature or with non-immune serum as control. The sections are then washed with phosphate-buffered saline and 1% bovine serum albumin before incubation with Protein-A (from *Staphylococcus aureus*) conjugated with 12–14 nm gold particles for 2 h at room temperature. After extensive washing with the saline and albumin buffer, sections are fixed with 3% glutaraldehyde in cacodylate buffer and post-fixed with osmium. After dehydration the material is embedded in Spurr resin and then sections cut for viewing in the electron microscope.

Silver enhancement

The gold labelling can be enhanced by precipitation of silver on the gold particles. The method gives up to 200-fold sensitivity. Silver staining is performed by immersion in a developer comprising a solution of acacia powder, hydroquinone and silver lactate adjusted to pH 3·5 by addition of citrate buffer.

Rotary shadowing

The powerful technique of rotary shadowing, refined in the 1970s in work on muscle proteins, has recently been revived in collagen research to study the complex collagenous molecules obtained from basement membranes. The appeal and power of the method is that, given a pure, homogeneous protein preparation, individual molecules may be readily visualised and their size and some aspects of their structure assessed. Glycerol is added to a 10–30 g ml^{-1} solution of pure collagen (or any other connective tissue macromolecule of interest) in 0·056M acetic acid to a final concentration of 50% (v/v). The solution is then sprayed directly onto freshly cleaved mica discs from 20–30 cm distance. The discs are immediately shadowed with platinum (at an angle of 9°) and then carbon (at 90°) in an evacuated chamber (1 × 10 Torr) by using an electron bombardment source and evaporation conditions of 40 kV/50 mA (for 0·2 × 5 cm platinum wire) and 40 kV/50–100 mA (for carbon). Carbon replicas can be cut to size and floated on to 400 mesh copper grids on water (Scheme 13.2). Specimens are then examined in the electron microscope at magnifications from 30 000–100 000 (see Fig. 6.7 and Fig. 13.1).

Scanning electron microscopy

Several methods are available for routine SEM. Tissue samples can be larger than for TEM and blocks are normally 1–5 mm^3. Fixation need only be carried out at the surface of the sample as this is the part to be viewed during SEM. Fixatives as described above are used at isotonic

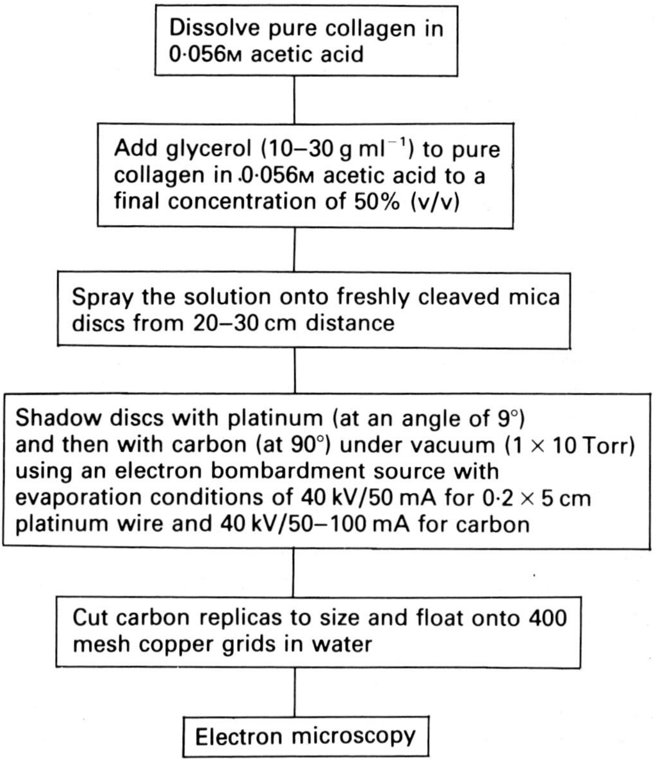

Scheme 13.2 Flow diagram showing procedures for the preparation of samples of meat connective tissue collagens for rotary shadowing electron microscopy.

osmolalities. Drying is preferable from organic solvents, freeze drying or by critical point drying (CPD). Freeze drying is a useful technique and introduces few artifacts as well as reducing shrinkage, but is not commonly used for muscle tissue.

For critical point drying (CPD) the block or sectioned sample on a glass slide is transferred from the last stage of dehydration with alcohols then amyl acetate to the CPD bomb with liquid CO_2. The bomb is then filled with liquid CO_2 and the temperature adjusted to about 36°C or until the pressure gauge reads 8–10 000 kPa. When the pressure and temperature are stabilised CPD is timed for about 10–20 min, at which time the pressure is released and the block or

Fig. 13.1 Electron micrograph of Type I collagen molecules prepared by the rotary shadowing technique.

slides removed from the bomb, coated with metal and stored in a desiccator prior to viewing in the SEM.

Further sophistications of the SEM technique may be used including freeze and dry fracture methods.

MECHANICAL METHODS—ISOMETRIC MEASUREMENTS, ULTIMATE TENSILE STRENGTH MEASUREMENTS, FRACTURE MECHANICS AND SHEAR TESTS

Over the last thirty years a considerable amount of research has been done in the field of the mechanical properties of meat. A variety of methods have been invented or modified from classical mechanics technology to allow testing of large and small samples of meat as well as partially purified meat connective tissues. Once again, it would be impossible to describe every method in detail and so only the most well known and one or two newer techniques will be covered.

In effect, three major approaches have been used: the investigation of the force developed in shrinking meat samples held at constant length during the heating process (isometric tension); the load needed, at a constant strain rate, to stretch or pull a piece of meat to breaking point (ultimate tensile strength measurements and, with modifications, fracture mechanics); and the load needed to shear a piece of meat under conditions of compressive force (shear or 'bite' tests).

Isometric tension tests

Isometric measurements are made on collagenous connective tissue samples in a specially built apparatus of the sort illustrated in Fig. 13.2. The sample is held between the fixed jaws, at its original length, in a bath of neutral buffer (normally 0·9% NaCl, 0·02M sodium phosphate buffer, pH 7·0) containing a stirring device to ensure rapid temperature equilibration in the bath. The buffer is heated at a rate of 1–3°C min^{-1}, through a linear cycle starting at 20°C and finishing, usually, at 100°C. The tension developed in the sample due to the shrinkage of the collagen in the tissue (whose length is held constant between the jaws) is measured and recorded on an X–Y plotter versus change in temperature.

Samples chosen for isometric measurements should be selected carefully. It is normal traditional practice to cite tension developed in terms of the cross-sectional area of the sample. This is not always

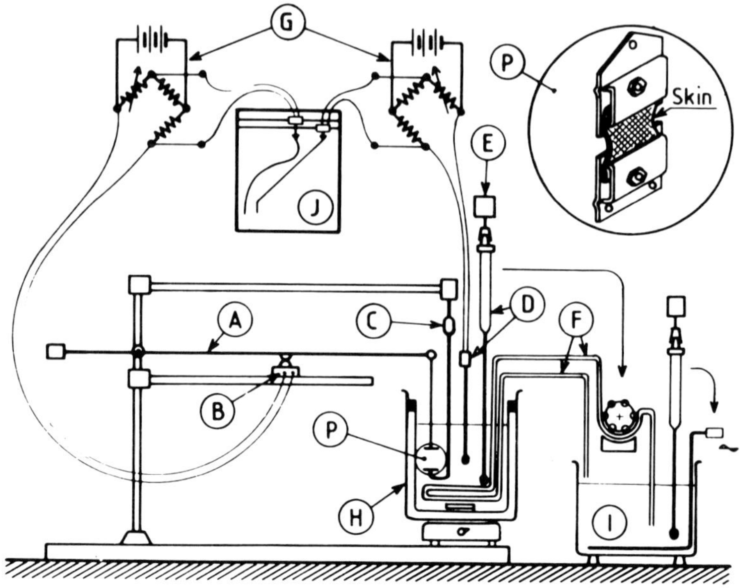

Fig. 13.2 Diagram showing the apparatus used for isometric tension studies of collagenous tissues.

possible in the case of meat connective tissues. Whilst it may be possible to take regular and relatively uniform samples of tendon for testing and the cross-sectional area of these may be measured reasonably easily, it is obvious that the less regular connective tissues of meat are not so amenable to such treatment. The most reproducible method for determining force developed per unit sample is on a basis of collagen content per unit dry weight of the sample determined by the methods described in Chapter 12. Thus, force can be expressed as units per milligram collagen in the sample.

Other versions of this technique have been reported. The total tension response can be measured by immersing the sample, pre-fixed in the jaws of the apparatus, in a bath of appropriate buffer (0·9% NaCl, 0·02M sodium phosphate buffer, pH 7·0) held at a constant temperature above T_s (70–90°C) and the isometric tension developed can be recorded against time over a period of 15–30 min.

It should be stressed that, like other mechanical methods, the interpretation of results from isometric tension measurements is dependent on some degree of mathematical manipulation. However, the method is a simple technique for qualitative comparisons of tissues.

Ultimate tensile strength (UTS) measurements

These methods rely, largely, on stretching a piece of meat between constantly widening jaws until it breaks and measuring the forces involved in the process. The method has been widely used in meat research and gives some indication of the variability in ultimate tensile strength (UTS) of meats from a variety of sources; for example, different muscles of varying eating quality from the same animal, the same muscle from animals of different ages, and samples from the same muscle cooked in different ways.

The problem with the technique, however, is that, like the shear tests discussed below, it measures a single parameter (UTS) which is based on a set of complex changes which occur in the meat on stretching. For example, the two major compartments of structure in the meat, the contractile apparatus within the cells and the connective tissue outside, may both be expected to contribute to the UTS in different ways. It is possible to produce data correlating differences in UTS with differences in connective tissue or collagen content but it is difficult to attribute the mechanical differences directly to the connective tissue differences without employing other methods. For this reason the UTS methods may be seen as useful ways in which overall strength and, therefore, toughness of a meat sample may be measured (and then, perhaps, correlated with connective tissue parameters), but not as primary methods for the investigation of the causative role of connective tissue in determining texture.

In brief, the methods utilise a standard mechanical testing rig, such as an Instron Universal Tensiometer (Fig. 13.3), in which the sample of meat, cut to a standardised size and pre-treated if necessary (e.g. cooked), is gripped between two sets of jaws. Glues and soft materials (e.g. rubber) set between the metal jaw and the sample may be used to enhance the grip and reduce the chance of the sample tearing or yielding at the jaw interface. Another method of avoiding this unwanted artifactual effect is to cut the meat sample with dumbbell-shaped or splayed ends for insertion in the grips. Samples should ideally be cut either along or perpendicular to the muscle fibre direction. Samples are then subjected to a constant force (strain rate) suitable for their size and bulk, e.g. 50–250 mm min^{-1} for samples with 30–50 mm length between jaws and 5–10 mm width of samples

(4–6 mm thick). The UTS is taken as the maximum observed load before rupture per unit cross-sectional area of the sample.

A great deal of information about the mechanical behaviour of linear connective tissues, e.g. tendons, has been obtained using the method and the typical stress–strain curve for tendon is well-known (Fig. 7.3). UTS and ultimate per cent elongation (UPE) for a wide range of connective tissues have been calculated under several different sets of conditions.

Shear tests

The use of machines which are designed to simulate the act of mastication of meat in the human mouth to measure the resistance of meat (i.e. 'toughness'), to such physical damage has been popular. Tests using such machines (often called tenderometers) are termed shear tests or bite tests, and the forces measured during such tests are generally referred to as shearing forces. Several devices are available for such tests, but perhaps the best described instrument is the

(a)

Fig. 13.3 (a) The Instron Universal Tensiometer; (b) the machine operating in the tensile strength mode.

(b)

Fig. 13.3—*contd*

Warner–Bratzler apparatus (Fig. 13.4). Modifications and improvements have been made to the original apparatus, as, for example, those made at the Meat Research Institute of New Zealand resulting in the MIRINZ tenderometer.

In general, the instruments consist of a blunt wedge, accurately loaded at a constant rate which compresses the meat sample against a base plate. The wedge shears through the meat to give a sharp yielding point which is determined by the toughness of the meat sample. The shear force can be measured on any of a range of scales but normally an arbitrary scale is used with units corresponding to 1·7 kPa. Toughness is measured as the force required to cleave a standard cross-section of cooked meat across the muscle fibres. The method does not always lead to complete cleavage of the sample in that connective tissue sometimes adheres to the wedge and remains undamaged at the end of the test. In spite of this, and mainly because the results of such tests correlate with taste panel data, the shear test is regarded as a meaningful and useful guide to toughness by many research workers.

Using standardised samples it is possible to correlate such parameters as connective tissue content with toughness and the degree of cooking with changes in connective tissue during the cooking process.

(a)

Fig. 13.4 (a) The Warner–Bratzler shear apparatus.

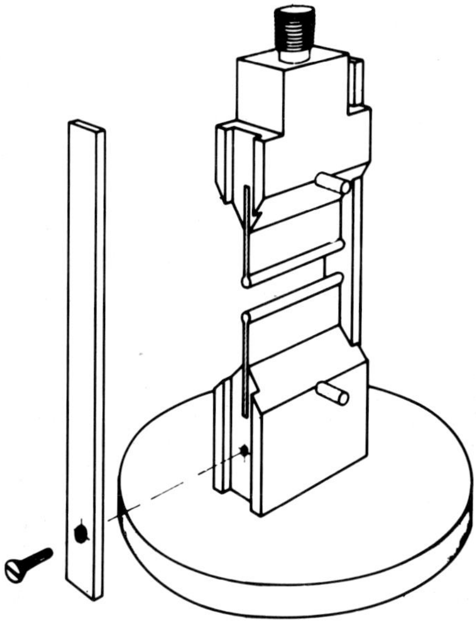

(b)

Fig. 13.4—*contd* (b) Volodkovitch apparatus.

However, it should be stressed that such methods should always be used in concert with other analytical techniques such as taste panelling, biochemical analysis of the tissues, and other mechanical tests. Shear tests are, after all, a rather crude way of investigating the bulk properties of meat tissue and, as such, provide rather complex results which may be extremely hard to interpret reliably in the absence of other data about the same samples.

Fracture mechanics

The study of fracture mechanics in meat has only recently begun although much has already been learned from the use of these techniques. The method is an application of the stress–strain/tensile strength type measurement discussed above and uses the same type of equipment. A break or weak point is introduced in the tissue sample and then a consistent force is applied such that the break or split spreads (Fig. 13.5). The break is referred to as the notch. Notches can be introduced into meat samples either parallel or perpendicular to the muscle fibre direction. The work of fracture is measured when the

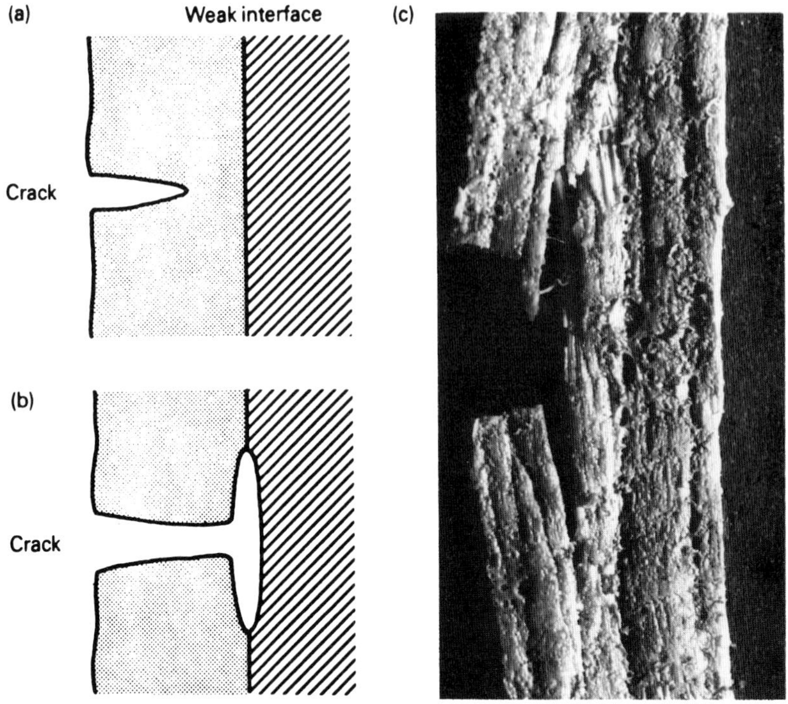

Fig. 13.5 Diagrammatic representation of a meat sample used for fracture mechanics study showing the introduction and propagation of a crack.

notch is caused to spread at a slow speed through the tissue by the application of a constant or variable force. Video apparatus can be used for qualitative assessments of the fracture behaviour of the material. After fracture specimens are often examined under both the light and scanning electron microscope in order to relate the structural changes in the sample to the observed mechanical data. Fracture methods, therefore, try to explain how and where fracture occurs in samples as well how much force is needed to cause the rupture.

There is as yet no indication that this type of method will be able to be used as a routine analytical test in meat samples. However, unlike other mechanical tests, it does seek to define the various roles of the different elements of meat as discussed in Chapter 7. In this respect the method should be seen as a primary research tool in the investigation of the causative role of connective tissue in determining meat quality.

TASTE PANEL TESTS

Taste panel tests on meat involve the subjective analysis of a meat sample by a taster using his or her own senses. Such testing obviously involves a great number of problems in standardising responses and understanding data obtained, but over the last thirty years much progress has been made in making such techniques scientifically useful. Taste panel tests are always carried out on meat as a whole and not on separated connective tissue and so, because meat is a complex tissue and connective tissue forms such a small part of it, taste panel results can only ever support and add to other data on the role of meat connective tissue in determining textural properties rather than acting as a primary source of data.

The two main methods of taste panel testing are the use of trained panellists for specific food items and the use of consumer panels. Only the first will be considered here. In this case, a panel of people is gathered together and each is assessed in terms of his or her ability to determine for texture, taste and smell with routine tests. People with problems in distinguishing texture, tastes and smells or who have difficulty in describing organoleptic quality accurately are rejected. The remaining people are assembled into a panel of at least six or more, and ideally ten.

Several types of test have been evolved to allow the discernment of certain characteristics or the difference testing of food products by taste panels. These include single sample tests, duo-trio tests, hedonic scaling, triangle tests and others. Several excellent text books have been written about these methods and so they will not be discussed here. However, perhaps one of the most powerful methods is that of quantitative descriptive analysis in which the panel uses a set of selected and agreed words or terms to describe the food product and allot scores to each of the characteristics or attributes those words or terms describe. This method can be used with repetitive testing to build up a score profile for the attributes of the product and sophisticated statistical computer packages can be applied to the quantitative data to both analyse the results and to test the efficacy and accuracy of the panellists themselves.

The selected panel is asked to consider the food item (i.e. the chosen meat sample or type) and the specific characteristics to be tested in that particular regime of tests and to discuss words which might adequately describe the main attributes of the meat to be assessed. In terms of texture such words might include chewiness, toughness, tenderness, resistance to the first bite, juiciness, etc. These words are called descriptors or attributes. This stage, with a full explanation of how to carry out the tests to be done, completes the training of the panel. The descriptors are listed on a scoring sheet each with its own scoring line with 'anchor words' at each end of the line to show the extremes of that particular attribute (for example, see Fig. 13.6).

Acceptability and Texture of Conditioned Meat

Place a vertical line across the horizontal line at the point representing your opinion of the characteristics described by the anchor words.

Appetising aroma	Unappetising aroma
Initially juicy	Initially dry
Moist after chewing	Dry after chewing
Not resistant to first bite	Resistant to first bite
Very easy to chew	Very difficult to chew
Very tender	Very tough
Expected meat colour	Unacceptable meat colour
Like extremely	Dislike extremely

Fig. 13.6 Typical example of a score sheet, showing horizontal score lines and anchor words, used for quantitative descriptive analysis (QDA) of meat samples.

Having agreed on the terms needed to describe the texture of the meat the panel now eat a selection of samples and score each attribute on a linear scale (Fig. 13.6). Multiple replicates of the same sample must always be done to allow statistical analysis to be performed on the data. The scores of the panellists may now be aggregated and averaged. In this way, the group of people acting as a trained taste panel may be viewed as a single instrument for testing the quality attributes of meats. Indeed, although the method seems highly qualitative and subjective, there is now good evidence to suggest that properly conducted tests of this kind lead to the production of highly reproducible and reliable data.

Although taste panels do not provide a primary source of data on connective tissue in meat they can give important information on the differences in quality of different meat samples about whose connective tissue much can be discovered by other means. This, then, allows the correlation of sensory analysis with hard data on connective tissue parameters. After all, the role of connective tissue in influencing meat texture is a question of paramount importance to meat scientists and therefore, to be able to quantitate meat quality reliably and reproducibly such that it can be correlated with biochemical and structural data is vital.

SUMMARY

The techniques of light microscopy including sample preparation, fixation and staining are discussed. The use of electron microscopy for the high power visualisation of structure in meat is also discussed, as are the specific methods of fixation, critical point drying and staining used in the preparation of samples for transmission electron microscopy (TEM) and scanning electron microscopy (SEM). The use of the SLS-crystallite method for the visualisation of groups of molecules has been superseded by rotary shadowing to visualise single molecules. More specific methods for the staining of collagen types with antibodies prepared from the sera of donor animals are discussed, both at the light and electron microscope levels. Mechanical methods are mentioned briefly in terms of isometric tension and ultimate tensile strength measurements, fracture mechanics and shear tests. Finally, the use of taste panels to assess subjectively the toughness/tenderness of meat is discussed and the several main methods used in such techniques are described.

FURTHER READING

1. Bailey, A. J., Restall, D. J., Sims, T. J. & Duance, V. C., Meat tenderness: immunofluorescent localisation of isomorphic forms of collagen in bovine muscles of varying texture. *J. Sci. Food Agric.*, **30,** 1979, 203–210.
2. Davey, C. L. & Winger, R. J., In *Fibrous Proteins: Scientific, Industrial and Medical Aspects,* ed. D. A. D. Parry & L. K. Creamer. Academic Press, London and New York, 1979, pp. 97–132.
3. Dodds, H. M. & Clark, G., Notes on Technic. An improved Van Gieson. *Stain Technol.*, **60,** 1985, 55.
4. Geissinger, H. D. & Stanley, D. W., Preparation of muscle samples for electron microscopy. *Scanning Electron Microscopy,* **III,** 1981, 415–426.
5. Gomori, G., The effect of certain factors on the results of silver impregnation for reticulum fibers. *Amer. J. Pathol.*, **15,** 1989, 493–495.
6. Humberstone, G. C. W. & Humberstone, F. C., An elastic tissue stain. *J. Med. Lab. Technol,* **26,** 1969, 99–101.
7. Kanwar, Y. S., Jakubowski, M. L. & Rosenweig, L. J., Distribution of sulfated glycosaminoglycans in the glomerular basement membrane and mesangial matrix. *Eur. J. Cell Biol.* **31,** 1983, 290–295.
8. Mallory, F. J., *Pathological Technique.* Saunders and Co., Philadelphia PA, 1938.
9. Margolena, L. A. & Dolnick, E. H., A differential staining method for elastic fibres, collagenic fibres, and keratin. *Stain Technol.*, **26,** 1951, 119–121.
10. Masson, P., Some histological methods: trichrome stainings and their preliminary techniques. *J. Tech. Meth.*, **12,** 1929, 75–90.
11. Purslow, P. P., The physical basis of meat texture: observations on the fracture behaviour of cooked bovine M. semitendinosus. *Meat Sci.*, **12,** 1985, 39–60.
12. Rowe, R. W. D., Elastin in bovine semitendinosus and longissimus dorsi muscles. *Meat Sci.*, **17,** 1986, 293–312.
13. Sims, T. J. & Bailey, A. J., In *Developments in Meat Science,* ed. R. E. Lawrie. Applied Science Publishers, London, 1981, pp. 29–59.
14. Sweat, F., Puchtler, H. & Rosenthal, S. J., Sirius red F36A as a stain for connective tissue. *Arch. Pathol.*, **78,** 1984, 69–72.
15. Van Gieson, J., Laboratory notes of technical methods for the nervous system. *N.Y. Med. J.*, **50,** 1889, 57–60.
16. Viidik, A., In *The Biology of Collagen,* ed. A. Viidik & J. Vuust. Academic Press, London & New York, 1980, pp. 237–255.

Chapter 14

Future Prospects

Throughout this book we have discussed the wealth of knowledge that has accrued over the last fifty years from the study of collagen, elastin, proteoglycans and other connective tissue components. In some areas the depth of our knowledge and understanding of these important macromolecules is extensive, whilst in others it is less convincing and complete. Indeed, although we know much about the general aspects of these molecules and their structure and function, most research has been done with model tissues rather than intramuscular collagen. Although we can apply much of this knowledge, the need to know more specifically about meat connective tissues is becoming ever more pressing.

Meat production is one of the most important sectors of agriculture. As the increasingly quality conscious consumer becomes more knowledgeable not only about the benefits of leaner meat but also the possibilities of producing meat that is both convenient to use and less variable, the producer will have to accept and accede to these new consumer demands. This reversal of roles has already begun and is having a dramatic effect on the attitude of the industry to quality control, from the production system to the product on the plate. Many aspects of quality have to be considered throughout the meat production chain including fat/lean ratio, colour, the requirements of meat specifically for processing, water-holding capacity and perhaps most important of all, texture. It is this last parameter that causes the main variation in fresh meat and is the major cause of rejection by the consumer. It is, therefore, becoming increasingly important to reduce this variation if meat sales are to be maintained.

The basic mechanisms determining the variation in texture are now fairly well understood to be due to poorly controlled refrigeration and the variation in the amount and quality of the intramuscular collagen. Refrigeration can be properly controlled and its effect on texture virtually eliminated. However, controlling the quality of collagen is not easy. As we have seen, rapid growth, particularly before slaughter, can reduce toughness ante-mortem, and treatment with enzymes can reduce toughness post-mortem. These are simplistic approaches, and what is now required is the application of our well established understanding of collagen to the modification of the intramuscular collagen and thereby its influence on meat quality.

In this concluding chapter we will examine some of the major problems to be solved in terms of our understanding of the role of connective tissue in determining aspects of meat eating quality. The areas of immediate concern are a detailed study of the correlation of the connective tissue structure at the molecular level and the texture of meat; the possibility of manipulating the properties of the collagen in line with above studies during growth and the possibility of controlled degradation of collagen post-mortem.

STRUCTURAL STUDIES

The role of collagen types

At the moment it is not clear why there are so many different collagen types associated with muscle and their individual functions remain to be elucidated. The minor collagens may regulate the fibre size of Types I and III collagen in different parts of the muscle. It is, therefore, important to establish the actual collagen content of meat in terms of genetic types. Are Types VI–XII collagen present in all muscles, and if so, where and in what quantities? The answer to these questions will provide the route to establishing the relationship of the various minor collagens to Types I and III, the main fibrous types, and the role of the minor types in, perhaps, linking various structures or various macromolecules.

The way these collagens interact is clearly very complex, but the use of modern molecular biology techniques and immunolabelling at the electron microscope level should permit resolution of this question in the near future. An important open question on the role of collagen in muscle is the mode of interaction of the myofibrils to the tendon or epimysial collagen which is crucial to the transmission of the muscle contraction to provide movement of the skeleton. The immuno-electron microscopy approach may reveal an answer to this long standing and vexed question. Whether these interactions play a role in meat quality remains to be seen.

The role of the epi-, peri- and endomysium

There is no doubt, from consideration of what we now know of the mechanical role of the epimysium, perimysium and endomysium in the swelling and shrinkage changes during salting and cooking and the final textural properties of the meat, that these structures play a key role in influencing changes in the eating quality of meats during processing. However, although their properties are consistent with their proposed role in determining the toughness of meat (Chapter 8), their relative contribution still needs to be assessed.

For example, the influence on mechanical properties of perimysial fibre size and orientation and fibre bundle size, the interaction of perimysial fibres with ground substance and, most importantly, the nature of the boundary between the perimysium and endomysium all need investigation. Detailed morphological studies of the shrinkage of endomysial and perimysial structures during cooking and the influence of the heating process on the perimysial–endomysial junction would also add to our understanding of the effects of these parameters on texture and quality. Physical and biochemical techniques are now being developed to address these problems and this important question should be resolved in the near future.

The mechanical properties of the endomysium also restrict the swelling of the myofibrils, as has been discussed earlier (Chapter 8). Mechanical or enzyme damage to the endomysium during conditioning and preparation of meat products is, therefore, critical to the ability of the meat to swell or take up water following salt treatment. Variation in water uptake by different muscles is a problem and further work should be carried out on the differences in the properties of the endomysium in different muscles. Alternatively it might be the activity of the degradative enzymes which bring about the endomysial damage that allows the swelling that is the determining factor in bringing about the observed differences. A better understanding of the role and mechanism of action of these enzymes and their effects on collagen and other connective tissue components during this process could provide better control during the processing of different starting materials.

Relationship with intramuscular fat

The perimysium contains most of the intramuscular fat, known as marbling, the amount depending on the overall fatness of the animal. The fat plays a role in the eating quality of the meat, probably by retaining fluids during cooking leading to increased juiciness. The nature of the collagenous tissue around these fat cells, together with that of the perimysium proper, could be important since excessive tension generated during cooking could lead to considerable fat loss and thereby a loss of juiciness. Very little is known about the relationship between fat and the connective tissue components, and there is clearly scope for a study of these interactions and their influence on meat quality.

NON-COLLAGENOUS COMPONENTS

The nature of the other connective tissue components has received scant attention, and although at present it appears unlikely that they play a major role in determining the texture of meat except in very special cases, further studies should be carried out on these components.

Proteoglycans

Without playing a direct role in textural determination, these physically minor components may have subtle interactions with collagen fibres which affect final quality. These interactions may be highly specific and dependent on the types of proteoglycans present. For example, the proteoglycans are believed to be involved in the binding and alignment of collagen fibres, and their removal or degradation could allow fibres to slide past each other under tension. Only by determining which proteoglycans are present and their distribution in various muscles can we begin to answer these questions.

Elastin

Many muscles contain relatively low amounts of elastin and it has always been assumed that this insoluble matrix molecule has little or no influence on texture. However, microscopy shows that small amounts of the protein are laid down as fibres within the perimysial structure of certain muscles (Chapter 7). This may be important, especially in the light of the fact that many mechanical measurements are made on the semitendinosus muscle due to its highly ordered muscle fibre orientation, despite the fact that this muscle is quite unusual in having the highest intraperimysial elastin content.

COLLAGEN TURNOVER AND GROWTH

As mentioned earlier, the amount and type of collagen in muscles can be influenced by the crude process of inducing rapid animal growth before slaughter. However, if ante-mortem influences on collagen in meat are to be properly controlled and used to the benefit of meat quality, then considerably more work is required. For example, studies are needed on the turnover of intramuscular collagen during growth, the differences in the turnover rate of collagen in different muscles, and the differences between these effects in fast and slow twitch myofibres. It is assumed that the total amount of collagen is related to hormone activity, which itself can be controlled, but this relationship needs to be firmly established to allow proper control. The use of synthetic growth agents from anabolic steroids to β-agonists clearly affects not only the overall rate of synthesis of collagen, but also the balance of synthesis and degradation, and thereby affects the quality of the collagen.

The future prospect of controlling the amount and type of collagen in muscle at the gene level cannot be dismissed. For example, recent studies indicate that there are promoter regions of the gene that control synthesis, and specific defects have been demonstrated to activate these promoter regions. There is some hope, therefore, that genetic engineering may provide some scope for the future. This seems far off at present and perhaps more faith should be put in the influence of more direct ante- and post-mortem factors.

Control of collagen cross-linking

The cross-linking of collagen is essential for a healthy animal, but the second stage of cross-linking involving the formation of multivalent mature cross-links is a point at which control might be productive. Clearly further studies on the nature, extent and location of these mature cross-links is crucial to any attempt to control the maturation of collagen. There is a pressing need to describe the types and structures of mature cross-links in collagens of various types and from various sources, notably in meat. There is also a need to provide the facility of rapid analytical methods to quantify these compounds in collagens from meat. Only in this way can the variable influence due to the different cross-links present in the whole range of carcass muscles be checked and ratified. Then the causative role of collagen in determining meat texture may more properly be understood.

EFFECTS OF CONDITIONING AND COOKING ON COLLAGEN

The enzymic degradation of both the perimysium and endomysial sheath during conditioning is crucial to their effective role in determining texture. More specifically, the degradative effects of enzymes on collagen in these structures is probably the key to many of the changes in the connective tissue that occur during conditioning and which, finally, influence the quality of meat after cooking. It seems likely that the cathepsins have a vital role in the degradation of collagens in both the endomysium and perimysium. However, the precise effect of these various enzymes on meat collagen needs to be accurately established to aid the control of the conditioning process. Similarly the identity and activity of the proteolytic enzymes involved need to be confirmed and determined before their activities can be controlled or artificially stimulated. Alternatively enzymes specific to collagen designed to work under conditions of low temperature and pH could be synthesised either by modification of known enzymes or by gene synthesis of new enzymes.

Meat products

In the future it may be possible not only to specify the quality of the natural collagen, but also to modify the properties of the collagen to suit a particular process. The technology of reformed meat and extension to restructured meat, a rapidly expanding area, all depend on the ability of the meat particles to bind together. The binding agent in the cooked state is primarily a myosin gel but as new binders are developed, e.g. alginates, other components of the meat may be important. The insolubility of collagen generally precludes it from playing a major role in binding and gel formation, but collagen is essential otherwise the products are soft and jelly-like. Modification of the properties of collagen could increase the binding capacity and the gel strength of meat binding proteins.

In conclusion, it can be stated quite clearly that connective tissues play a crucial role in the sensation of eating quality of meat and meat products. Better control of the quality of meat demands a better understanding of the constituent components of meat. Connective tissues have been particularly intractable in providing ready answers, but the ground work has now been laid in the case of the major component, collagen, and a greater effort will be readily rewarded in our understanding of the role of this ubiquitous protein.

Index